Horatio Bryan Donkin

The Diseases of Childhood

Medical

Horatio Bryan Donkin

The Diseases of Childhood
Medical

ISBN/EAN: 9783337369941

Printed in Europe, USA, Canada, Australia, Japan

Cover: Foto ©berggeist007 / pixelio.de

More available books at **www.hansebooks.com**

THE

DISEASES OF CHILDHOOD

(MEDICAL)

BY

H. BRYAN DONKIN, M.D. Oxon., F.R.C.P.

PHYSICIAN TO THE WESTMINSTER HOSPITAL AND TO THE EAST LONDON HOSPITAL FOR
CHILDREN AT SHADWELL; JOINT-LECTURER ON MEDICINE AND CLINICAL MEDICINE
AT WESTMINSTER HOSPITAL MEDICAL SCHOOL.

NEW YORK:
WILLIAM WOOD & COMPANY.
1893.

PREFACE.

THIS book is based to a great extent on the records and recollections of nearly twenty years' experience at the East London Hospital for Children and elsewhere, and includes the substance of some lectures given at Westminster Hospital and of a few contributions to the Westminster Hospital Reports.

Bound by prescribed limits of space as well as by the publishers' requirement of a clinical work for practitioners and senior students, I have assumed the reader's general knowledge of the diseases discussed, and emphasised only the points pertaining to childhood. Projected chapters on mental disorders and on affections of the skin have, moreover, been abandoned, and notices of 'variola' and some other maladies, of which my personal experience has been inconsiderable, have been omitted.

To the many writers and teachers whom I have studied and followed but few references have been made. My debts to them are great and conspicuous; but I have striven throughout to avoid repetition of statements howsoever trite on authority howsoever good without recourse to the records of my own observations.

My sincere thanks are due to many Hospital Residents, especially at Shadwell, who have materially aided me in the collection of cases from the note-books. From three successive medical officers there—Mr. Scott Battams, Dr. Hastings, and Dr. Ware—I have, further, received important help, both critical and clerical. To Dr. Hastings a special expression of my gratitude is offered for his friendly assistance in the work of proof-correction and for much valuable advice.

I am deeply obliged to several other friends and colleagues:—to Mr. J. L. Hague for having kindly relieved me of the laborious task of index-making; to Dr. J. A. Coutts for the critical reading

of many chapters; to Dr. R. G. Hebb for his ever-ready help in
matters pathological; to Dr. W. A. Wills for assistance in the
collection of cases from the Westminster Hospital Records; and
to Dr. Sturges and Dr. Eustace Smith for permission of free access
to their respective wards at Westminster and Shadwell.

While the final sheets of this volume were passing through the
press a valuable Report was issued by the Clinical Society of
London on the duration of the periods of Incubation and Contagiousness in certain of the commoner infectious diseases. Seeing
that this Report is founded on material which was, for the greater
part, either entirely new or inaccessible to earlier writers, and that
it has been drawn up with great care under the direction of a
Committee of the Society, I have deemed it well to record its
main conclusions in the form of an Appendix to Section III.
For this abstract I am wholly indebted to the kindness of my
colleague, Dr. Dawson Williams, whose highly important part in
the preparation of the work is specially acknowledged by the
above-mentioned Committee.

H. BRYAN DONKIN.

LONDON, *June* 1893.

CONTENTS.

INTRODUCTION.

SECTION I.—DISORDERS OF THE ALIMENTARY TRACT AND OF THE ABDOMEN.

CHAPTER I.

INFANTILE WASTING.

CHAPTER II.

AFFECTIONS OF THE MOUTH.

CHAPTER III.

AFFECTIONS OF THE FAUCES.

CHAPTER IV.

GASTRO-INTESTINAL DISORDERS.

CHAPTER V.

GASTRIC AND INTESTINAL DISEASE.

CHAPTER VI.
CONSTIPATION.

CHAPTER VII.
INTESTINAL OBSTRUCTION.

CHAPTER VIII.
PERITYPHLITIS AND TYPHLITIS.

CHAPTER IX.
PERITONITIS AND ABDOMINAL TUBERCLE.

CHAPTER X.
ASCITES, JAUNDICE, AND DISEASES OF THE LIVER.

CHAPTER XI.
ENLARGEMENT OF THE SPLEEN.

CHAPTER XII.
URINARY DISORDERS.

CHAPTER XIII.
ANASARCA AND KIDNEY DISEASE.

CHAPTER XIV.

SECTION II.—GENERAL DISEASES.

CHAPTER I.

RICKETS.

CHAPTER II.

SYPHILIS.

CHAPTER III.

SCROFULOSIS OR STRUMA.

CHAPTER IV.

TUBERCULOSIS.

CHAPTER V.

ANÆMIA, PURPURA, AND SCURVY.

SECTION III.—ACUTE FEBRILE DISEASES.

CHAPTER I.

PYREXIA.

CHAPTER II.

DIPHTHERIA.

SECTION IV.

DISORDERS OF THE NERVOUS SYSTEM.

CHAPTER I.

SPASMODIC DISORDERS.

CHAPTER II.

THE PARALYSES OF CHILDHOOD.

CHAPTER III.

ACUTE DISEASES OF THE BRAIN.

CHAPTER IV.

CHRONIC DISEASES OF THE BRAIN.

CHAPTER V.

CHOREA.

CHAPTER VI.

HYSTERIA AND FUNCTIONAL NERVOUS DISORDER.

CHAPTER V.

ACUTE BRONCHITIS AND BRONCHO-PNEUMONIA.

CHAPTER VI.

PNEUMONIA.

CHAPTER VII.

PLEURISY.

CHAPTER VIII

ON PHTHISIS AND MEDIASTINAL GLAND DISEASE.

SECTION VI.

DISORDERS OF THE HEART AND CIRCULATION.

CHAPTER I.

CONGENITAL HEART-DISEASE.

CHAPTER II.

CARDIAC INFLAMMATION AND VALVE-DISEASE.

CHAPTER III.

PERICARDITIS.

CHAPTER IV.

RAYNAUD'S DISEASE.

DISEASES OF CHILDHOOD.

(MEDICAL.)

INTRODUCTION.

WITHOUT personal experience and careful study of disease as it appears in early life the practitioner meets with many difficulties of diagnosis and treatment which are not to be coped with by the knowledge gained from the practice of ordinary hospitals or the study of the general text-books of medicine. There is, therefore, a clinical reason for the existence of hospitals and even books devoted to the disorders of children. Doubt-less, however, a mindful consideration of the anatomy, and still more of the physiology, of infancy and childhood will usefully direct the studies of those who are well acquainted with general pathology. Strictly speaking, the subject of disease in children is no speciality, and it may be safely assumed that however great the practical advantages may be of special hospitals and facilities for this branch of clinical study the best diagnosis and treatment will be accomplished by those who have concurrent experience of disease in general at all times of life. It is mainly in the practical application of that knowledge which should be common property that assistance is wanted for the clinical study of such disorders as are prominently incident on childhood or assume a special aspect at that period.

A very large proportion of the maladies of childhood are the direct outcome of imperfect and at the same time rapidly progressive growth and development. Apart from congenital malformations there is a con-stant liability to injury from external stresses, as shown so markedly in the disorders of the digestive and nervous systems in infancy. Rickets is a salient instance in point, and it will be seen that a large majority of alimentary and nervous disturbances at this time are due not to any primary change in the organs concerned but to untoward strain put upon them by influences from without. Gastric and intestinal disorders evidenced by vomiting, diarrhœa, pain or other symptoms are very often directly referable to ingesta which are in quantity or quality unsuitable to the infantile organs, and, unless long neglected, are frequently curable

A*

by obedience to simple physiological rules. Again, the peripheral nervous system of the infant is in developmental advance of the spinal, and both are far ahead of the higher organs of brain control. Hence we meet with numerous instances of disorder arising from what are seemingly the slightest and often undiscoverable external causes, as is evidenced among other affections by convulsion, readily excited febrile disturbance, and disordered breathing. Under this heading, too, we may probably rank much of the great liability of young children to catarrh of mucous membranes arising from the action of external causes, such as changes of temperature, on the nervous periphery, and the grave results of such catarrh in the respiratory tract leading to a rapid collapse of the lung from the imperfectly organised nervo-muscular mechanism of breathing.

Many affections which are not confined to children occur with so much greater frequency in early life that they are practically diseases of childhood. Of such are many of the exanthemata, whooping-cough, membranous and other forms of laryngitis, certain inflammatory affections of the brain and cord, and, among other intestinal disorders, intus-susception. Still other diseases, such as tuberculosis, rheumatic fever, enteric fever and lung-inflammations, have certain peculiarities more or less marked in childhood which give them a claim to be included in a special work, and familiarity with such affections in some of their important aspects can only be gained from extensive experience among children. Besides considerations such as these, most writers on the subject advance a further apology for themselves by urging the special difficulties which beset the investigation of diseases in infancy through the inability of the sufferers to give account of their feelings. Not to dwell too long on the matter of the examination of sick children, which is ably and amply and sometimes rather tediously treated by authors, I shall here give but a brief sketch of what seems to me of most importance, leaving it to be more or less filled up in the progress of the work and by the knowledge or common sense of the reader.

In all cases of illness in children, who react readily to but very slight disturbances, the most careful and complete examination possible should be made of all parts and organs, and all deviations from normal function noted. For this it is necessary that the observer should be familiar with the healthy appearance and habits of children from birth onward. The normal changes of colour in the skin of the new-born infant must be kept in mind, the characters of the excreta studied, and the nature of the movements, especially of the eyes and limbs, observed. Want of intimate acquaintance with the normal infant may lead to a mistaken diagnosis, healthy and morbid signs being often confused. Full attention should always be paid, with due critical reserve, to anything that is reported as abnormal by mother or nurse, for though frequently coloured by fancy or fright their statements may be very helpful to both the experienced and the inexperienced doctor. In examining a child it is well to postpone until the last those procedures which are likely to be felt most irksome, but no general routine need be prescribed. As a rule the abdomen should be carefully palpated and percussed when the child's attention is distracted and before it is stripped for that general inspection which is usually necessary on a first visit, if not afterwards as well; for even if the child do not cry the reflex contractions of the muscles are so easily excited by handling that it is often very difficult to learn much of the size and condition of the abdominal contents. A little practice and patience, even if the child be restless or crying, will generally insure fairly complete auscultation of the lungs with the unaided ear, or with the stethoscope, which, in the case of less disciplined private patients, should for preference be a flexible one; and the heart can be listened to with some success under similar conditions. The time for percussion must be awaited, as a rule, until the child be quiet, for the results of this method of examination in young children are even in the most favourable circumstances far more difficult of interpretation than those of auscultation. The examination of the mouth, and especially of the fauces (which should never be omitted, and is especially valuable in many obscure cases of pyrexia) is often difficult and must frequently be forcible. For this it is well to be prompt, causing the child's head and limbs to be firmly held, and pushing the spatula or spoon well back on the tongue. Much time is wasted and much annoyance given to the child and its attendants by hesitating and feeble attempts to inspect the throat. The use of both the ophthalmoscope and the laryngoscope requires much practice when the patient is restless or too young to co-operate with the observer, and is often impossible. For important practical purposes, however, neither of these instruments is often necessary in the case of infants, and in many instances of cerebral symptoms which call for the diagnostic aid of the ophthalmoscope the torpid condition of the patient renders its use comparatively easy.

Thermometric observation, to be of real value, should not be made in the mouth, but either in the rectum, the child's body being kept still by one hand placed on the abdomen, or by carefully holding the instrument in the axilla or groin for five minutes or more, according to its sensitiveness. If there be reason to suspect pyrexia from a rash or other symptoms and the axillary temperature appear normal a second observation should always be taken in the rectum.

Details of what to observe, of the deviations from the normal in external appearance, and other physical signs of organic disease will be given when the various groups of symptoms pointing to general or local disorder are considered. I will only add here that in dealing with children the doctor should be as gentle and as natural as possible in manner and strive to drop entirely, if he have ever acquired, that grave professional style which is often deemed necessary with adult patients. During the whole of the visit, and specially if it be the first, alertness of eye and ear will often give us valuable information, and we may learn much before the child be touched or even aware that it is noticed.

The period of years included here under the term "childhood" extends from birth to the usual time of the establishment of puberty which may be roughly set down to the fifteenth year.

"Infancy" for practical purposes is to be understood as extending to the end of the second year when the first dentition is usually complete. In accordance with the usage of my colleague, Dr. Eustace Smith, I apply the term "early childhood" to the next two years, for there is a marked falling off in the incidence of several disorders, both nervous, alimentary and pulmonary, before or about the end of the fourth year.

CHAPTER I.

INFANTILE WASTING.

Wasting in infancy as the result of insufficient or improper food is the subject-matter of the present chapter. Several local and general diseases, however, are most prominently evidenced by loss of flesh, and we must never be content with the diagnosis of simple atrophy from dietetic causes without excluding as far as possible, after careful examination and inquiry, the antecedence or concurrence of other mischief. As examples of some among many affections which I have often seen diagnosed as simple wasting in infancy and early childhood I may mention not only tuberculosis and syphilis, but also undiscovered empyema without marked symptoms or pyrexia, and the atrophic sequelæ of enteric and other fevers. A very large number, however, of the deaths of children in their first year is due to starvation, whether caused by insufficient food or by ingesta which, being to a great extent indigestible or unassimilable, have little or no nutritive value. Innumerable cases, too, of rickets and other disorders are either largely the direct and often disastrous results of improper feeding, or are favoured in their development by the draining and starvation of the body arising from the frequently consequent diarrhœa and vomiting. The most ordinary and glaring fault in feeding infants is the substitution of most or all of the requisite milk by farinaceous material, which, if not properly cooked or malted, is no food at all, and at the best is half-starvation diet. It contains practically no fat, and lacks that indefinite antiscorbutic quality which is essential to health.

In the case of an infant healthy at birth and suckled by a mother sound in body and mind at intervals of two hours gradually increased to three hours until weaning-time, there is the least to fear from serious alimentary trouble leading to wasting. The mother's milk contains, of course, all that is necessary to nutrition, and in such a form that

5

the scanty salivary secretion of young infants and the small power of the pancreatic juice, during the early months, of turning starch into sugar or acting upon fat have as yet no physiological import. The sugar is ready formed in the milk which contains also the necessary salts, the fat is easily assimilable, and the gastric juice of the infant is in full activity to deal with the proteid casein. That the infant requires much more fat and less carbo-hydrate in proportion to proteid food than the adult, and a greater proportion of water, is shown by the following figures which I quote from the averages drawn by Dr. Cheadle from standard analyses of human milk and from estimations of the physiological requirements of adult diet. These data may be regarded as sufficiently accurate for the establishment of the above-mentioned conclusions, but partake too much of the artificial nature of all physiological averages to be rigidly applied as a standard in individual cases.

In human milk, to which of course all imitation- or substitution-diets for infants should conform as far as possible, the percentage of the elements is as follows :—

Proteid .	3.500
Fat .	3.000
Carbo-hydrates	4.000
Salts .	.138
Water .	89.362
	100.000

The standard requirements for adults are thus proportioned :—

Proteid .	5.00
Fat .	3.00
Carbo-hydrates	15.00
Salts .	1.15
Water	75.85
	100.00

The remembrance of the above formula for infantile diet, together with that of the fact that about a pint of fluid so constituted is the daily amount necessary for an average infant during the first month, will remind us of the theoretically proper though, as we shall see, not always practicable dilution of the various articles of food which we may have to use in substitution of mother's milk. After the age of one month until the sixth is reached the quantity of this nourishment should be gradually increased to about two pints, and two and a half pints or even more may be taken after the tenth month. The intervals of feeding should at first be about every two hours, and later somewhat longer, according always to those requirements of the individual case which can be learnt by experience alone. In this context the following Table, taken from an article by Dr. Emmett Holt in Keating's *Cyclopædia of the Diseases of Children*, and based on the ascertained weight of a number of healthy infants before and after nursing, is valuable as indicating the approximate quantity for the meals of average infants, although, according to my experience, children over six months can very often take more with advantage.

Age.	One Feeding.	Number of Feedings.	Daily Amount.
	Oz.		Oz.
2 weeks	2	8	16
1 month	3	8	24
2 months	4	7	28
4 ,,	5	6	30
6 ,,	5½-6	6	33-36
9 ,,	7-7½	5	35-38
12 ,,	8-9	5	40

Certain conditions of the mother's milk, as regards quantity or quality, may render it unfit or insufficient food for the child, and either a wet-nurse or artificial feeding must be resorted to if a short trial of suckling result in nutritive failure as shown by wasting or much discomfort, or still more by persistent vomiting, unhealthy motions or diarrhœa. If, however, the breast-milk be deficient in quantity only, or the mother be otherwise unable to suckle the child sufficiently often, alternate suckling and artificial feeding are far better than weaning.

An infant is not likely to thrive well on the milk of a mother who is suffering from syphilis or tuberculosis. Apart from the discovery of definite ill-effects we are always justified in advising that the children of such mothers should be otherwise fed. Repeated emotional disturbances or febrile illnesses on the part of the mother are frequently accompanied by some modification of the milk which causes digestive disorder in the child, and there are cases where an analysis of the milk shows that the various elements are in too great or too small proportion for due nutrition, thus accounting for alimentary disorder or wasting and pointing to a change of diet.

The most cases by far of simple infantile wasting are met with in the poorest classes and are largely due either to chronic under-feeding and weakness of mothers who, from their very poverty, persist in nursing their infants, or to the calls of daily work which hinder regular nursing and necessitate substituted or additional diets that are usually of improper quality and practically innutritious. Among the more fortunate classes mothers who do not nurse their children frequently engage a wet-nurse or avail themselves of the best instructions as to artificial feeding. The labouring classes are thus forced to supply a very large contingent to the population of starved and diseased children, for wet-nursing, of course, and often even the purchase of enough good milk for approximately efficient nutrition are out of their power. Wasting cases, however, occur not only when the mothers cannot but also when they will not suckle their infants, and the children of many well-to-do people fail under the stress of artificial feeding ignorantly or carelessly directed.

Loss of weight and fretfulness are the leading *symptoms* of atrophy due to insufficiency of proper food. The normal child should increase in its first month by about one-third, in half a year by more than double, and in one year by treble its original weight ; but, without weighing, wasting is usually apparent from the more or less rapid disappearance of fat from all parts of the body. The face becomes peaked and wears an old and anxious appearance ; the fontanelle is depressed ; the child is constantly crying and sucking its fingers or anything on which it can lay hold ; the skin is pale, dry, and wrinkled, loses its elasticity, and may show fine desquamation. Some infants may thus continue wasting without further marked symptoms until drowsiness sets in upon extreme weakness, and with all the wheels of being running slower and slower they may die unnoticed in their sleep. Sometimes almost sudden death, as observed by Mr. Scott Battams, may result from collapsed lung, many children having thus succumbed at Shadwell Hospital during the small hours of the morning when vitality is normally at its lowest ebb. Much more often, however, further disturbances are observed. Progressive anæmia and weakness cause indigestion of the little food taken, and vomiting accelerates the process of the malady. If the diet be milk it is ejected in unchanged curds, and when indigestible food has been given, as is the case in so many instances of hand-fed children, the symptoms of irritation of the alimentary canal are seen earlier and persist. The tongue is furred ; vomiting is frequent or constant after food, the ejecta often having an excess of mucus and a sour offensive smell ; the belly is frequently distended with gases and somewhat tender, and there are facial signs of abdominal distress. Diarrhœa, too, is very common, the motions losing their healthy consistency and yellow colour and becoming watery and often green, with sour smell and highly acid reaction. With these dyspeptic symptoms the appetite sometimes fails or ceases altogether ; the temperature tends to fall, and in some extreme cases may be below 90° F. ; the heart's beats become fewer and the respiration shallower ; thrush may appear in the mouth ; the extremities grow cold and blue, and the child dies, sometimes in a convulsion. In many cases, however, the appetite may be ravenous while the wasting progresses. The ingesta remaining long in the stomach and intestines there is acid fermentation and consequent flatus with much pain which seems to be temporarily relieved by renewed feeding.

It is unnecessary here to give a more detailed account of the varying symptoms and course of untreated or maltreated wasting which depend so much on the individual powers of resistance and the nature of the ingesta with which attempts at nourishment are made. The subject will be further dealt with under the clinical headings of Diarrhœa and Vomiting. If a child be fed on milk which is poor or insufficient in quantity, as so often happens, or on any diet which is deficient in the fatty element, there may be no vomiting or diarrhœa, as we have already seen, or at all events not for long, and there is frequently constipation. Irritating substances, such as unchanged starch which is one of the commonest, or masses of any food which the individual stomach cannot deal with will cause vomiting and, probably, diarrhœa. The less vomiting or diarrhœa there has been the better is the prognosis when the case comes under treatment, but in many of even the apparently worst instances with almost a skeletal appearance there is considerable ground for hope, and, in the absence of evidence of other disease, we are often justified in soon making a highly favourable forecast.

Post-mortem examination shows in many instances nothing but universal wasting and anæmia even when there has been much diarrhœa and vomiting; in some protracted cases, however, signs of chronic intestinal catarrh or occasionally even small superficial ulcers may be found. Taking into consideration the very frequent absence of any notable change in the stomach or intestines and the fact that many severe cases of wasting, even with long-continued diarrhœa and vomiting, often start on the road to recovery in the course not of weeks, but of days, when properly treated, I have long been of opinion that catarrh of the stomach or intestines plays a much less important part than is often taught in the cases we are considering, and that when it does occur it is the result of continued irritation from undigested and fermenting substances in the alimentary canal. On this point I am therefore in accord with Professor Henoch, who has recourse to the older chemical theory of acid fermentation to explain the common characteristics of the vomit and fæces in cases such as these. A certain degree of catarrhal flux may result from the irritation of the gastric mucosa by indigestible material, and this alkaline fluid may assist in neutralising the gastric juice and thus favour fermentation; but it is greatly surprising to the inexperienced how very frequently the simple administration of diluted milk in small quantities, or the substitution of a little whey or dissolved white of egg, or indeed the ingestion of scarcely anything but water for a day or two will quickly remove all untoward symptoms and prepare the child to receive its duly nutritive diet. A chronic gastric catarrh—and it is with chronic cases that we are dealing now—could scarcely be cured so simply or so soon.

There is one very common cause of the dyspeptic disturbances and consequent wasting of hand-fed children, apart from any obviously bad quality of the food given. Want of cleanliness of the bottle or feeding-tubes (which may but contain the stale remains of milk) often occasions disturbance by allowing the direct entrance of deleterious germs into the stomach. A certain number of cases, of less intensity than those generally known as summer diarrhœa or " cholera infantum," which are in all probability due to the action of morbific organisms can be rapidly cured by attention to this point.

Such are the leading symptoms and causes of atrophy from insufficient nutrition without antecedent disease. It must never be forgotten that in a large proportion of wasting infants among the poor the deprivation of sunlight and of wholesome air and many positive evils besides contribute greatly through lowered general vitality to the digestive and assimilative failure of these victims of heredity and environment, and hinder or prevent their response, even in the absence of special gastric affections, to the most approved and careful treatment.

The question of the **treatment** of these cases of wasting mainly resolves itself into that of correcting the child's diet according to physiological principles either by way of supplementing its natural diet or substituting an artificial regimen.

I shall leave for subsequent notice those temporary digestive disturbances, whether catarrhal or otherwise, which from time to time, apart from apparent dietetic errors, may interfere with nutrition, only emphasising here the fact that in the first year such disturbing causes are not very frequent in children duly fed and cared for.

When a systematically suckled infant with no sign of other disease fails to thrive we must consider what can be done to secure success before giving the order to wean, for the natural food should never be lightly forbidden. If the mother be healthy and her milk plentiful and given with due intervals there may be some excess or defect in the nitrogenous or fatty elements of the milk. This may be ascertained by analysis, and at any rate intermediate meals of cow's milk properly diluted as hereafter to be described may be tried for a while, or artificial feeding or possibly wet-nursing altogether may be temporarily resorted to. A doctrinaire dieting of the mother with a view to modifying her milk has seldom a successful and often a bad result, by interfering with her digestion, and I would incidentally remark that all attempts at medication of the infant by dosing the mother are to be condemned as useless or pernicious trifling. An increase of the fat in the milk may be attained in some cases by a richer nitrogenous diet on the mother's part. The best way for the mother to secure a good milk-secretion is to indulge in a plentiful mixed diet with

no stint of fluid and no more alcohol than what she may take in modera-
tion at other times; to take plenty of exercise in the fresh air; and to
obey generally the known laws of health. If a woman cannot so order
her habits and be unable or unwilling to prefer the function of nursing
before all things else for the allotted time it were better for the child
to run the risk of hand-feeding from its start in life. If the breast-milk
be deficient in fat a certain amount of cream may be given to the child
at intervals, and, should the albuminous elements be in excess, and thus
cause difficult digestion, the method devised by Dr. Eustace Smith may
be tried of giving the child a diluent draught of barley- or lime-water or,
I would add, of water alone, just before putting it to the breast. With
this aid the otherwise too indigestible curd may often be successfully
dealt with. When the mother's milk is altogether too poor and scanty,
in spite of careful diet and other hygienic measures, and especially when
even the few breast-meals that the child may have in addition to other
milk appear to disagree as well as to be unsatisfying, wet-nursing or
weaning should be at once established.

Most cases of atrophy, however, occur when appropriate breast-feeding
is out of the question owing to the mother's inability or disinclina-
tion to suckle her child entirely. Wet-nursing doubtless gives a child
the best chance of progress without drawbacks, but there are, in my
opinion, so many difficulties and objections both medical and socio-
logical attaching to this question, that with an expression of general
disapproval of this method I shall pass on to the matter of artificial
feeding.

Cow's milk, with of course the cream, duly diluted with a regard
to both general physiological principles and individual possibilities is
practically the only basis for hand-rearing owing to its very ready
supply as well as to its generally best qualities in spite of certain faults.
Goat's milk is richer in fat than cow's milk and therefore good food for
healthy children and possibly, for some cases of weakly infants, but it
has no advantage over cow's milk as regards the indigestibility of its
curd in comparison with that of human milk. Ass's milk has a very
easily digestible curd, but, being probably no richer in nutritive qualities
than cow's milk diluted with two parts of water and much poorer than
human milk, can only be of temporary use in cases of feeble digestion
and is, further, very expensive.

Exceptionally robust infants may occasionally do well on almost or
quite undiluted cow's milk provided no cream be abstracted, but the
average baby will fail to digest it for any length of time. Although
the amount of fat is practically the same in cow's and human milk the
former is richer in proteids, poorer in sugar, and has a slightly acid
instead of a slightly alkaline reaction. Its casein, too, coagulates in much

larger curds, causing difficulty in digestion.   Besides this, cow's milk as usually obtained several hours after leaving the udder contains (and especially in the warmer season) many bacteria, some of them probably more or less morbific, and may in certain circumstances be infected with the germs of tubercle, scarlatina, enteric fever or other specific diseases.

To adapt cow's milk, therefore, to the infantile needs, we must (1) dilute to reduce the amount of proteids, (2) endeavour to promote the digestibility of the curd, (3) add some sugar, (4) boil or otherwise sterilise, and (5) render it of slightly alkaline reaction.

The two points of *dilution* and *curd-digestibility* may be considered together.   It will be seen at once that dilution which may duly reduce the proportion of casein will unduly reduce that of the fat and other solids, and further, that even if the reduction of the fatty element be compensated by the addition of cream a much larger quantity of fluid than the necessary quantity of mother's milk must be ingested in order to obtain the due amount of albuminoid food.   Now the difficulty here raised seems very great, and it is further, I think, magnified by those who rely for their clinical data on hard and fast analytical statements as to the normal proportions of the various ingredients in human and cow's milk respectively, omitting perhaps to remember the greatly varying relative quantities of these ingredients in human and cow's milk, and also in individual women and individual cows, as well as the fact that the milk of both women and cows is much richer in solids, and especially in fat, at the end of the milking process and when the intervals of milking are short.   In illustration of the great theoretical difficulty that all students must meet with in this matter of dilution I will but refer to two standards given respectively as a basis for action by Dr. Cheadle in London and Dr. Rotch in Boston, U.S.A.   The former quotes the relative proportions of proteids in cow's and woman's milk as 5.404 to 3.924, while the latter gives it as 4 to 2 or 1.   It is plain that here as elsewhere biologico-chemical averages, owing to the complexity of their data, are not of paramount value for practical application to individual cases.   My own experience fully corroborates this and justifies me in saying that as a rule it is not necessary to weaken the best cow's milk so much as by two parts of water or other diluent except, perhaps, in some few cases in the first weeks of life, and that even when such dilution is found necessary, a slightly increased quantity at all meals and shorter intervals between them (as compared with what natural suckling would indicate) go far towards overcoming the objection of giving to the infant a considerably larger amount of fluid as a whole than it might take at the breast in twenty-four hours.   If cows and women and milk and babies were all constant quantities our confessed difficulty in the hand-rearing of human young on cow's milk would probably be as great in practice as it seems to be on paper. The comparatively indigestible nature of the larger curds of cow's milk is perhaps the most important factor after all in this problem of dilution. In some cases indeed of wasting and alimentary disturbance the difficulty is almost insoluble and among other causes often necessitates a substitution-diet. It is usually taught that if lime-water or barley-water be used in varying proportions with pure water as diluents the curd becomes finer and therefore more digestible; but from repeated experiments I have found, as Dr. Rotch found, that outside the body there is practically no difference in the appearance of the clots of milk formed by acetic acid whether in mixture with water alone or with barley- or lime-water, and I have never known, even in cases when continued and careful observation was possible, that a child who persistently vomited hard curd when fed on milk well diluted by water ceased to do so when, without any other change in treatment, lime- or barley-water was wholly or partially substituted for the original diluent. When the digestibility of the curd is not attained by simple dilution pancreatised milk may be used with advantage. To prepare this, a pint of milk, according to Sir William Roberts, is diluted with a quarter of a pint of water and heated to about 140° F. Two teaspoonfuls of Benger's "liquor pancreaticus" with twenty grains of bicarbonate of soda are then added and the mixture is poured into a covered jug which is placed in a warm situation to keep up the heat. After an hour or an hour and a half the produce is raised to the boiling-point to prevent further action of the ferment. The milk can then be used, further diluted or not, according to circumstances. Another method, often very useful, of promoting digestibility of the curd is to mix with the milk a small quantity of well-baked flour or of one of the best prepared farinaceous foods. About a teaspoonful of such an addition to five or six ounces of milk is frequently found to have good effect. I have said that it is unpractical to lay down very hard and fast rules as to the quantity of milk to be taken in twenty-four hours by any given child in view of the variability of requirements and digestive power, but have practically found that from three-quarters of a pint to a pint of cow's milk diluted to about twice its bulk will usually suit the youngest babies, if given in divided doses about every two hours or sometimes oftener, although such meals may be larger in quantity than is naturally indicated. Both the quantity and dilution may have to be frequently altered in either direction, and each case must be carefully studied on its own merits, especially at the outset. By these means much more success will be attained than by the prescription of any tabulated routine however elaborately worked out on the system of averages. Regurgitation of milk immediately after a meal means that the stomach has been overfilled, and usually indicates less abundant meals with probably shorter intervals. When a healthy child has arrived at the age of about three months it is almost always able to digest milk diluted by one-third only of water, an approximately perfect diet.

The deficiency of sugar in cow's milk is theoretically best compensated by the addition of sugar of milk, but a small quantity of cane-sugar in each bottle (just sufficient to give a slight sweet taste) usually answers all practical purposes. Pure cow's milk is at least 2 per cent. lower in sugar than human milk, and, diluted, of course still lower.

To prevent decomposition by destroying germs all milk given to infants should be sterilised by boiling or steaming. The milk should be thus treated as soon as received twice a day, and kept in a receptacle carefully stoppered with cotton wool. If the steaming process be adopted it is well to place the feeding-bottles containing the milk in the steamer, the nipples being protected by an indiarubber cap, thus preventing the necessity of pouring the milk from one vessel to another before using it. A convenient method of steaming milk is detailed by Dr. Rotch in an article in Keating's *Cyclopædia* on "Infant Feeding," to which I would refer the reader on this and other points. His steamer consists of a tin pail eight or nine inches in diameter and nineteen or twenty inches deep, raised on three legs sufficiently high to allow a Bunsen gas-burner (or spirit-lamp) to stand under it. Four inches from the bottom of the cylinder is a perforated tin diaphragm on which the milk-receptacle or the feeding-bottles stand while being sterilised. There is a small vent in the cover for the escape of steam. Water is placed in the steamer to the depth of about one inch. After the water has been boiling a few minutes the vessels containing the milk are put into the steamer and allowed to remain for twenty minutes. A readier form of steriliser, Dr. Rotch adds, is a simple colander, with a lid, placed on a kettle. If the milk be sterilised in a vessel containing a larger quantity than is enough for one meal the receptacle should be carefully stoppered with cotton wool and kept cool, and the feeding-bottles thoroughly scalded out immediately before use. Another method of sterilising is immersion of the feeding-bottles containing the milk in boiling-water for thirty or forty minutes.

A complete apparatus for sterilising, with feeding-bottles, as devised by Professor Soxhlet of Munich, can be obtained in the market, but by following the above directions the desired object will be satisfactorily attained. There is in my opinion no objection to boiled milk other than its taste which to many children accustomed to unboiled milk is somewhat repellent. With most young infants, however, this point is inconsiderable. The scum is thicker and more coherent than that which forms on milk steamed for an equal time, a small proportion of the albumen being thereby coagulated and lost, but from my experience at the hospital for children where boiled milk is always used, and from knowledge of many infants reared entirely in this manner, I believe that boiled milk is practically as nutritious as fresh milk and at least equally digestible. Steam-sterilised milk, when used at once, has, I think, no drawback at all, and, judging from some samples I have tasted after being kept for some months, is nearly if not quite as palatable as fresh milk. When it is kept long in bottles the cream separates in large masses, but all that is requisite then before using it is to thoroughly shake or whip it up in order to diffuse the cream. More prolonged and repeated sterilisation is necessary when the milk is to be kept for an indefinite time, and this, in our lack of complete knowledge of the possible changes induced by such a process, is a point in favour of sterilising milk at home for daily use as above recommended. All things considered, however, it would seem that an abundant and cheap supply, under effective supervision, of thoroughly good milk in a sterilised form would be a great national boon, for although boiling is effective for sterilisation, there is in the houses of the poor much risk of fresh contamination, and there is no practicable guarantee of the proper quality of the milk supplied to the masses.

Before leaving this subject I would shortly notice the allegations of some that boiled or otherwise sterilised milk is lacking in the antiscorbutic qualities of fresh milk. In answer to those who quote isolated cases with scorbutic symptoms which have disappeared when fresh milk was substituted for sterilised I instance, besides the prevalent and increasing experience in all countries of the apparently harmless use of boiled or steam-sterilised milk in both hospitals and private families, the following evidence kindly supplied me by men who have had much greater personal experience of infants exclusively reared on such milk than any English physician can probably claim. Dr. Emmett Holt of New York has observed during the past three or four years large numbers of normal infants at several institutions who were thus fed, many of them being under notice until they were three years old or over, as well as several others under a similar régime during a hospital residence of three months. Dr. Holt informs me that in all his experience of the last eight years at least he has met with only two marked cases of scorbutus, the one in a hospital infant who from inability to digest milk in any form was kept exclusively on a malted food for over three months, the other in an infant seen in private practice who had been fed from birth on condensed milk and on a malted food of a brand highly approved and widely used in England and elsewhere. He adds that in his experience and that of other physicians in New York scorbutus is a very rare disease, while the practice of sterilising milk apart from cases in hospitals is well-nigh universal among well-to-do people. I have been informed also by Dr. Glaevecke, of the University of Kiel, that, while there is a very widespread use in Germany of boiled and steam-sterilised milk in hospitals and asylums for infants as well as in the families of both the well-to-do and the labouring classes, he has never seen a case of infantile scorbutus and that this disease is extremely rare altogether in Germany. He adds that, in his experience, since the use of sterilised milk has become so common, cases of thrush (*Soor*) have greatly diminished in frequency and severity.

To render cow's milk sufficiently *alkaline* the addition of lime-water is the best method. For this purpose about one and a half or two teaspoonfuls to a pint of milk is enough.

I believe that in a very large number of cases healthy children can be reared successfully from the first on the principles above sketched, but doubtless there are many instances, especially in weakly infants and those who have been brought very low by bad attempts at artificial nourishment, where a more accurate adaptation of cow's milk to infantile requirements is necessary and should always be tried before recourse to any substitution-diet. I would, however, once more emphasise here that the commercial abstraction of cream from milk, so commonly practised to a greater or less degree, is to be credited with a considerable proportion of the alleged nutritive failures of cow's milk. This is probably evidenced by the frequent and rapid recovery in hospital practice of much-wasted infants, treated by good cow's milk alone, who had nevertheless been previously well cared for and fed on milk according to proper instructions.

It is owing to the frequent failure of cow's milk, given more or less in accordance with the above directions whether with or without lime- or barley-water, to agree with or nourish a child that condensed milk and a host of patent and other foods invite the attention of the physician and the public. I have now for a long time almost entirely discontinued the use, in the case of young infants, of such proffered aids after many past trials. Doubtless some babies thrive well on condensed milk or on perhaps any of the more carefully prepared predigested or malted farinaceous foods, and a few indeed appear to thrive on one or other of the worst and most widely advertised articles in the medical market. The abundant professional testimony, therefore, which favours these preparations and causes them to exist and endure, is not perhaps to be credited wholly to the commercial spirit, but partly to that uncritical experience which is always so shy of the "instantia negativa" in scientific evidence.

Absolutely no sound result can be attained by the "trial" of such foods on any scale at hospitals for children, for the clinical material is too complicated and fluctuating, and individual differences too great, for any true comparison to be made. Very many of these foods are certainly well taken by many children over seven or eight months old when the organism is becoming ready to deal with other than a milk diet ; but the elaboration of these manufactures is then proportionally unnecessary, or may be to some extent objectionable, as tending to diminish natural function by anticipating it. It must be remembered that these preparations must really be criticised as the first diet of infants. The only practical, as the only logical proof of the value of such foods, would be that they satisfactorily nourish the youngest infants who fail to respond to their natural food or the closest possible imitation of it. The main objections to these articles are admirably summarised by Dr. Rotch in his article on "Infant Feeding," above quoted, and, since his positive advice on this subject is no less to the point than his criticism, I shall add here a short abstract of his conclusions. There is no guarantee, says Dr. Rotch in effect, that even if the published analysis be genuine the subsequent preparation shall correspond to the samples, and there is every inducement to the contrary. In many cases the foods are demonstrably not what they claim to be, the analysis which takes the physician's fancy being usually made regardless of the ultimately necessary dilution, so that what enters the child's stomach is not correspondent to the concept of it in the medical mind. The addition of starch, even if changed, as alleged, into glucose, is not only unnecessary, as the sugar might be added from the first, but also generally erroneous, as the natural sugar is that of milk which should be converted into glucose by the natural functions hereby allowed to fall into disuse, and there is a similar but still greater objection to the predigestion of albuminoids by peptonised foods with healthy children, in that their stomachs are capable of dealing efficiently with such material and should not be prevented from exercising their function. This last objection, however, of Dr. Rotch is not applicable to the temporary use of pancreatised or peptonised foods to tide over a bad time in many cases of disordered digestion from various causes, but I can thoroughly endorse his general criticism, having found that in most cases of simple atrophy, and still more with normal infants who for different reasons must be hand-fed from birth, the use of all patent and predigested foods can usually be dispensed with, and that substantially good results are but seldom obtained with them as compared with the simpler methods above indicated.

Of condensed milk I would say, again in accordance with Dr. Rotch and others, that it is often very useful, especially for the children of the poor who are mostly unable to obtain or properly adapt a sufficiency of good fresh milk. It is tolerably uniform, conveniently portable and easily digestible when duly diluted with about twelve to fifteen or more parts of water. But as a food it is largely deficient in fat from its original composition and in solids generally from its necessary dilution, and there is often an undue excess of cane-sugar. The deficiency of fat, however, can be remedied by adding from one-eighth to one-tenth part of cream to the diluted milk. Without this addition some children will apparently thrive for a while, but most will soon fail, and some plainly waste. Healthy infants fed from the first on condensed milk and digesting it, often seem to grow even fatter than suckled ones, but their fat vanishes rapidly under feverish and other morbid conditions. It is further believed by many, on apparently good grounds, that condensed and, still more, desiccated milk-foods are deficient in whatever confers upon fresh milk its well-known antiscorbutic quality. Some evidence of varying degrees of probably scorbutic symptoms in infants thus fed has been given by Dr. Cheadle, and since his first publication on this subject I have from time to time met with infants deprived of fresh milk and fed either on farina or in a few instances on condensed milk alone who suffered from ulcerated gums and wasting, but soon recovered when their diet was duly regulated.

Once more I quote from Dr. Rotch in brief the directions for the preparation of what appears to me one of the best substitutes for mother's milk, the result of the researches and practice of Dr. Meigs and himself. The ingredients in eight parts are *milk* one part, *cream* two parts, *sugar-water* (of the strength of 18 drachms of milk-sugar to one pint of water) three parts, the remaining two parts consisting of a mixture of *plain water* and *lime-water* in the relative proportions of three to one. The mixture should be made (with the exception of the lime-water) as soon as the milk and cream are received, and then boiled or steamed in a sterilising apparatus, the lime-water being added after the mixture, carefully stoppered with cotton wool, has partially cooled. The food should be prepared in one or two quantities for the daily use, and should be kept similarly stoppered in a cool place or on ice. In default of success with the rougher method at first mentioned this plan is to be highly recommended. Another method of adapting milk to weak digestions is the preparation of what is known as artificial human milk. The cream is first skimmed off, and the milk then divided into two equal parts. From one half the casein is removed by rennet, and the resulting whey and the whole of the cream are added to the other half. Dr. Cheadle urges that this should never be kept long for fear of clotting, and should as a rule be made at home.

In dealing thus with the subject of artificial feeding the most important part of the therapeutics of simple atrophy has been covered. Just as it is necessary with a healthy child to decrease gradually the dilution of the milk, so in atrophic cases we must often begin with children of several months old as with a normal infant at birth, with subsequent modifications according to progress. The healthy infant should be

suckled or fed as above described for at least nine months as a rule,
but here the law is not rigid, many thriving well for a year without
any change and others doing better with earlier additions to their diet.
After this age a certain proportion of farinaceous food and meat-juice
should be given, but as my subject is disease I cannot here describe
appropriate diets for advancing childhood. Such details find place in
special works, and notably in the practical book of Dr. Eustace Smith
on the "Wasting Diseases of Children." The great importance of a
free supply of fat in some form in the food of early childhood cannot be
too much insisted on, especially in view of the fact that most young
children object to the fat of meat, which it is useless to force upon them,
and of the great likelihood that the deprivation of fatty food is largely
contributory to the production of rickets. It is probable that the fat
necessary for general tissue-formation must be supplied as such from
without, and hence when ordinary animal fats are refused cream or
butter should be supplied. Cod-liver oil is specially useful to the
children of the poor. Many such show failure in nutrition without any
digestive disturbance, and rapidly improve when cod-liver oil is added
to their diet.

Of the various disorders of digestion from improper feeding which
cause continued wasting and are evidenced by obstinate gastric and
intestinal disturbances I shall subsequently treat more in detail. Suffice
it to say here that in hospital practice especially, as well instanced in my
experience of the poor working population of the East of London, there
are numerous children with no definite disease who either from mere
half-starvation or, what amounts to the same thing or worse, from sharing
their parents' diet, "eating anything," are quite unable to digest milk.
In such cases it is often quite useless to give either medicines or even
artificial digestives with milk, although sometimes from the latter class
of substances considerable help can be derived. Pancreatised milk, for
instance, may be usefully given for a time, and the child thus nourished
while its organs are recovering from previous ill-treatment. But it is
sometimes necessary to give almost absolute rest to the stomach for a
period which, however, must be but short. In such cases we order whey
and water or whey and barley-water in equal parts or some liquid preparation of meat, such as veal-broth or weak beef-tea, with a few drops,
frequently repeated, of brandy. There is no advantage in this temporary
diet other than that of securing gastric rest and slight stimulation to the
organism, and it must not be continued longer than a few days. Infants
who are fed for several days in this manner make no progress in nutrition,
and are very apt to suffer from œdema of the lower extremities owing
in all probability to the anæmia of partial starvation. An attempt therefore at return to a duly directed milk diet, at first perhaps wholly or

partially pancreatised, must be soon made, and made again after each intermission.

In the case of children who remain unable to digest milk in any form or in any quantity adequate for nutrition, as shown by persistent vomiting with or without diarrhœa and by marked wasting, we must substitute for the whole or a part of the milk some other diet containing due proportions of the essential elements. Such a diet may be supplied by varying proportions of cooked farina, such as bread-jelly or some kind of malted flour, of raw meat-juice, and of cream. In the absence of milk fresh meat-juice, containing, as it does, the necessary proteids and salts, is absolutely necessary as an antiscorbutic, for young infants cannot take fresh vegetables. I have frequently found that babies who could not digest milk do very well for a short time on a mixture of raw meat-juice with some form of malted farina and a little sugar, cod-liver oil being given in small quantities to supply the necessary fat. As an excellent formula for the basis of a complete but temporary substitution-diet for milk I would mention Dr. Cheadle's combination of bread-jelly, raw meat-juice, cream and sugar, the proportion of the meat-juice being that suitable to a weakly infant, and to be raised gradually to double as digestive power increases. The bread-jelly is prepared by soaking in cold water for six or eight hours four ounces of stale bread, made preferably of "seconds" flour. After being well squeezed the pulp is boiled in fresh water for an hour and a half, strained and rubbed through a fine hair sieve, and allowed to cool into a jelly. A tablespoonful of the jelly is to be mixed with eight ounces of warm water previously boiled. To five teaspoonfuls of the solution six teaspoonfuls of raw meat-juice, two teaspoonfuls of cream and about half a teaspoonful of white sugar are added. Both the jelly and the raw meat-juice must be prepared twice in the twenty-four hours, the cream taken in night and morning, and the meat-juice must not be added to the food when hot. From two to three ounces of raw meat-juice may thus be given in the twenty-four hours. The meat-juice is prepared by finely mincing the best rump-steak and adding one part of cold water to four of the meat, and the mixture should be stirred and left to soak for half an hour. The juice should then be forcibly expressed by twisting through muslin.

As an intermediate diet between milk and the above about three tablespoonfuls of milk either pancreatised or not may be added to the bread-jelly solution instead of the six teaspoonfuls of meat-juice.

The *constipation* sometimes seen in wasting infants may be often successfully treated by frequent and long-continued kneading of the abdomen with the oiled hand; by small enemata of warm water or soap and water, or containing a drachm or two of castor oil; or with a few drops of castor oil, occasionally repeated, by the mouth. Other drugs, such as senna, aloes, or cascara may be tried. As soon as a proper diet is well taken this medication may almost always be discontinued.

Colic and flatulence as evidenced by the signs of abdominal pain, and also *vomiting* may often disappear with regulated diet. These symptoms are usually much relieved by bicarbonate of soda with syrup of ginger or spirit of chloroform. Bismuth, too, may be given with good effect, peppermint-, caraway-, or cinnamon-water being a useful excipient.

Diarrhœa is to be treated at first in much the same way. Like the other symptoms, unless it be chronic, or marked intestinal catarrh have been set up, or there be any undiscovered and independent disorder, it will often subside with proper diet. While undigested food is passing away in the motions astringents are out of place ; rather should we then give a purge or two ; but when numerous fluid evacuations of whatever nature, with or without much mucus, are once established it is always desirable, in my opinion, to endeavour to control them according to the methods to be referred to under the heading of diarrhœa.

In cases of atrophy, when there is a tendency to diarrhœa, cod-liver oil, otherwise often so useful, should as a general rule not be given.

CHAPTER II.

AFFECTIONS OF THE MOUTH.

Dentition.

The period of eruption of the milk-teeth, which lasts as a rule from about the sixth to the twenty-fourth month, is one of generally rapid organic advance. The glandular and higher nervous organs show at this time an especially active development in strong contrast to their condition in the first few months of life. Such a period is therefore liable to be accompanied by disturbances in predisposed children. Controversy is abundant as to whether any or all of such disturbances are caused by or merely coincident with the process of dentition, but the discussion appears to be but scholastic or at best unpractical. A large number of children, indeed a great majority of healthy sucklings, cut their teeth without sign of trouble, and there are thus no symptoms exclusively consequent on dentition. On the other hand, many infants, especially weakly ones, suffer from marked but irregular pyrexia coincident with the eruption of some or all of their teeth and disappearing with the accomplishment of that process. It is my opinion, based on much

observation and carefully-noted hospital cases, that this pyrexia occasioned
by the local irritation is often the only morbid symptom. The child
may be restless but does not seem very ill, and recovery is rapid and
complete. The temperature may rise to 103° or 104° F. Frequently
there are obvious signs of local irritation in redness and swelling of the
gum, sometimes ulceration, and occasionally aphthous stomatitis. It
seems useless to attempt to classify these cases further, or to discuss how
far slight restlessness and pyrexia during the cutting of a tooth is to be
regarded as literally pathological. We have no reason to ascribe to dentition as a cause any of the more widespread disturbances of the organism
which must frequently coincide with it, for when such affections occur
they are as a rule easily referable to other sources. The hypothesis of
the causation of otitis and of several nervous or other phenomena by
reflex irritation from the dental branch of the fifth nerve is as gratuitous
as it is at first sight plausible, for such phenomena are frequent enough
in infancy without any dental irritation at all. Such simple applications
of the anatomical primer have often in medical literature the credit of
scientific acumen.

The gastric and intestinal derangements, such as vomiting and diarrhœa,
occur mostly in artificially-reared infants, and the convulsive attacks so
often set down to dentition are nearly always attributable to rickets or
some ulterior diathetic condition. All the pulmonary and cutaneous
troubles which are treated by writers as due to dentition, either directly,
or indirectly through a liability to disorder induced by the pyrexia, are
so entirely of the same nature as those which are familiar at this age
apart from the eruption of individual teeth that they may safely be
regarded as not causally connected with teething. The practical lesson
to be learned from the study of disorders which accompany or seem to
result from dentition is—never to make the diagnosis of "teething" until
after a most exhaustive examination and inquiry and never to be satisfied
with it before the child is well. I have said that I believe considerable
pyrexia may result from teething alone, and we may be often right,
when we find evidence of local irritation, in making this diagnosis; but
we must never overlook the possible existence of cerebral, pulmonary,
intestinal, faucial or other disorders, and should carefully watch for their
signs and symptoms.

I have never seen any benefit, local or remote, from lancing the gums,
except in cases where hyperæmic tension was great, and would recommend that pyrexia, gastro-intestinal trouble, and all other ailments which
may be coincident with dental eruptions should always be treated on
their own merits, and that medicine should not be given unless indicated
otherwise than by the supposed cause of "teething." The disorders
which are frequently thus named are temporary and insignificant, and

thus have an appearance of yielding to almost any remedy. No better instance of the inconsequent reasoning of some therapeutists can be found than in the constantly recurring statements in journals of attacks of convulsions, sickness, pyrexia, abdominal pain, or of numerous other symptoms disappearing with marvellous certainty after a few days' use of purges, or other drugs of widely different or negative properties.

While any symptom of illness exists in an infant let us beware of basing our prognosis on the agreeable hypothesis of dental causation, which is a rife cause of bad blundering. I have three times seen cases of meningitis which had been positively pronounced, even after several days' observation, to be merely disorders of teething. It need scarcely be said that a medical reputation is sorely damaged by thus mistaking the early irritable, restless, and whining stage of brain disease.

Stomatitis.

Ulceration of the mucosa of the mouth is of very common occurrence in children, especially in the poorer classes. There are certain appearances besides the usually well-known redness and dryness of the buccal mucosa in young infants which are familiar to those who have experience with children, but are sometimes liable to be mistaken as morbid. Henoch has rightly drawn attention to certain small yellowish-white patches with a red border which are often seen symmetrically placed on either side of the palate just behind the alveolar arch of the jaw, and may readily bleed when touched. They occur in healthy children and are sometimes erroneously called syphilitic. Henoch's explanation that they are due to friction with the tongue in sucking seems to be right. He states that in badly-nourished children they may occasionally and even deeply ulcerate. They usually disappear of themselves, but should they spread they should be locally treated by touching with a solution of zinc sulphate or silver nitrate. The various common kinds of stomatitis may be classed as follows :—

1. **Aphthous.**[1]—This occurs very frequently in infants in the form of small gray or yellowish papules or perhaps vesicles situated anywhere on the buccal mucous membrane, but generally first observed inside the lower lip and at the tip and edges of the tongue. They rupture or become flat, and a superficial small round ulcer appears. In a large number of cases occurring in apparently healthy children these little ulcers give

---

[1] It is well to confine the term *aphthæ* to this form of initially discrete ulceration of the buccal mucosa, in accordance with most authorities here and abroad. Others use the term as synonymous with "thrush." There is no special fitness in the word for denoting either of these different affections, for it was used originally by Hippokrates to mark, from its meaning of "kindling," the fiery inflammation of erysipelas or *sacer ignis*.

no trouble other than a local soreness, and rapidly disappear without treatment. They are in all probability due to a germ from decomposing food, though there is no positive evidence on this point. Occasionally several members of one family may be affected. Not seldom aphthous stomatitis runs a more severe course, especially in weakly children who are suffering from alimentary trouble and appear to afford a special nidus for the disorder. The ulcers may spread and become confluent, covering almost the whole buccal cavity, and there may be swelling of the sub-maxillary glands and of the face with considerable rise of temperature. The breath is foul and the saliva profuse. In these cases it is plain that the general condition must attract our main attention, for this underlies the severity of the buccal ulceration. But the local discomfort is great, and the pain felt on touching the ulcers, especially on the tongue, is often so severe as to interfere much with feeding. This should be *treated* by the application of salol or borax and glycerine, and by chlorate of potash in small doses given internally for a short time. This latter drug should not, in my opinion, be omitted, comparatively slight though its modern credit may be. The local use of it as a mouth wash is not equally effective. In extensive ulceration the application of nitrate of silver may be found necessary.

The practical import of this affection may be gathered from the general condition of the child. It is insignificant *per se*, but when obstinate and spreading in a case of severe illness it is one more indication of the gravity of the case.

2. Ulcerative Stomatitis or Stomacace.—This is a very common form of mouth-disease, especially in hospital patients. It begins as a rule in the gums, which are at first red, swollen, and spongy, bleeding readily, and soon afterwards take on a yellowish-grey ulcerated surface. This ulceration may be temporary, or may spread and endure for weeks or more. The breath is foul, the saliva abundant and often bloody, and there is pain with all movements of the mouth. The teeth may become loose and fall out, and the ulceration frequently spreads to the lips and cheek and to the tongue where it is in contact with the gums or lining of the mouth. The ulcerated patches are of various size, and sometimes very extensive. The glands beneath the jaw are tender and swollen, and there is more or less pyrexia.

This disorder, whether slight or severe, is most common in children over two years old, and is probably the result of bad nutrition or a bad constitution, or both combined. It is seen in rickets and tuberculosis, has been observed in connection with congenital heart disease, and is doubtless due in some cases to a scorbutic condition. There is, however, as far as I have observed, no essential difference between the most severe cases of ulcerative stomatitis which are certainly not scorbutic, as far as their history can tell us, and those far less frequent ones whose diet might occasion scurvy. That scurvy, however, may account for some cases of this disease in children is not to be denied; for a history of the absence of milk, fresh meat and vegetable food from the dietary is sometimes clearly to be obtained, and more or less marked instances are sometimes seen after a diet of desiccated or even condensed milk, or when antiscorbutic food has been very deficient in quantity. Several micro-organisms, some of which have been stated to be inoculable, have been found in some cases of ulcerative stomatitis, and these may possibly be the proximate cause of the affection. Many severe examples of this malady are seen from time to time in epidemics of measles, the ulcers being frequently covered with a membranous exudation the removal of which leaves a bleeding surface. The question of the alternative diagnosis of diphtheria may thus be raised, especially when there is concomitant faucial inflammation with exudation. Several cases of this kind occurred in the East of London in the autumn of 1890, many showing also severe broncho-pneumonia and laryngitis, but none being followed by the sequelæ of diphtheria.

Dr. J. F. Payne[1] has made the observation that a pustular eruption on the lips and hands often accompanies ulcerative stomatitis, and most aptly suggests that both eruptions arise from the same virus, the different lesions being due to their different locale.

The disease is certainly very rare in children, however weakly, who are properly fed, cleaned, and cared for. I have nevertheless occasionally seen ulceration and sloughing neither deep nor lasting of the mucosa of the gums and cheek in apparently healthy and cleanly children, where no cause could be assigned. It is by no means rare to see such ulceration as the direct effect of a jagged tooth. In this context may be mentioned the ulcer near the frænum linguæ which is observed in some cases of whooping-cough from the friction of the tongue upon the lower incisors. Henoch attributes ulcerative stomatitis in some cases to the processes of the second dentition, but both the proof and disproof of this theory seem equally difficult to establish.

The discovery of diphtheria as evidenced by pharyngeal affection is sometimes preceded by that of a stomatitis which is not easily if at all distinguishable from the simple disorder. It may be questioned whether such stomatitis be diphtheritic in nature, or merely a favourable soil for the specific poison.

Chlorate of potash or soda, taken internally, appears generally to control this kind of stomatitis, unless the bone be involved, and the application of salol in glycerine, of the strength of one drachm to the ounce, to the parts affected by means of a brush is very effective, recovery

[1] St. Thomas' Hospital Reports, 1883.

often taking place with great rapidity. It may be necessary in the severer cases to apply some astringent, such as alum, or a solution of sulphate of zinc or sulphate of copper (each 10–12 grains to the ounce), or nitrate of silver (2–5 grains to the ounce), to the ulcerated surface, but in many of the milder instances careful cleansing of the mouth with a weak solution of permanganate of potash or some other antiseptic will be all that is required. In all the numerous cases where there is cachexia of any kind general hygienic and nutritive measures are indispensable, the diet must be specially attended to when scurvy is in question, and any co-existent disease must be expressly treated. Bone-affection must be dealt with by surgical measures.

3. Gangrenous Stomatitis (*Noma, Cancrum Oris*). — The essential characters of this disease, which has only clinical rank as a separate affection, are that the ulceration rapidly becomes gangrenous, invades the deeper tissues of the cheek or lips, perforating and destroying them in their whole thickness, and mostly causes death. Clinically, the first usually observed symptom is swelling of the face, most often of the cheek, but it may involve the lips and chin and the lower eyelid. The most marked part of the swelling is found by careful feeling to be hard, but is generally neither tender, nor, at first, red. If the inside of the mouth be examined a gangrenous ulcer is seen, usually opposite the hard part of the swelling, and the breath is almost always offensively foul. There is much salivation and swelling of the neighbouring glands and of the surrounding connective tissue. A certain degree of pyrexia always exists, sometimes very slight, sometimes high ; and although some cases die very early from exhaustion before perforation occurs, others linger on for two or three weeks. Some seem to suffer no pain, and even appear lively long after extensive destruction of the face, including the bones, has taken place. Before perforation the swelling becomes hard and centrally red and shiny, and then turns black. Occasionally the whole process, from the first observed symptom to perforation, is but of a few days' duration. In some fatal cases there is no external lesion other than swelling, and the patient dies from pulmonary inflammation or gangrene owing to septic inhalation from the sloughing mucosa of the mouth. In one case of mine death occurred thus after the buccal ulcer had ceased to spread and had shown signs of the healing process.

Diarrhœa has been frequently observed, attributable probably to the swallowing of putrid matter from the ulcer, and the *broncho-pneumonia* which occurs, and is fatal in so many instances, may be referred to inhalation of the same. Gangrene of the lung is frequently found post-mortem. The few cases which recover suffer deformity from contraction of the face, adhesion of the cheek to the jaw, or turning out of the lower eyelid.

This disease has not been proved to result from the action of a specific germ, although cases have been reported where micro-organisms were found in the blood or tissues. It occurs most often in cachectic children past infancy, after the exanthemata, especially measles and less commonly enteric fever, and sometimes after affections of the lungs or bowels. Tubercular children are said to be liable to it. Henoch quotes two cases where noma seemed to arise out of an ordinary ulcerative stomatitis. I have myself seen one probable example of this, and Eustace Smith relates the cases of a cachectic brother and sister suffering respectively from these diseases. Henoch further gives one case, which recovered, of gangrenous perforation of the cheek, which began as an abscess without involvement of the buccal mucosa. It would thus appear that this disease has no ætiological claim at present to separate nosological rank.

I add an abstract of a case of my own, illustrating this point among others. A boy of three years old began to have "ulcerated mouth" in form of "small blisters" six weeks before admission to hospital. Five weeks before admission he had measles, and the mouth became much worse. He was found to have marked bony signs of rickets on admission. The mucous membrane inside the left angle of the mouth was gangrenous, and the alveolar ridge of the lower jaw was carious. Eight teeth had recently dropped out. The mucous membrane on the inner surface of the lower jaw and under the tongue was also gangrenous, and there was disgusting fœtor. The glands under the jaw were much enlarged and the tissues around infiltrated, causing the appearance of a huge double chin. The child could swallow well. There were no signs of broncho-pneumonia. The diseased surfaces were touched with strong nitric acid, and chlorate of potash and iron given. Eight days after admission, the child growing weaker and the temperature ranging from 99° to 102°, bronchitic signs were heard, and on the ninth day there were signs of consolidation. There was a slough as large as a crown piece on the external and lower surface of the left lower jaw. The child died on the tenth day with a temperature of 106°.

To prevent, if it be ever possible to prevent by art, the usually fatal issue, the parts involved should be actively *treated* by cautery or exci-sion as soon as observed. Literature records but few recoveries, and fewer, if any, cures. My own few cases, but five or six in all, have been fatal, although locally treated, but none were seen in the earliest stage. Before the ulceration has involved the deeper tissues some hope may be entertained of arresting it. Ordinary caustics are probably useless. Henoch recommends the thermo-cautery.

In the following case of a boy, æt. 3½, admitted with enteric fever, local treatment of the disease by strong nitric acid seems to have had some good effect, although the child died from inflammation of the lungs.

On admission there was increased salivary secretion, the gums bled easily, and the breath smelt very foul. Ten days subsequently, rose-spots still existing, the left side of the face was found to be swollen and red, and the breath was exceedingly fœtid. A black slough the size of a shilling was seen under the cheek. The child did not seem in any way worse, and the temperature was 100°. Ether was administered, the slough scraped away with a sharp spoon, two upper molar teeth with a necrosed piece of jaw were removed, and the healthy mucous membrane being covered with oil the raw surface was cauterised with fuming nitric acid. On the following day the fœtor had disappeared, and there was no sign of the disease spreading. In the evening pulmonary symptoms came on, with physical signs of bronchitis only, and the child died at midnight. Post-mortem there was no sign of spreading disease in the mouth. There was early broncho-pneumonia with incomplete but widely-dispersed consolidation of both lungs.

The parts should be kept throughout as clean as possible by syringing with washes of permanganate of potash or chlorinated soda. Every effort should be made, in spite of the great local difficulties, to give concentrated nutriment, such as the most approved preparations of meat-juice, at frequent intervals, and the patient should be freely stimulated with alcohol. It may be necessary to feed the patient through the nasal tube, or by enemata which, owing to the absorption of their watery vehicle, are far preferable to suppositories. Opium should be given when there is pain, and I would recommend it in all early cases in view of the possible action of its healing power.

For the sake of the attendants the mouth of the patient should be covered with cotton-wool or a compress soaked in solution of thymol or some other disinfectant, and a constant spray of carbolic acid or creasote should be maintained close to the bed. The child should be kept as much as possible in the prone position, to lessen the swallowing and inhalation of the putrid discharges.

My colleague, Mr. L. A. Dunn, recommends, after a preliminary tracheotomy, free cauterisation of the diseased surface, and, plugging the mouth antiseptically, would treat it as a closed cavity, all food being given by the nasal tube.

4. Thrush is the name given to an affection of the mucous membrane, especially of the mouth, characterised by numerous milk-white specks and patches of various sizes. The mucosa of the mouth is dry and red and sometimes painful, and the white patches have a tendency to run together, especially over the palatal arch. This disease is due to the presence of a fungus which, formerly classed as an Oïdium and allied with the fungus of favus, is now referred to the group of yeast fungi. The growth, according to Plaut, is characterised by abundant mycelium, produces fermentation, and when purely cultivated causes distinct thrush in the crop of fowls.

Thrush in its simplest form attacks young infants, and mostly, if not entirely, those who are weakly or are artificially fed. It is not necessarily accompanied by any inflammation. It may be conveyed by means of infected articles, such as bottle-tubes, but probably not to a perfectly healthy mucous membrane. In the milder cases a regulated diet and careful removal of the patches by wiping out the mouth with some such soft substance as lint will quickly cure the disease, with or without the use of the glycerine of borax to the parts involved. In many instances, however, the affection is much more severe, the fungus growing rapidly, occupying the whole mouth and pharynx, and causing superficial ulceration which is seen on the forcible removal of the white substance. For this development there must probably be antecedent disorder of the mucous membrane and an acid state of the secretions, and there is often evidence of concomitant derangement of the whole alimentary canal. Diarrhœa may be marked and fatal. It is in cases of wasting from bad feeding and disease that this severe form is usually or always met with. In adults thrush occurs almost exclusively with extreme exhaustion from disease, such as typhoid fever or phthisis, and generally warrants the gravest prognosis. The fungus is found post-mortem in the œsophagus and stomach, and occasionally in the intestine, but not in the ciliated epithelial regions of the nose or larynx. Very frequently in the graver cases there is a red eruption from a superficial dermatitis spreading from the anus to the nates, due to the irritation of acrid discharges from the bowel. This is often seen, among other diseases, in connection with congenital syphilis. It is to the antecedent disorder which favours the luxuriant vegetation of the fungus that the serious symptoms must be ascribed, and the extent of the thrush itself must be regarded rather in the light of an index to the gravity of such disorder. Even in the mildest cases some decomposition of the food in the mouth is to be assumed as forming the nidus for the fungous growth.

The remains of milk are sometimes mistaken for thrush in the mouth, having much the same appearance, but they are readily removable when touched, and microscopical examination will decide the diagnosis. Occasionally also a patch of desquamation of the epithelium of the tongue may be confused with thrush on superficial examination. The *treatment* of thrush in mild cases consists in removing the patches as above mentioned ; in the more severe cases the general condition calls first for attention, but the local trouble should be relieved by washing or syringing the mouth with weak alkaline solutions, such as carbonate of potash or soda, or by the use of glycerine of borax. In some cases the application of a solution of nitrate of silver (1–2 grains to the ounce) to the mucous membrane of the whole mouth will prevent further development of the affection. In any case, with especial regard to the co-existent stomatitis, the chlorate of potash may be given internally, or a few grains of borax may be advantageously taken three or four times a day. Strict cleanliness must be observed in the feeding-apparatus of patients with thrush, and starchy and saccharine articles of diet must be given sparingly or not at all. Alcoholic stimulation will be sometimes necessary.

CHAPTER III.

AFFECTIONS OF THE FAUCES.

Acute Catarrh of the Pharynx.

VERY young infants are but little liable to catarrhal inflammation of the fauces, but sore throat from such a cause is frequent after this period. Some attacks begin with feverishness and evidence of soreness on swallowing, and may be soon followed by nasal catarrh. The pharynx is seen to be red, the tonsils and faucial pillars slightly swollen, and the uvula elongated. In other cases the temperature is high from the first, there may be slight rigors or even convulsions or vomiting, and the child may appear very ill with furred tongue, anorexia and headache. Pain in the throat may not be prominent although tenderness may always be demonstrated by pressure in the direction of the tonsils, and inspection will show a pharyngitis with more or less tonsillar swelling. Such an affection, with or without pain, often ushers in the exanthemata, especially scarlatina and sometimes enteric fever; and this association should always be borne in mind when we are making our diagnosis.

I would lay stress on the absence of complaint or evidence of discomfort about the throat in many of these cases in childhood, for omission from this cause to examine the fauces frequently mystifies a case of febrile nature which would be otherwise clear. Both these sets of cases are apparently similar in origin, the difference of symptoms depending probably on the child's idiosyncrasy, and the affection usually passes off in a few days. The tonsillar swelling may be more or less prominent than the general inflammatory condition of the fauces. Some at least of these cases seem to be due to chill, but the clinical group is probably of no single ætiological origin.

Such cases as these require but little or no active treatment. Warmth and confinement to bed, with a saline mixture containing chlorate of potash, are the best remedies; and if there be much pain or restlessness a few grains of Dover's powder, according to age, may well be given. The temperature may run high, as in many other slight disorders of childhood, but does not require any antidotal treatment. I strongly deprecate the meddling administrations of the now fashionable "antipyretics," such as antipyrin and aconite, not without past experience of them. They are never indicated, nor, when operative, always harmless.

Acute Tonsillitis.

Under this heading we have again to recognise a somewhat heterogeneous class of cases which, owing to the prominence of tonsillar inflammation from the first and the absence of general catarrhal symptoms, deserve to be considered apart for practical purposes, although ætiologically multiform. Many cases are marked by a tendency to recur and by the frequent presence of patches of whitish or yellowish inspissated exudations on the swollen tonsil. Pain and dysphagia may be great, but on the other hand are often insignificant, and the tonsillitis is sometimes recognised only by inspection during the routine inquiry as to the cause of any given case of feverishness, or, it may be, of but slight malaise. We see tonsillitis with all degrees of local severity and constitutional disturbance, and have often to remark on the apparent disproportion between acuteness of general symptoms and slight local discomfort, although severe tonsillitis is always attended by marked febrile movement. Probably the affection is referable to several causes. Unhygienic surroundings, such as bad drainage or foul air, are often supposed to be probable sources, and sometimes with apparent reason, but the affection may arise suddenly in a seemingly healthy child in presumably good sanitary conditions. As in the adult, so perhaps still more in the child, there is a frequent connection between tonsillitis and *rheumatism*. Tonsillitis, indeed, besides being a frequent accompaniment of rheumatic fever, may sometimes be the first indication of rheumatism evidenced afterwards by less equivocal signs of that disease. In some instances, and especially in older children, the inflammation rapidly goes on to suppuration, and the abscess in one or sometimes in both tonsils may burst, the case ending then in rapid recovery. Exactly the same kind of tonsillitis, whether exudative or suppurative, is seen in adults, and there is nothing special in respect to its clinical course in childhood. The frequent difficulty of diagnosis, however, between exudative tonsillitis and other diseases, especially incident on childhood, in which tonsillitis plays a prominent part, renders this subject of great importance. In practice we often have to decide whether a given case of tonsillitis is non-contagious or a symptom of diphtheria or scarlatina, and all experienced practitioners know that there are certain cases where this diagnostic question, even after the most careful observation and inquiry, is at first insoluble. The more or less essential signs of scarlatinous and diphtheritic throats will be dealt with in their proper place; but it must be noted here that the appearance of the throat and tongue in many cases of scarlatina is but slightly if at all characteristic, and that when the bright redness of the scarlatinous pharynx is absent the swelling or patches of ulceration of the tonsils may be almost or quite indistinguishable from that of the independent affection we are now considering. We must here remember how frequently most acute diseases lack their typical signs, and should therefore hesitate for a while before positively excluding scarlatina in any case of suddenly occurring tonsillitis, especially when there has been vomiting. The early appearance of the rash in scarlatina will as a rule soon remove doubt, although in some undoubted cases the scarlatinous sore throat is followed by no observed rash. With regard to diphtheria the difficulty is greater, of longer duration, and, apart from the possible discovery of the probably specific bacillus, sometimes insuperable. There are cases proved diphtheritic by their sequel or by their association with marked instances of the disease which show but little or no membranous appearance, no nasal symptoms, and very slight, if any, constitutional disturbance; while in non-diphtheritic tonsillitis the exudation frequently covers the tonsils as with a membrane. In most doubtful instances we should wait for two or three days before pronouncing on the nature of the case, by which time the exudation, if it be not diphtheritic, will usually be markedly diminished. Having erred and seen many others of long experience err in the too hasty diagnosis of such cases I would emphasise strongly the danger of overmuch confidence in typical descriptions or in one's own clinical acumen. It were better to be confronted with our mistaken caution in a case where the sequel, or several subsequent attacks, may disprove diphtheria, than to see or hear of a child's death from heart-failure or other untoward event after we have pronounced the diagnosis of simple tonsillitis. The existence or recent antecedence of scarlatina or diphtheria in the house or close neighbourhood of a case in question may give us some diagnostic aid, as also might the very rare possibility of excluding all likely sources of such infections.

The usual characteristics of this form of tonsillitis are, however, enough in most instances to prevent much diagnostic difficulty, at least after a day or two. There are frequently well-marked ulcers at the onset quite unlike diphtheria. The white or yellow or yellowish-white exudation, differing from the greyish membrane seen after a few days in most diphtheritic cases, occurs at first in small patches which neither coalesce nor rapidly spread, and leaves the soft palate, uvula and back of the pharynx unaffected. It usually occupies, moreover, one tonsil only, or one before the other.

Glandular enlargement under the lower jaw does not help us much in diagnosis, for it frequently occurs in tonsillitis while it may be but slight in diphtheria, and we must remember that the typical "ashen-grey" appearance of diphtheritic membrane is not generally seen at the early time when the diagnostic question is at once most difficult and most important. The constitutional symptoms and fever, sometimes very high, are generally prominent in tonsillitis, and there may be rigors; while in diphtheria there may be but slight or no marked symptoms of illness, and often very little fever. The urine in tonsillitis is usually scanty and high-coloured, and not seldom slightly albuminous.

If we are satisfied as to the non-contagious nature of a given case of tonsillitis, or at least that it is neither scarlatina nor diphtheria, as after due experience and careful observation we can perhaps often be, the **prognosis** may be said to be nearly always good. There is certainly good reason to believe from many recorded instances that there is an epidemic form of tonsillitis which is neither scarlet fever nor diphtheria, but the greatest caution is necessary before allowing this probability to influence our diagnosis or prognosis.

In the most acute cases the temperature may rise as high as 105° or more, and sometimes after five or six days there is a critical fall. The affection very often recurs without any apparent exciting cause. Doubtless in some cases this recurrence causes chronic hypertrophy of the tonsils, but on the other hand, as we shall presently see, the origin of chronic enlargement of the tonsils is often very early and insidious, and this condition induces a great liability to acute attacks. In those cases of non-suppurative tonsillitis where patches of whitish or yellowish exudation are seen the question of insanitary conditions as a possible cause must be thought of, and the patients, if necessary, removed, not only for their immediate good, but also in view of the undoubted fact that any continuing morbid condition of the fauces is favourable soil for the reception of the diphtheritic poison.

The **treatment** of tonsillitis is mainly symptomatic, but there may be some causal indications. For the local trouble, hot poultices or compresses frequently renewed should be applied to the throat, and steam should be inhaled either with an ordinary apparatus, or, in the case of young children, by means of a steam-kettle at the foot of a tent-bed. In some very acute cases the use of ice both internally and externally is advantageous. In suppurative tonsillitis with severe pain the abscess should be punctured. Opium, either as laudanum or Dover's powder, should be always given when there is much pain, and is often very useful when there is none. A saline mixture with antimonial wine and chlorate of potash seems to be frequently beneficial. Salicylate of soda is said by many to be of value, but, as the rapidity of the improve-

ment which follows its use is never comparable to the evident action
of this drug in acute rheumatism, there is no proof of its specific effect
where an affection tends to spontaneous recovery in a few days. I
am quite in accord with many authorities in recommending an initial
aperient in this as in many other febrile states where there is often constipation at the outset and where at least a feeling of relief frequently
follows on purgation. The diet should be liquid and as concentrated
as possible. Alcohol is unnecessary unless indicated by the general
conditions.

Chronic Enlargement of the Tonsils.

This affection is common in children beyond infancy, especially after
the age of five or six years. It may sometimes result from repeated
attacks of acute tonsillitis, but in my opinion much more often begins
gradually apart from any acute attack and is rarely evidenced by local
pain. It is by far most frequent in children of the so-called "scrofulous" constitution as indicated by tendency to glandular enlargement and
ready inflammatory reaction to injuries of skin, mucous membranes,
joints, and other structures, but is not limited to such cases. The
symptoms are snoring in sleep and sometimes audible breathing in
waking hours ; an altered quality of voice commonly called "speaking
through the nose," and often a considerable degree of deafness referable
to obstruction or occlusion of the Eustachian orifices. There is frequently
concurrent nasal catarrh, and sometimes the tonsils are so large that
they meet with the uvula and almost block up the pharynx. A further
symptom is a great liability to acute attacks of tonsillitis which then are
very distressing. Swallowing is not always affected, and is often performed with ease in these chronic cases. I believe with Henoch that the
somewhat frequent nervous symptoms in these children, such as night-
mare and sudden startings on being awakened from sleep, are connected
with obstruction to breathing, and not referable to catarrh of the alimentary canal, for in many there is no sign of gastric trouble, and most if
not all symptoms often disappear after a complete excision of the tonsils.
Adenoid growths in the pharynx, especially in the posterior nasal tract,
again to be referred to in connection with respiratory diseases, often
concur with enlarged tonsils. It is this complication especially which
leads to the great obstruction to breathing sometimes observed, with
marked retraction of the lower end of the sternum causing somewhat
of the appearance of pigeon-breast. Cough is frequent in these cases,
and there is especially loud and sonorous expiration. With the imperfect
aeration of the blood arising from this condition there is sometimes
considerable impairment of body-growth. According to Dr. Colcott

Fox, psoriasis and chronic tonsillitis, with or without adenoid growths, are often associated in childhood.

The **treatment** of chronic tonsillitis consists in endeavouring first to improve the child's general condition which is often at fault, and then to diminish or prevent the local trouble. Often much is gained by giving cod-liver oil, iron, arsenic or strychnine, the tonsils becoming sometimes gradually smaller, or more often, with improved general nutrition, ceasing to enlarge or to suffer recurrent inflammation. But in all marked cases I unhesitatingly recommend excision of the tonsils which, when carefully performed by a competent surgeon, always markedly relieves and very often works a complete cure. The best that can be said of the protracted methods of cauterisation so much in vogue is that they are scarcely so successful as even the frequently incomplete removal of the glands which has brought the radical operation into the undeserved disrepute so much insisted on by some specialist adherents of the slower methods of treatment. There are few better instances of the invaluable aid of surgery to medicine than the many cases of chronic suffering which may be rescued from chronic and expensive maltreatment by this single operation, involving neither previous "training" of the patient nor any considerable after-care. I am convinced further that there is no objection to the operation even when unnecessarily performed.

Retro-pharyngeal Abscess.

It is of great importance to remember and search for this disease in all cases of throat trouble in young children, especially in those under two years old. I do not doubt that many cases run their favourable or it may be fatal course undetected, not only from forgetfulness to explore the pharynx with the finger, but also from the frequent situation of the abscess either in the naso-pharynx or behind the larynx. In many instances the first noticeable symptom is dyspnœa, when the abscess is low down behind the larynx, and is then apt to be erroneously referred to primary laryngeal mischief, especially if there be any cough or hoarseness. If on the other hand the abscess be in the naso-pharynx, there is obstruction to breathing through the nose, which, when accompanied, as it frequently is, by a thick nasal discharge and glandular and areolar swelling in the neck, may closely simulate at first sight either scarlatinal or diphtheritic disease. The simplest cases both for diagnosis and treatment are those where the abscess is in the intermediate part of the pharynx within easy reach of the finger and gives rise to some dysphagia as the first and most prominent symptom.

The affection consists in inflammation leading to abscess in the submucous tissue of the pharynx in front of the spine, and is by far most often seen in the first few years of life.   The origin of most cases is very doubtful.   My own experience is in accord with that of Professor Henoch, who, quoting from the large number of sixty-five cases observed by himself, very strongly contradicts the prevalent belief that either spinal caries or the scrofulous condition is a common cause, and shows that the majority of sufferers are previously healthy.   A certain number have their origin undoubtedly in measles or in scarlatina, instances of both of which I have seen.   In the unexplained majority undiscovered traumatism from swallowing pieces of crust, bone, or other hard substances may possibly play a greater part than can ever be demonstrated.

If a retro-pharyngeal abscess be undiscovered and therefore untreated it frequently bursts with the disappearance of all symptoms, but not seldom the pus burrows in various directions.   It may cause suffocation by pressure on the larynx or by discharging into it, may rupture into the auditory meatus, or may largely invade the areolar tissue of the neck, giving rise to great swelling and sometimes pointing externally. Occasionally the pus extends to the mediastinum and may open into the œsophagus or the pleura.   Many cases come first under observation with much cervical swelling on one or both sides according to the situation of the abscess.   This should always excite a suspicion of the probable cause when accompanied by dysphagia or dyspnœa.   In perhaps a majority of cases the abscess is not literally retro-pharyngeal in situation, but forms on one side of the pharynx, causing often great unilateral swelling.

There are rarely any notable symptoms of onset, but there may be some fever with or without slight rigors.   Some pain and tenderness are probably constant, but seldom early detected, and there may be stiffness of the neck with the head kept back or on one side.   When the abscess is low down the dyspnœa and stridor from pressure on the larynx may be great, although dysphagia may not be very prominent.   The stridor, however, unlike that of most laryngeal mischief except extensive membranous blocking, accompanies expiration as well as inspiration, and although both cough and hoarseness of voice or cry may be present they are very often absent, and never approach in intensity to the similar symptoms of primary laryngeal disease.   Swelling and especially fluctuation in the neck in cases of this kind are an additional help to the right diagnosis.   Mr. Scott Battams informs me that he has been twice able by observation of these points to relieve completely and at once by external incision cases of extremely urgent dyspnœa to which he had been summoned for the performance of tracheotomy.

In cases where the abscess is high up in the naso-pharynx and there is impeded nasal breathing with perhaps nasal discharge, but no laryngeal difficulty, scarlatinous and diphtheritic affection of the fauces may usually be excluded by the absence of their proper signs on inspection of the pharynx, and, if we still suspect diphtheria limited in its local expression to the naso-pharynx, the presence of much external swelling will go very far to exclude it and to corroborate the diagnosis of retro-pharyngeal abscess. A careful examination with the finger passed up into the naso-pharynx behind the soft palate will usually detect the swelling and thickening caused by an abscess, but in one case of my own arising after measles in which, from much cervical swelling, impeded nasal breathing and absence of faucial or laryngeal signs, I thought of abscess, nothing was detected by repeated digital examination. The child soon died, apparently with septicæmia, a dusky rash being observed the day before death, and an unburst abscess was found very high up in the naso-pharynx with great inflammatory thickening of the tissues. Although the cervical swelling in this case, mainly on the right side, was not fluctuant I greatly regret that I did not at once order free incision.

It should be remembered that a small abscess at the back of the visible pharynx which but slowly increases may be evidenced by nothing else for some time than by the infant's crying while sucking or otherwise feeding, or by its taking very little at a time. Something appears to ail the child who may waste somewhat and become flabby. I have seen a case in point where an infant without any noted sign of local pain had been treated dietetically for about ten days with all perseverance but no success. A sudden suspicion that there might be some oral or pharyngeal trouble led me to inspect the mouth and explore the fauces with the finger, and thus to discover at once a small retro-pharyngeal abscess which was quickly relieved by puncture.

In all cases, therefore, of doubtful trouble in the throat of young children a thorough exploration of the pharynx should be made. If an abscess be found it should be opened at once with the proper surgical precautions of guarded bistoury and medianly-directed incision, and the child's head should be quickly bent forwards after this operation. Aspiration of the abscess, recommended by some, is open to the objection of possibly necessary repetition. Much external swelling when the internal abscess is out of sight and reach, and especially fluctuant swelling, indicates external opening which is often followed by immediate relief and rapid subsidence of symptoms.

CHAPTER IV.

GASTRO-INTESTINAL DISORDERS.

In the heading of this chapter, which will be mainly concerned with the common and prominent troubles of vomiting and diarrhœa in infancy for the most part referable to gastric dyspepsia, I have for practical purposes departed from the systematic or anatomical classification of disorders of the alimentary tract according to which the affections of the stomach and of the intestines would be separately treated from the point of view of structural disease.

Although such morbid conditions of this tract as catarrhal inflammation and even some degree of ulceration are more or less frequently associated with the affections to be discussed and may play in turn a symptomatic part of their own, yet they are very often absent, as evidenced by post-mortem examination in even very chronic cases, and must mainly be regarded when present as the result rather than the source of the disorders they accompany. That a catarrhal condition of the stomach or intestines may be sometimes inferred with much probability from the nature of the vomit or fæces, and that in certain cases ulceration of the bowel may be strongly indicated by the passage of blood from the anus is not to be questioned, but I would emphatically teach that in the gastro-intestinal disorders of infants which especially concern us now the less we picture " catarrh " as a substantive condition in our ætiological diagnosis of the cases before us, and the more we think of the functions of digestion and absorption and the subtler changes underlying their disorder, the better our hygienic, dietetic and medicinal treatment will be.

First considering, therefore, the general subject of vomiting and diarrhœa, with their frequently associated symptoms of abdominal distress or general disturbance, as pointing chiefly to gastric dyspepsia, I shall subsequently deal with the different morbid conditions of the stomach and intestines which, whether strictly primary or not, may be indicated by special symptoms or found post-mortem.

Vomiting.

The import of vomiting as an indication of chronic gastric dyspepsia depends mainly on consideration of the condition of its subjects, the nature of the food taken, and of the clinical association of other symptoms, especially diarrhœa and wasting. I have already spoken generally of this symptom of stomach trouble from dietetic causes and of its treatment in connection with the subject of infantile atrophy, and must further allude to it under the head of diarrhœa with which in its most serious forms it is far most frequently associated. It is therefore sufficient here to treat of vomiting mainly from the diagnostic point of view in its various clinical relationships.

In young infants otherwise healthy vomiting is frequently seen as the result of overfilling the stomach, the small size of which is often forgotten in feeding them either from the breast or the bottle. Such regurgitation of food is as a rule not excessive, and is immediately sequent upon sucking. In the absence of failure of nutrition or any other sign of ill-health there is no cause for anxiety even although this vomiting be often recurrent, but smaller and sometimes perhaps more frequent meals are certainly indicated. When, however, vomiting is persistent and copious, and still more when it does not follow immediately on feeding, it should at once be regarded as morbid and indicative of dyspepsia with its possible train of accompaniments and sequelæ. Before therefore any further symptoms arise the general condition of the child and the quality of its diet must be carefully considered. Sudden acute vomiting in a healthy child, accompanied by eructations and by evidence of abdominal pain or discomfort, often points at once to undigested food as the cause, but it must be remembered that although usually some diarrhœal and other symptoms soon follow on continued dyspeptic vomiting yet chronic vomiting may often lead to rapid and serious wasting without other definite symptoms, at least for some time, and even with constipation. In such cases the meals are ejected before any considerable absorption can take place; thirst is excessive, and the infant will perhaps take greedily any food that is offered it; the tongue may be furred or later become red and dry; and thrush may appear in the mouth. Death may result from pure exhaustion, or may be immediately due to pulmonary collapse or inflammation or to sinus-thrombosis with coma or convulsions.

In such cases as these unaccompanied by diarrhœa there is very often, and in my experience mostly, some further cause than improper feeding which, however, doubtless often plays a very large part. In some instances neither careful feeding nor any medicine gives material help, and we usually find that just in proportion as the food is apparently appropriate there are indications in the child of previous ill-health or evidence of anæmia or of some other general malady. Not to mention several cases of this kind in children who were the subjects of scrofula and therefore perhaps of gastric catarrh as part of the catarrhal disorder so frequent in that affection, I have seen many instances of obstinate vomiting and fatal wasting in infants who had previously suffered from syphilitic symptoms, and have indeed been often led to a suspicion of syphilis, confirmed by subsequent inquiry, from the mere existence of vomiting and wasting recalcitrant to all treatment. To syphilitic wasting, however, often without either diarrhœa or vomiting, I shall allude further in its proper place.   Rickets and the conditions of low vitality from defective hygiene so generally incident on poverty must also be reckoned with, besides strictly dietetic improprieties, as important factors in the production of dyspepsia with vomiting and deficient absorption.

In many cases of chronic vomiting gastric catarrh, as the result of irritant or undigested food, probably contributes to persistence of the symptom, and is evidenced by the ejection of much mucus.   Constipation and flatulence are here often observed, but usually there is diarrhœa.   I must, however, repeat that it is the exception to find any post-mortem evidence of affection of the gastric mucosa, even in the most chronic cases of vomiting with or without diarrhœa.

Before attributing vomiting, whether often repeated or not, to the above-mentioned group of causes we must exclude as far as possible other affections of which it may be symptomatic.   It may be the earliest or the most prominent mark of *cerebral disease*, such as tumour or meningitis, but it is then sooner or later accompanied by headache, cardiac irregularity, or some of the many other signs indicative of its true cause, and is usually unattended by other ordinary signs of gastric disorder such as furred tongue, gaseous eructation, abdominal pain or diarrhœa.   It is often however directly occasioned in cerebral cases by the taking of food.   Vomiting is also very often the first observed symptom of *acute febrile diseases*, among which the exanthemata—especially scarlet fever,—pneumonia and tonsillitis are prominent, and it is frequent in pertussis as well as in many cases of pulmonary disorder with cough.   The accompaniment of fever and the sequence of some of the characteristic symptoms and signs of the above-mentioned affections will usually clear up doubt.   Febrile attacks however with gastric symptoms in children mostly beyond infancy not infrequently occur, and may be mentioned here although they are often probably not of gastric origin.   The onset may be sudden with vomiting, high temperature, and even some delirium.   In some cases there is diarrhœa, in others constipation, and there may be cough or hurried breathing with or without harsh respiratory or dry crepitant sounds.   These attacks are usually of but a day or two's duration, but are apt to recur with various intervals free from gastric symptoms, and are often unassociated with any discoverable dietetic error.   A nervous origin has been suggested for these symptoms in some cases with much apparent reason owing to their sudden appearance after excitement, to their frequently sudden subsidence, and to the tendency of their subjects to suffer from other nervous disorders. In some instances these attacks usually, be it noted, interpreted as arising from "the stomach," or in medical language from "acute gastric catarrh," are proved by the sequel or by careful examination to be connected with unsuspected tonsillitis or pneumonia. Sometimes their origin is by no means apparent, and in the absence of all evidence of dietetic causes I prefer no diagnosis to the theory of gastric catarrh arising idiopathically or occasioned by a conveniently hypothetical chill.

Whenever vomiting occurs with pyrexia, especially in children beyond infancy, we should think of the possibility of enteric fever before deciding on purgative treatment which in many instances may be beneficial. An emetic in such cases is mostly advisable and always innocent even while we still stand in doubt, and a single purge, if generally indicated, will probably do no harm whatever the illness prove to be.

Intestinal obstruction, and especially *intussusception*, must also be remembered as a cause of vomiting which may be the first symptom noticed. It is probably only in children of over six or seven years of age that *nervous vomiting*, which is by no means rare, need be thought of. Such vomiting may or may not be excited by food, is mostly accompanied by other signs of nervous disturbance, and is often prominent in plainly hysterical patients.

The **treatment** of that most important form of persistent vomiting in infants which is primarily due to gastric derangement and mostly associated with diarrhœa is chiefly dealt with under the headings of Infantile Wasting and Diarrhœa. It need therefore only be said here that dyspeptic vomiting must be treated mainly on dietetic principles, medicines being regarded as adjuncts. When vomiting is urgent the greatest rest possible for the stomach is necessary, and much benefit in recent cases can be derived from giving nothing but water for a day or even somewhat longer with an occasional small dose of diluted alcohol. In more severe cases indicating longer rest to the stomach nutrient enemata of pancreatised or peptonised milk, or of meat peptones, in bulk not exceeding from half to one fluid ounce at a time, should be tried. The rectum must be carefully cleaned out by a water-injection before the nutrient enema is given. As a merely symptomatic remedy, if nothing else, for dyspeptic vomiting especially when accompanied, as it is in most acute and many chronic cases, by epigastric pain and eructation due to acid fermentation in the stomach, an alkali such as sodium bicarbonate in an aromatic water, with or without a minim or two of syrup of ginger or compound tincture of chloroform, is certainly most valuable. It is true, as urged by Henoch and others, that an alkali only neutralises the acid formed by the fermenting contents of the stomach, and does not touch the morbid process itself; but while deferring the question of the best possible causal treatment of gastric dyspepsia I would express my conviction that not only does far more effectual relief to the vomiting and pain result from the alkaline treatment than from the hydrochloric acid, calomel, salicylates, creasote and other remedies given for their supposed anti-fermentative or anti-septic action, but that the stomach being thus kept at comparative rest for a while in many cases soon becomes able to respond to the curative effect of an appropriate diet which prevents fermentative processes. Bismuth subnitrate with or without small doses of the aromatic powder of chalk and opium often gives valuable additional help. When cases of this kind are taken early under treatment appropriate diet alone after a preliminary purgation by castor-oil or calomel will often cause rapid improvement. It is at this time also that the anti-fermentative class of remedies I have mentioned might be expected to be useful; but in obstinate cases which respond but little or not at all to dietetic treatment I must confess that my experience of these medicines has ended mostly in disappointment. When urgent symptoms have subsided in any case of gastric vomiting, leaving the child very weak, some of the peptonised, pancreatised or malted foods, substituting artificial for natural digestion, are often of much temporary value. To papain as a help to digestion my experience has not been favourable; in chronic and otherwise obstinate cases it is I think quite useless. The all-important dietetic treatment should be undertaken and patiently persevered with on the general principles laid down in the chapter on atrophy. I believe with Dr. Goodhart that those who use the simplest methods, with due ingenuity as regards detail, with the aid of a faithful and intelligent nurse and with complete disregard of the reported success or failure of any previous treatment, are the most likely to effect the desired cure.

Acute vomiting and empty retching may often be relieved or checked by the application of a light mustard poultice or hot stupes to the abdomen. In some chronic and unyielding cases of vomiting even in quite young infants occasional or daily washing out of the stomach with warm water as recommended by Epstein and others and largely practised by Seibert of New York is of undoubted value. Food should be withheld for at least two hours after the operation and then given in very small quantities. It is often better to administer nourishment for a while through a tube passed into the stomach by the mouth or nose, this method apparently favouring the retention of food by avoidance of the reflex effects of swallowing until such time as the nervous irritability of the stomach be allayed.

Diarrhœa.

Under this clinical term I include all cases where the prominent symptom is frequent liquid discharge from the bowels whatever its character and whether accompanied or not by undigested food, much mucus, blood, or other morbid material. Although the nature and conditions of diarrhœal stools may often indicate some more or less definite morbid state of the intestines or stomach there is no possibility in most cases of making an ætiological or anatomical diagnosis from this standpoint, severe diarrhœa often occurring with no special post-mortem signs visible to the naked eye in the alimentary canal, extensive ulceration being sometimes observed when there has been no diarrhœa, and cases clinically similar being very frequent with widely differing intestinal lesions. The diagnosis therefore of the probable cause which is the chief basis for the proper treatment of diarrhœa must depend mainly on clinical considerations of the conditions, concomitant symptoms and history of each case or group of cases.

The clinical division of diarrhœa in childhood into **Acute** and **Chronic** is at once salient and practical, and it may be said, although with important exceptions hereafter to be noticed, that cases of the first class are not as a rule referable to gross lesions of the alimentary canal as their immediate cause whereas in many of the latter class such lesions may be inferred or, in fatal instances, are discoverable to a greater or less extent. Chronic diarrhœa too, with or without marked inflammatory or ulcerative signs in intestine is often the result of acute diarrhœa from whatever cause arising.

Acute Diarrhœa.—Diarrhœa beginning more or less suddenly, whether attended or not by gastric symptoms or by fever, and whether transient or ingravescent in character, is by far most common in children under two years of age, is ultimably referable in a large majority of cases to improprieties in the quality or quantity of food, and is greatly favoured by warmth of season and bad hygienic surroundings of all kinds. Children fed exclusively at the breast supply but a small contingent to the diarrhœa total even among the poor of large cities, but since in these classes, on whom the incidence of infantile diarrhœa is by far the heaviest, this manner of feeding at least according to hospital experience in London is decidedly rare even during the first six months of life, we find no great immunity from the worst forms of diarrhœa among infants of the suckling age. The majority of cases, however, in my experience as in that of others, such as Dr. Emmett Holt's in New York, who deals with large numbers in Infant Asylums, is met with in the second half year of life, a period during which nearly all the children of the poor and some of the well-to-do are from necessity or ignorance miscellaneously and improperly fed. Both the incidence and mortality of diarrhœa though considerably less than at the last-mentioned period continues enormously high until the age of two years after which it diminishes with great rapidity. Although space forbids me to quote evidence for the conclusion that in a large majority of cases of infantile diarrhœa the probable cause is the poisonous activity, in promoting decomposition or fermentation in the alimentary canal, of bacteria introduced with the food and especially in milk, I would state at the outset that—after careful consideration of the conditions in which diarrhœa arises—such a doctrine seems to me practically established, and this in spite of the fact that no constant or demonstrably specific form of infective micro-organism has as yet been discovered in the dejecta of any group of cases with however characteristically similar symptoms. Many and elaborate bacteriological researches have been made into this question, and I would especially refer to those of Drs. Booker and Jeffries published in the *Transactions of the American Pediatric Society* for 1889. Dr. Jeffries' observations go far to prove that the large class of summer diarrhœas is the result of the products of bacterial growth in the food and in the alimentary canal, and indicate the paramount necessity of sterilising all milk that is given to infants in almost all cases. In this matter of vital importance the line of treatment thus pointed out should be followed without waiting for the more certain ætiological and diagnostic knowledge which may probably be hopefully looked for. In the light of analogy from other infective diseases we may infer, from the acuteness of so many of these cases of infantile diarrhœa and the favourable conditions to germ-development in food or otherwise among which they arise, that they are largely of microbic, and probably of multifariously microbic origin.

Before, however, describing in more detail this important and extensive class of cases which occur mostly in the summer months and are often of epidemic nature I would call attention to the also common forms of diarrhœa of generally less dangerous character which, although possessing no very distinctive symptoms and in many instances not perhaps to be denied bacterial origin, seem at least to be primarily caused by demonstrable improprieties of diet and to be often relieved or soon cured by due attention to feeding and general hygiene. Some of these cases are merely temporary, characterized mainly by increased peristaltic action and fluid secretions from intestine, and due to undigested or irritating material passing from the stomach to the bowels. Beyond some abdominal pain in those which begin suddenly and the prostration which results from all diarrhœas these cases may have no other marked symptoms, neither vomiting nor fever being necessarily present. Rest, a preliminary purge followed by bismuth and opium if the discharges be not very soon controlled, and the blandest or almost starvation diet for a short time is the necessary treatment for this class of cases. With predisposed and weakly infants, and in bad hygienic and other favouring conditions such attacks may however rapidly become grave or chronic.

In many other instances we have to deal with acute diarrhœas, generally in infants artificially fed, which are accompanied by undoubted signs of gastric disturbance such as vomiting of milk-curds or other undigested matter, evidence of abdominal pain and much furring of the tongue, and are often though by no means always marked by varying degrees of pyrexia. There is frequently much thirst, and the motions are copious and of either a dark-brown or greenish colour with or without visible undigested material. After a while fever, if present, may disappear, and pain and vomiting may cease, leaving only diarrhœa, in the form of frequent and foul-smelling liquid motions, to mark the case. If not soon checked by appropriate dietetic and medicinal treatment some or all of these symptoms may rapidly increase, and wasting with all the dangerous phenomena of exhaustion may ensue as in the typical instances of summer diarrhœas, presently to be noticed, from which indeed these cases are not to be too strictly distinguished. I only mention them separately to emphasize the great importance of thinking at once of dietetic causes in all diarrhœas however slight or severe they may be and of promptly treating them from this point of view. It is especially in such attacks where there is evidence of initial gastric disturbance, of improper food or of undigested matter in the fæces, whether or no there be accompanying fever, that prompt treatment may be rapidly curative and stay the progress of the case towards gastro intestinal catarrh or inflammation of either acute or chronic kind. Many of these patients who escape the common event of early death first come under our notice as well-marked instances of entero-colitis with abundance of mucus and often more or less blood in the discharges from the bowels.

From the milder instances, then, of infantile diarrhœa with but little gastric disturbance and little fever we may find all grades up to the acutest kind which from its greater prevalence in the warm season is known as *summer diarrhœa* or, under its gravest aspects, as "cholera infantum." The onset and course of these cases, which are so frequent in cities and crowd the wards and out-patient rooms of hospitals for children in London from July to August and often later, are very various, and we are frequently unable to distinguish practically as regards forecast between those which begin gradually with but few symptoms other than diarrhœa, and those where the attack is sudden with vomiting and much fever. In many there is of course a history of bad feeding with a longer or shorter period of malaise and wasting previous to the special symptoms complained of; in many, equally of course, there is evidence of syphilis, rickets, a history of previous whooping-cough or measles or other affections favouring or causing diarrhœal symptoms and intestinal catarrh; but, also in many, impropriety of diet, other than that connected with bacterial contamination, can probably or certainly be excluded, and previously healthy and well-cared-for infants frequently suffer and not seldom succumb. It is in this latter class of patients and especially when there is a high febrile onset with vomiting that a bacterial origin specially forces itself upon our minds. With regard to the significance of fever in these cases as a whole I would say that many which run a favourable course under appropriate treatment begin with gastric symptoms and some fever; that many which become rapidly worse and some with even the symptoms known as "cholera infantum" may have no fever at the beginning and little or perhaps none at the fatal end; but that continued fever is always of the gravest prognosis and when over 103° F. almost always indicates death. Continued vomiting is also a very dangerous symptom, and is usually present with continued fever. On the whole it may be said, in spite of the many fatal cases I have seen without fever, that a persistently normal temperature, other things being equal, contributes to a favourable prognosis.

In proportion to the amount of the diarrhœa there is wasting and prostration which may be excessive even in cases which recover. The stools usually lose their normal yellow colour almost from the beginning, becoming green, brown or gray, are frequently accompanied by flatus, and have a foul odour. In some cases they are excessively profuse and watery with at first sickening and later but little smell. With this condition especially the symptoms may be those of rapid collapse and practically indistinguishable from those of Asiatic cholera. Rapid and early collapse may however take place with profuse liquid motions while the stools are still fæculent and in such cases too we observe the apathetic or drowsy condition, the dry skin, the extreme pallor or cyanosis, the cutaneous chilliness, the sunken eyes, the depressed fontanelle, the disordered breathing, the rapid irregular action of the heart, and the weakness or impalpability of the peripheral pulses which together form a salient example of the conditions often described as "hydrocephaloid." The head may also be markedly retracted as in other cases of profound exhaustion, and the secretion of urine is diminished or arrested.

In most of the large number of cases which run a slower course either to recovery, to more or less chronicity, or to death the infants continue restless and are apparently in abdominal pain, with distended belly and drawn-up legs, until in the worst instances symptoms of exhaustion become predominant and drowsy apathy supervenes. The tongue is coated or more often dry and red, and the thirst is excessive.

Broncho-pneumonia and collapse of lung often follow or concur with the abdominal symptoms and may be the immediate cause of death. Convulsions may accompany the onset of the disease in some cases marked by much pyrexia, and are frequent just before the fatal end.

In a certain number of cases, after the subsidence of gastric symptoms and more or less diminution of the fever which may have been present, obstinate but slighter diarrhœa remains with markedly mucous or bloody stools indicating probably the existence of intestinal inflammation, but I would record here the warning not to regard bloody stools, even if frequent and plentiful, which last but a few days as necessarily or usually the result of ulceration. In many severe examples of gastro-intestinal disturbance, both in children and adults, poisonous or acutely irritating articles of diet may cause so-called dysenteric stools in marked degree which are doubtless the result of intense intestinal congestion. In the frequent cases where, after the subsidence of acute symptoms and of all vomiting, diarrhœa alone remains in the form of frequent, small and foul-smelling liquid motions, and wasting progresses in spite of careful feeding and symptomatic medicinal treatment, the irremediable fault seems to lie in the failure of intestinal absorption. In many such cases no catarrh or coarse lesion is found after death, although usually the intestines are much thinned and the glands atrophied. The instances of profuse diarrhœa with symptoms of collapse which may be characterized as "cholera infantum" are not, according to my experience, to be practically distinguished otherwise than by these salient marks and a mostly fatal tendency from the great class of acute diarrhœas which we regard as due to decomposition of food or fermentation within the alimentary canal as the probable result of bacterial poisoning. Nor are such instances, though confessedly not very frequent, quite so rare here as, according to Dr. Emmett Holt, they would appear to be in New York. Although most cases of this nature die in spite of all treatment, including those who have been healthy and in good condition up to the time of the acute seizure, yet some recover after sinking into a state of deep collapse. Prognosis is worst with a very high temperature and in very young infants with a history of untoward surroundings and of previous disease, especially diarrhœa.

In many instances these choleraic symptoms are not observed from the first but follow on a stage of previous diarrhœa with or without marked evidence of intestinal inflammation. After death, however, in cases of rapid course anatomical signs of intestinal mischief other than microscopical evidence of loss of epithelium are usually absent.

The early *diagnosis* of the nature and import of cases of acute diarrhœa, especially in children beyond earliest infancy, involves some practical considerations other than those of merely the state of the alimentary canal.   It is not of much moment as regards treatment, although it may be ultimately as regards prognosis, to decide whether or no there is evidence of intestinal catarrh or inflammation in any given case.   The more violent the symptoms and the more continuous the pyrexia the more likely such a condition becomes, but we must remember that in many cases of acute onset which may last even for weeks no special appearances are noted post-mortem.

Rather should we trouble ourselves to exclude other affections which may be marked at their onset by fever, vomiting and diarrhœa, or may be frequently accompanied by diarrhœa as the result of impaired digestion.   Thus in many acute diseases such as measles, pneumonia, scarlatina, and others there may be sometimes much initial diarrhœa, and in every case with fever we should make a complete physical examination for local signs including inspection of the throat, and wait for the period of distinctive symptoms of the exanthemata before finally pronouncing our diagnosis.

Acute attacks of diarrhœa again are especially liable to be developed without discoverable dietetic errors in infants who are the subjects of rickets or syphilis.   They may possibly be due to chill and certainly often accompany attacks of bronchial or nasal catarrh ; but marked diarrhœa as an isolated symptom is in my opinion but rarely attributable with any plausibility to intestinal catarrh as the result of cold alone.   It is well to remember that occasionally obstruction of the bowel by intussusception may be symptomatically ushered in by vomiting with liquid evacuations from the parts below the lesion, and that sometimes even the later symptom of melæna may possibly be confused with the result of enteritis.   Increased vomiting, however, with arrested intestinal flux and other signs of intussusception will, as a rule, soon resolve doubt. The acute diarrhœas of children beyond two years old are generally more or less demonstrably due to dietetic errors when not merely a part of more general disease, acute or chronic, although some may possibly be due to cold.   With persistent pyrexia and distended and tender belly inflammation or ulceration of the intestines, especially of the colon, may be suspected.   Tubercular diarrhœa is as a rule chronic, and in enteric fever initial diarrhœa is but rarely prominent.

The naked-eye and other characteristics of the stools in cases of diarrhœa should always be noted whether minute microscopical or chemical research into their nature be possible or not.   Examination of the stools may be of considerable practical use in acute cases, while in chronic it is very often of high importance.

In a healthy suckled infant the stools are of a more or less bright yellow colour, of pasty consistency, acid reaction and slightly sour smell. Their frequency varies from three to five or six during the first few weeks, after which time they may be from one to three until weaning time. With a more mixed diet the stools gradually assume the adult characters until, after two or three years old, they are in no way specialised.

Very watery stools are always a bad sign, occurring when small and frequent in chronic cases, and being both frequent and profuse in the so-called cholera infantum. In the latter the reaction is generally neutral or alkaline.

Green stools are very common in acute diarrhœa and may persist even in some chronic cases. According to the best evidence it appears that this colour is due to biliverdin and implies an alkaline condition in some part of the alimentary canal which is abnormal in infants wholly fed by the breast.

Stools may be of very dark colour from various causes. Besides alimentary disorder raw meat-juice and certain drugs such as bismuth and iron will turn them dark brown or black, as also will blood from the upper part of the canal. Much visible mucus, or mucus and red blood in the stools point often to an inflammatory or ulcerative condition of the colon or rectum. According to Emmett Holt's observations the more intimately mixed this mucus is with the stool the higher is its source in the intestine. The mucous shreds he mentions as sometimes resembling false membrane and distinguishable from it by being very readily broken down by a stream of water I have seen from time to time. Definite false membranes as the possible result of colitis are scarcely ever seen in infancy and but rarely in childhood. I have had no experience among children of this condition which according to many observers is most frequent in young adult women of neurotic tendency in connection with chronic gastro-intestinal disorder.

Masses due to undigested food in the stools are common, but are more especially important in chronic cases. Lumps of casein are frequently seen in the diarrhœal stools of milk-fed infants, and round masses of fat are often abundant. Starch may occur in large quantities, revealed by microscopic examination or by the iodine test.

Morbid anatomical signs in the intestines of cases of infantile diarrhœa of the class we are considering are often, as I have said, practically absent, and when present vary considerably with no certain relationship to the symptoms observed during life. It is however roughly true that the more severe and febrile the case, and the longer its duration, the more likely we are to find some naked-eye evidence of intestinal catarrh or inflammation such as mucous coating, or more especially, distinct vascular congestion of the lining membrane of the bowels, chiefly in the colon and ileum. Much swelling of the solitary glands or of Peyer's patches is also often seen, and sometimes superficial destruction of the mucosa over irregular areas as the result of catarrhal inflammation may be found

D

if carefully looked for.   Ulceration of these glands, usually known as
"follicular" enteritis, is observed in some of the subacute cases which
run a somewhat protracted course towards death without much or any
fever.   These ulcers are usually small and superficial but sometimes
penetrate beneath the mucosa; they are mostly situated in the colon
though sometimes seen above the cæcal valve.   Neither melæna nor any
other clinical symptom distinguishes this kind of ulceration, which, as
Dr. Emmett Holt observes judging from the excessively rare occurrence of
healed ulcers in the intestines of infants, is probably nearly always fatal.
The lesson I have learned from post-mortem examination in cases of acute
and subacute infantile diarrhœa is that the conception of catarrh or of
any other morbid condition of the intestines is not very helpful towards
treatment, and I am quite in accord with Holt's conclusion, based on
much observation both macroscopical and microscopical, that attempts at
anatomical subdivision of this class of cases are of no clinical value.   I
would however refer the reader both on this and other points to Dr.
Holt's article on "The Diarrhœal Diseases" of children in vol. iii. of
Keating's *Cyclopædia,* as the most masterly and thoroughly practical
account of the subject within my knowledge.

The post-mortem fact of severe entero-colitis in some cases of infantile
diarrhœa and vomiting will be referred to for the sake of convenience in
the next chapter.   Entero-colitis is indeed to be regarded as for the most
part secondary to more or less prolonged attacks of diarrhœal disturbance,
and has no claim to a separate ætiological position.

The *treatment* of cases of acute and subacute infantile diarrhœa consists mainly in removing or antagonizing its chief cause by the institution
of appropriate diet; in giving medicines to check the intestinal flux and,
if it be possible, to arrest morbid processes in the alimentary canal; and
in securing the best hygienic and other conditions for recovery.

Rest in bed with avoidance of chill and as much fresh air and light
as possible are necessary for all the worst and youngest cases, while those
beyond early infancy and of less acute character may be carried out of
doors for a while in favourable weather if they be kept strictly at rest.
Absolute cleanliness should be observed, all soiled linen being quickly
removed, and daily or more frequent tepid bathing is advisable.

In respect of diet, the acuter the diarrhœa and vomiting with or without fever, the more important it is to keep the stomach at rest, and in
cases seen at the outset nothing should be given for several hours or
perhaps a day or more besides frequent small quantities of cold water
which has been boiled, or barley-water, to allay the thirst which is
always present and to compensate for the loss of fluid by the bowel.
Repeated small doses of brandy will however generally be needed.
Very small quantities of milk or other food appropriately diluted or

otherwise treated, according to the methods described in the chapter on wasting, and not without lime-water, should then be tried, after one preliminary purge of from half a drachm to a drachm of castor-oil or half a grain of calomel for a child from six months to a year old. At this early stage I give, as already stated under the heading of vomiting, 4 or 5 grain doses of bicarbonate of soda in some aromatic water every four or six hours, believing it to be practically useful in allaying discomfort and lessening the tendency to vomit. Should the symptoms continue or increase, or the diarrhœa be great, we must check the flux if possible, and among other remedies which may be tried in various cases 10 grains or more of bismuth subnitrate combined with 5 grains of the aromatic powder of chalk and opium is generally the most efficacious. In all severe and continuing attacks of acute diarrhœa, with or without vomiting, astringent treatment is necessary, and I believe that opium in some form, carefully proportioned to the case and the age, its safe action being judged of by its observed effects, is almost indispensable.

The milk which, as I have said, we should first try in very small quantities in quite early cases, after granting some hours' rest to the stomach, is exceedingly often rejected ; we must then give no more until the acute symptoms have subsided. We may try for a while small quantities of raw or nearly raw meat-juice with or without a little sugar, or of some of the peptonised preparations of meat-juice, or may tide over a day or two with beef-tea or veal-broth, or with white of egg dissolved in water which has been boiled, of the strength of one or two eggs to a pint,—all in small quantities at a time,—having recourse again to milk, with lime-water, as soon as marked improvement has set in. During all this time, and indeed in most cases from the very first, repeated doses of alcohol are necessary, children of under a year old often taking from 20 minims to half a drachm every two hours or even more with advantage. The necessity for this drug, however, and its quantity depend of course on the general indications in each individual case. It is usually valuable in direct proportion to the amount of diarrhœal flux.

It is impossible here, even were it useful, to enter into detail as regards the minutiæ and alternations of diet necessary in many cases of this kind ; a part, however, of this ground has been covered in a general way by what I have already said concerning wasting. We have always to bear in mind that the less the contents of the stomach are, compatibly with the ingestion of sufficient nutriment to sustain life, the more likely it is that the morbid processes in the alimentary canal, due as they so often are to bacterial action, will be discouraged. We must therefore be satisfied to give but little at a time, and to leave if possible two hours' interval between each feeding. The eminently rational therapeutic object of antagonizing by medicines the morbid processes of fermentation resulting from bacterial action can scarcely be said to have been as yet attained in any satisfactory degree. Were it practicable, this method of treatment should be undertaken in all cases from the very outset. I have myself been mostly disappointed, after many trials, with the results of repeated doses of calomel, of the hydrochloric and lactic acids, the salicylates, carbolic acid, creasote, resorcin, and other drugs which have been recommended for this purpose. In the quite early cases which, when not excessively severe, will often soon recover if treated at once by the methods above sketched I have learned nothing from the administration of antiseptic drugs, while in the much larger number of hospital cases, which are usually of some standing before they come under observation, I have but very rarely had much reason to believe that any improvement was due to this class of medicines. I have nearly always been forced to have recourse at last to the bismuth and opium or other astringent treatment. In established cases indeed we may have but little theoretical reason to hope for much success from germicidal drugs, for our knowledge of this subject at present points to the probability of rapid absorption of the poisons generated by bacterial action. Naphthol, however, now largely used in practice for its antiseptic properties may possibly prove to be of value. Its action is seen in the reduction of the frequency and quantity of the stools and the destruction of their offensive odour, and the drug is apparently useful in this direction in proportion to its early administration. It is insoluble or nearly so in water, and of disagreeable taste ; but when once swallowed seems usually to cause no further disturbance digestive or otherwise.

Naphthalene, hitherto more often given as an internal remedy than naphthol, has appeared to me from several recent trials to be quite as efficacious as naphthol in either of its forms and is considerably cheaper. The dose of either drug for infants is from 2 to 4 grains repeated two or three times in the day. It must be mentioned here that bismuth in the form of the subnitrate or the carbonate, not to mention the more modern and now much-used preparation of the salicylate, has been stated to possess antiseptic properties, and it is said by Dr. Emmett Holt to be among the best of this class of remedies and superior to naphthalene. It is possible, of course, that the recognised usefulness of bismuth in diarrhœa may be accounted for by its antiseptic action. If so it should be administered as soon as possible in all cases. The antiseptic treatment, however, of diarrhœa and vomiting in infants, especially in the acute epidemic cases during the summer, is still on its trial. The most difficult and unsatisfactory cases to treat are those where diarrhœa of some kind persists, often in the form of but small though offensive liquid motions, after vomiting has ceased, and where the child takes food readily or even greedily. Most of these cases, though they may linger long, ultimately die, but some few improve with continuous use of bismuth and some, it would seem, although my experience here has as yet been comparatively small, with naphthalene or naphthol, after the failure of other treatment. These drugs, being insoluble in the stomach, may act, as some think, on the lower bowel which is the chief seat of such lesions as may be found in fatal cases. In still other instances, which in my experience are somewhat more numerous than those which apparently respond to drug treatment, gradual improvement may be shown after all medicines have been stopped and ultimate recovery may take place after a long period of careful nursing and varied diet with no special regard to theoretical considerations. With respect to opium, which is in my opinion necessary in most acute cases of diarrhœa, with or without vomiting, mainly for the purpose of arresting peristaltic action and thus diminishing the flux which in itself is dangerous or fatal, I believe that this drug may be safely administered, with due precaution and observation of its effects, to children of almost any age. To no child under a year old should more than $\frac{1}{4}$ to $\frac{1}{2}$ a minim of tincture of opium be given as a first dose. The effect should be carefully watched, and subsequent doses regulated accordingly. Many infants require more than these doses for any appreciable result. In this context I may remark that just as adults may differ in their reaction to powerful drugs such as opium and belladonna so do young children, and that it is not in my opinion quite justifiable to affirm that children are more or less susceptible than adults to certain drugs in doses proportionate to their age. It may be true that many children can take ten-minim doses or more of tincture of belladonna without harm, but I have seen several instances in young subjects where distinct physiological symptoms have occurred after the administration of but half these doses for no very long time, and I feel certain from several trials that the usual ten-minim dose of tincture of belladonna is too small for usefulness in adults who can generally take in health three times this quantity without appreciable effect. I know of no good evidence for the oft-repeated statement that young children as a class are able to take proportionately larger doses of any poisons than adults; nor do I think that babies are much more susceptible than their elders to proportionately smaller doses of opium.

Stomach-washing should be employed in cases when vomiting is obstinate, and irrigation of the lower bowel, as described under the next heading, will often be found very useful in acute cases. With proper care there is but little to fear from this operation. In the acutest form of diarrhœa and vomiting known as "cholera infantum" and characterized by great general irritability, by innumerable and copious watery motions soon becoming neutral or alkaline, and by rapidly ensuing collapse and wasting, with sunken fontanelle, stupor, coma or convulsions, we must at first give nothing but frequent small doses of brandy and plenty of cold water or barley-water to assuage thirst, and should endeavour at once to arrest the vomiting and diarrhœa by repeated subcutaneous injections of morphia, beginning in the case of children under a year old with not more than $\frac{1}{100}$ grain.   If vomiting still continue brandy or ether must be injected subcutaneously.   Hot mustard baths are to be ordered in the stages of collapse.   When in the early irritable stage the temperature runs high, a warm bath gradually cooled down to about $85°$ should be given, and repeated with subsequent accesses of fever.   To conclude these remarks on treatment which I am conscious are no more than very inadequate hints to be supplemented by the practitioner's ingenuity, resource, and, above all, patient perseverance in each individual case, I quote, from the article by Dr. Holt above referred to, the following words of guidance which may serve to remind us of broad principles often neglected :—" No matter," he says, " how strongly we may be convinced of the value of any drugs or combination of drugs, if they continue to disturb the stomach they are worse than useless.   The use of all drugs is of very minor importance as compared with dietetic and hygienic treatment.   In the management of any single (acute) case the important points are thorough evacuation of the stomach and bowels, and then rest to these organs again for from twelve to twenty-four hours.   No cases do worse than those whose mothers cannot appreciate the value of starvation and insist upon giving milk in violation of the rules laid down."

Chronic Diarrhœa.—Chronic diarrhœa in infancy and early childhood may be divided into two main groups.   The *first* consists of those cases which are most often not referable to any previous acute gastro-intestinal attack, but are all the same the result of continued improper feeding, generally defective hygiene, inherent weakness, or a combination of these factors. Whether or no there be accompanying intestinal catarrh in some degree marked and definite inflammatory changes are absent, and improvement or recovery usually follows if the removal of the causal conditions be effected before extreme emaciation have been established.   The *second* group contains most of the graver cases which are either certainly or very probably referable to definite lesions, such as entero-colitis as the not very infrequent result of more or less acute attacks of infantile diarrhœa and vomiting, or intestinal tuberculosis.

Instances of the *first* group are very common in hand-fed infants, and in those who are the subjects of rickets, syphilis, scrofulosis, or other general conditions involving nutritive disorder.   Many cases seem to date from whooping-cough, from measles and other exanthemata, or from broncho-pneumonia, and some occur in tubercular children without tubercular intestinal disease. The more marked and definite the cachexia of the child and its consequent predisposition to react more readily to the exciting cause of improper feeding which is common to most of this class of cases, the less hope there is of recovery from attention to dietetic and hygienic measures alone. Many, however, of the symptomatically graver cases with much wasting and enlarged belly, often described even now by the terms "tabes mesenterica," or "consumption of the bowels" according to the social status of the patient, recover completely and rapidly when properly fed and cared for, the diarrhœa being thus clearly due to faulty diet with frequently bad hygienic conditions.

Besides the diarrhœa for which these children are usually brought, either before or after marked wasting has set in, there is no salient symptom of gastric disorder, the appetite being often good. With progressive diarrhœa both anæmia and wasting become prominent. In established cases the motions are frequent, foul, discoloured and watery, but usually contain masses of undigested material. There is no fever, although acute attacks of gastro-intestinal disorder are apt to set in after but slight exciting causes and especially in the warmer season. Neglected diarrhœa of this kind supplies a large contingent to the cases of chronic infantile wasting already described. Extreme emaciation may ensue, which may ultimately defy all reparative treatment; extensive dermatitis of an "impetiginous" or ecthymatous character often occurs; thrush appears in the mouth, and broncho-pneumonia, often with masked symptoms, or an acute attack of diarrhœa frequently ends the scene with or without convulsion or retracted head.

It is often possible to elicit a history of the passage of frequent, large, offensive, and semi-solid stools before actual diarrhœa or marked wasting has set in. If these important symptoms of indigestion be early observed careful dietetic treatment will probably preclude the establishment of chronic diarrhœa.

In older children chronic diarrhœa unconnected with marked intestinal disease or tuberculosis is almost always due to errors of diet and is apt to be very obstinate to treatment although rarely fatal. The motions, containing undigested matter, are not so copious as in infants, and wasting is less marked, but irritability, sleeplessness, abdominal discomfort and anæmia are prominent. Much mucus probably indicates catarrh of the lower bowel. Symptomatically much the same as this form of dyspeptic diarrhœa is that which is not referable to dietetic errors, and has been described by Trousseau and others as "nervous diarrhœa." I fully recognise the frequency of this affection in children, as also in adults. It would seem that in these cases almost any food, liquid or solid, except perhaps in the most minute quantities, causes undue peristalsis of stomach and bowels, and that the fault lies in individual nervous irritability. The effect on nutrition owing to the prevention of digestion and absorption is of course great. There may or may not be much mucus in the stools the frequency of which varies considerably. Of chronic diarrhœa as the alleged result of exposure to cold or insufficient clothing I can say nothing more than that I regard such exposure as mainly an exciting, although perhaps a frequently exciting, cause in children otherwise predisposed. While therefore much impressed with the multitudes of half-clothed children who never suffer from diarrhœa I am none the less convinced that in many cases chronic diarrhœa is much alleviated by due attention to the preservation of body-warmth.

The *second* group of chronic diarrhœas are those connected with demonstrable intestinal lesions, whether simply inflammatory, and then usually the result of acute diarrhœa and vomiting, or due to tubercular disease. Clinically these cases are marked by prevailing recalcitrance to both dietetic and medicinal treatment, progressive and excessive wasting, a dry red tongue, great tendency to thrush or to aphthous stomatitis, and much mucus with sometimes blood in the stools. There is very often tenderness on pressure on the abdomen which is usually distended but may be flat. With all this there is no vomiting and appetite may be ravenous. Apart from tubercular disease of the intestines, to be dealt with subsequently, which can often be diagnosed, even in the absence of pyrexia, by the evidence of tubercle elsewhere, there is no absolutely positive means, at least in the earlier stages, of diagnosing this class of chronic diarrhœas from the group already described; but attention to the above-mentioned points will usually guide us aright and aid us in making a prognosis. The presence of pus in the stools as revealed by the microscope is a valuable indication of intestinal inflammation or ulceration, as also is the discovery, which however is mostly prevented by abdominal tumefaction, of enlarged mesenteric glands. These cases are mostly fatal, their duration being usually measurable by a few months.

In that minority of severe cases which recovers we may suspect, although scarcely prove, that there has been considerable catarrhal or follicular inflammation of the bowel, especially of the colon, which has not proceeded to ulceration. The more acute and recurrent the symptoms of the primary attack or attacks of gastric disorder which so often usher in this form of chronic diarrhœa have been, the graver the intestinal lesions probably are, and the less the hope of ultimate recovery. The lesions other than tubercular found post-mortem, on which these cases of chronic diarrhœa depend, are sometimes severe catarrhal inflammation of the colon and lower part of the ileum but more often follicular ulceration of varying degrees, and sometimes patches of lymph are seen. The ulcers are often very small, and it may be remarked here that in most cases of chronic ulcerative diarrhœa there is no blood in the stools. Enlargement of the mesenteric glands is frequently to be noted.

In a certain number of cases of severe intestinal inflammation with prolonged diarrhœa a parenchymatous nephritis characterized by more or less swelling and granular opacity of the cortical epithelium has been observed by some and attributed to the intestinal condition. Occasional slight albuminuria with or without œdema and, more especially, a dry inelastic state of the skin are said to be clinical indications of this complication. It would seem, however, that the connection of these phenomena is not easy to define, for these symptoms occur in very different disorders and are by no means constant in chronic diarrhœa however severe and prolonged. The same morbid appearance in the kidney is moreover very often seen in association with high temperature as, for instance, in fatal cases of pneumonia or enteric fever, and marks the first stage of the well-known scarlatinal nephritis.

The *treatment* of chronic diarrhœa need regard but little the question of the nature of the intestinal lesions on which it sometimes depends, for we have seen that in proportion as the diarrhœa can be referred to definite enteritis the less recoverable it is. Practically we are here almost entirely concerned with dietetic and hygienic measures.

The body-warmth must be maintained by proper clothing, and with infants and young children, especially when there is a history of previous delicacy, a flannel binder round the belly is to be advised. All attention, too, should be given to ventilation, and the maximum of fresh air and sunlight should be secured by changing surroundings where they are unfavourable.

The general dietetic treatment is to be ordered according to the principles laid down under the headings of wasting and gastro-intestinal disorders. When an infant is brought for advice for diarrhœa of some weeks' duration our methods will of course vary somewhat, according to what we can definitely ascertain of its previous treatment. In most cases, however, a properly regulated diet suitable in quality for a healthy infant, though as a rule considerably less in quantity, should be instituted anew under our own eyes, whatever we may have been told, and this alone will frequently cause rapid improvement. The wasting cases with a long history of diarrhœa which are successfully treated by good nursing and dieting alone are very numerous. In some severe instances, where we are satisfied that appropriate diet for healthy children cannot be taken, and remedies such as peptonised milk and others directed to gastric digestion have failed, we must resort first to the smallest quantities of milk and lime-water, and then, for a short while, to whey with meat-broth, also in very small quantities. If, after a day or so, the diarrhœa is no less when milk is given we must try raw meat-juice, prepared according to the method before mentioned, or better, seeing that it is here the sole diet for a time, by merely pounding some raw mutton in a mortar, pressing the pulp through a fine sieve, and scraping it off. Only about a teaspoonful of this should be given with a little sugar or in some aromatic confection three or four times a day. One great advantage of this food is its small bulk, which is but slightly provocative of peristalsis. As an adjunct to this method of treatment brandy in 10 or 15 minim doses diluted to a drachm with water is often useful, and, certainly, whenever the intestinal flux is considerable, bismuth subnitrate in powder with the aromatic powder of chalk and opium, in 10 to 15 grain doses of the former and 5 grain doses of the latter, should be given two or three times a day. This is in my opinion the best astringent, and may be ordered with advantage in nearly all cases. When marked improvement sets in this medicine and the brandy should be discontinued, and a gradual return to normal diet attempted. If the child even at its worst be very thirsty it is better to give it water to satisfaction occasionally than in smaller quantities frequently repeated. When we have reason to suspect the existence of such a lesion as colitis underlying the continuous diarrhœa we must proceed in the same way both as to diet and drugs, and may try in addition the effect of irrigation and astringent enemata. If ulceration be found by examination in the rectum small enemas of starch and opium, not exceeding half an ounce or perhaps an ounce in bulk, should be used, and I would further recommend them in all cases with tenesmus and frequent motions, whether mixed or not with blood, where local irritation may be suspected though not proved. But in most cases where the supposed lesion is higher up irrigation and astringent local treatment are not satisfactory. Copious irrigation with warm water through a large-sized flexible catheter or a medium-sized stomach-pump tube previously well warmed may be performed daily or every other day, and high injections subsequently made of 5 or 6 ounces of a tannic acid solution of the strength of 20 grains to the ounce or of some other astringent. The tube should, in either case, be carefully passed 6 or 8 inches beyond the anus. In obstinate cases this method, strongly recommended by many high authorities, should be tried, although its theoretical promise seems greater than its practical performance. Both the irrigation and injection must be carried out very slowly, with careful attention to the child's general condition. Large and high injections seem possibly capable of causing much reflex shock. Once in my experience, perhaps not as a mere coincidence, collapse, convulsion and death followed soon after the operation, in a very puny and chronically wasted infant. With older children, owing partly to their great caprice of appetite, dietetic treatment in most established cases is not so soon followed by improvement as with infants. The staple diet to be aimed at in all cases is milk, but it should be supplemented by different kinds of meat-juices. When milk disagrees very small quantities of pounded lean meat may be given, or occasionally the yolk of an egg beaten up with a little brandy or sherry, or small quantities of bread and butter. Malted biscuits or rusks as more easy of digestion are recommended by Dr. Eustace Smith and others, and pepsin or peptonised foods may be useful in those cases where continuing gastric dyspepsia maintains the intestinal irritability. In all cases of chronic diarrhœa the greater and more direct the part played by gastric dyspepsia from improper feeding the more tractable are the symptoms, and the more simple dietetic treatment fails the greater is the probability of incurable lesion.

When convalescence is established hygienic and dietetic prophylaxis should not be relaxed, for these children are prone to relapse from slight causes. Iron, arsenic, and cod-liver oil are each or all of frequently good service here.

In the cases above alluded to of "nervous diarrhœa" which, at least in older children, are not rare and can often, after a while, be diagnosed with considerable confidence, dietetic treatment is of small value. Other nervous symptoms are usually present, and we cannot then rely chiefly on drugs for cure. But in all such cases small doses of opium, according to age, repeated about three times a day will tend to control materially the bad gastro-intestinal habit, and sometimes are alone curative. For the rest, change of scene, outdoor life and the nervine tonics are highly important. For a short time in some cases the sedative influence of the bromides is very helpful.

CHAPTER V.

GASTRIC AND INTESTINAL DISEASE.

In the preceding chapter the greater part of the subject of gastro-intestinal disorder has been dealt with from the clinical point of view, and we have seen that such disorder bears as a rule no certain relationship to discoverable morbid states. It remains for us now to notice shortly some of the diseased conditions of the stomach and bowels which may be inferred with all probability or are found post-mortem, and are often, though by no means always, evidenced by more or less definite symptoms. The subject of typhlitis and perityphlitis will be treated in a subsequent chapter.

Affections of the Stomach.

Substantive disease of the stomach figures but slightly in childhood.
Ulceration is rare, and for the most part tubercular in connection with
general tuberculosis.   It has been observed in very young infants.   The
symptoms are the same as in the adult affection, but small ulcerations
both single and multiple may occur with no symptoms at all.   Dr.
Goodhart [1] reports a case in an infant two days old where a small gastric
ulcer was found to be the cause of fatal hæmorrhage from the mouth and
anus.   I may here incidentally refer to other causes of hæmatemesis
with or without melæna in young children, such as ulceration of the
cardiac end of the œsophagus of which Henoch gives an instance, and
capillary hæmorrhage which has been seen in several cases and inferred
in others and may perhaps be sometimes attributable to increase of venous
pressure from great respiratory obstruction.   The hæmorrhagic diathesis
again, or purpura, may cause hæmatemesis, which in these cases is often
accompanied by cutaneous eruption or bleeding from other mucous membranes.   Epistaxis, oral or faucial ulceration, and the sucking of sore
nipples may also be the source of blood in vomit, and gastro-intestinal
hæmorrhage is occasionally seen in the malignant forms of the exanthemata, especially of small-pox.   The treatment of gastric hæmorrhage,
whether due to ulceration or not, involves no special consideration in
childhood.

With regard to **gastric cancer** I need only say that a very few cases
have been reported.

Of **gastric catarrh** I must speak somewhat more in detail, for although
its presence is as a rule only inferable from symptoms, and post-mortem
examinations even in cases of long-standing gastric disorder give little
colour to the teaching that it is to be regarded as a primary affection,
yet it is doubtless frequent in greater or less degree as the result of
irritation by indigestible food or of poisonous decomposition of food
either before or after ingestion, and tends to aggravate and prolong the
disorder out of which it arises.   Whether gastric catarrh is a common
source of disorder apart from the above-mentioned causes or from some
more widespread disease, such as the strumous constitution in which
the glandular system is specially apt to suffer, is a point on which
authorities differ much.   In my own opinion there is not much clinical
evidence in favour of giving it a much more independent nosological
position in children than in adults, and it is usually admitted that in
adults both the severe and acute as well as the chronic forms are
mainly due to some recognisable irritant or to plain indiscretions in

[1] *Pathological Society's Transactions*, 1881, vol. xxxii.

diet or drink. A chill either general or arising from undue exposure of the abdomen is frequently stated to be a cause of gastric catarrh in young children, as presumably evidenced not only by feverishness, vomiting or bowel derangement either in the direction of diarrhœa or constipation, but also, and even apart from such symptoms, by sallow complexion, disturbed sleep, irritability, abdominal pain, syncope or convulsions. I believe that most or nearly all of such cases are capable of 'further ætiological analysis, and that they can usually be referred either to demonstrable dietetic causes or to nervous or some other general disorder. In some of these so-called gastric attacks which begin suddenly and are attended by fever there is, as I mention again under the heading of Pyrexia, a concurrent affection of breathing with more or less evidence of bronchitis which is as temporary as the fever and the gastric symptoms. I do not deny that more children than adults may suffer from even tolerably severe gastric catarrh as a part of a general catarrhal condition evidenced by the state of the nasal or bronchial passages, or that in scrofulous and rickety children this affection is probably excited with great readiness; but I fail to recognise the frequency of gastric catarrh, febrile or non-febrile, acute or chronic, however indicated, affecting otherwise healthy children as the mere result of "cold." This view however by no means weakens the great importance of attention to warmth of clothing and other hygienic measures in the treatment of all nutritive disorders in infants and young children.

Gastric catarrh is practically important from the therapeutical point of view as the substantive cause of symptoms in proportion to the suddenness and acuteness of those symptoms and their ultimate referability to some definite source of gastric irritation. We may infer the existence of gastric catarrh with approximate certainty, after the exclusion of other disease, when there is anorexia, headache, abdominal discomfort and vomiting of food, followed by retching and vomiting of much mucus which persists after the stomach is empty. In cases which are accompanied by pyrexia, especially when this is persistent, we must ever be on the look-out for the characteristic signs of other and more general disorders, and whether the temperature is normal or not we must not lose sight of the possibility of cerebral disease. In the acuter forms of gastric catarrh diarrhœa is often prominent.

The more chronic forms of gastric catarrh as the result of continued ingestion of improper food depend on less certain inference for their diagnosis. Practically chronic catarrh of the stomach is not of much importance either in diagnosis or treatment, for we are constantly met with the fact that, however well-marked the symptoms may be which many authorities confidently attribute to gastric catarrh as a cause, post-mortem examination in such cases reveals but little evidence of this condition.   A chronic gastric catarrh which is not discoverable post-mortem need not be the central object of our attention during life, and it must be confessed that none of the classical symptoms so often referred to chronic gastric catarrh are in any way characteristic of this state, and that they all may occur when it is practically absent.   I would remark here that many cases of short duration, marked slightly if at all by feverishness or urgent stomach symptoms but rather by nausea and flatulence or occasional vomiting, with furred tongue, irregular bowel action, headache, pallor and other signs of disturbed circulation such as cold extremities or even syncope, which are often put down to gastric catarrh, are really of nervous origin and are very frequently connected with a definite family history of various neuroses and followed in the individual by established migraine or other nervous disorder in later life. I have seen instances of marked epilepsy treated as "gastric catarrh," which has in the terminology of some authorities an indefinitely compre-hensive pathogeny.   Some of the above-mentioned attacks may indeed be excited by dietetic errors and relieved by appropriate treatment, but a large number are quite independent of such origin and yield, if they yield to treatment at all, to measures directed to the care of the nervous system.

The more chronic cases with like symptoms in older children, marked by alternating constipation and diarrhœa, attacks of abdominal pain, pallor, languor, wasting, variable appetite and often a dry cough are usually, as is well known, referred to chronic gastro-intestinal catarrh. From my experience however of these cases, which are quite common, I agree with Dr. Goodhart in recognising in them a generally strong neurotic relationship.   They are apt to begin suddenly without ascer-tainable cause; they usually respond but little to merely dietetic or medicinal treatment, but often readily improve with general hygienic measures and change of scene and surroundings; and they are very frequently characterized by prominent symptoms and history of neurotic disorder.

In the *treatment* of cases of acute gastric catarrh from whatever cause arising absolute rest to the stomach is before all things indicated at first, and even for some days only small quantities of liquid food should be given.   An initial emetic is often very useful, and if nausea or vomiting be still prominent the stomach should be thoroughly washed out with warm water.   Alkalies, such as bicarbonate of soda or potash in some aromatic vehicle are often valuable as aids in treatment.   All farinaceous substances and sweets should be forbidden, and a very gradual return be made to the normal diet.

In suspected instances of chronic gastric catarrh such dietetic and medicinal treatment .as indicated in the chapter on gastro-intestinal disorders should be followed without undue consideration of any hypothetical condition of the stomach. It is especially however in those cases where there is excess of mucus in the vomit that repeated washing out of the stomach will be found useful.

Intestinal Catarrh and Enteritis.

I have already treated of clinical symptoms often seen in connection with intestinal catarrh. From constant irritation of undigested and decomposing food a chronic intestinal catarrh is often set up which in spite of all treatment may result in a severe or ulcerative colitis. The acute forms of colitis are mostly evidenced by fever, great restlessness succeeded later on by an apathetic condition, dryness and redness of tongue with frequent stomatitis, distension and tenderness of the abdomen, or by more or less diarrhœa with mucus in excess and not seldom blood in varying quantity. On the other hand we find from time to time in the post-mortem examination of children who have died from various diseases and without diarrhœa a marked intestinal catarrh and sometimes severe enteritis with extravasations or fibrinous exudation. I have seen hæmorrhagic enteritis of the ileum and colon in a case of febrile wasting with no other marked symptoms where tubercle was suspected but nowhere found. Henoch reports two cases of marked gastro-enteritis without bowel symptoms in connection with chronic nephritis, in one of which the post-mortem examination showed the ileum to be covered in places at its lower part with a coherent fibrinous membrane, and Goodhart quotes one of severe "diphtheritic" inflammation of the colon and rectum where there was only towards the end some watery diarrhœa with slight melæna. In most instances where definite inflammation or ulceration of the colon and lower part of the ileum is found post-mortem there is a history of a previous attack of gastro-intestinal disorder, but in some the intestinal symptoms are alone observed from the outset, beginning either gradually or suddenly with much fever and constitutional disturbance. When there is much blood in this latter class of cases the term dysentery is often used, but I would subscribe to the opinion of Dr. Emmett Holt that the mere appearance of considerable blood in the stools is insufficient to establish a special form of the affection. It must be remembered that blood is often seen in the stools in many cases of chronic diarrhœa which recover, and does not necessarily imply ulceration, while in follicular colitis there may be numerous small ulcers where melæna has never occurred.

I have seen some marked examples of true *dysentery* in young children who began to suffer after returning to England from the East

Indies where they had recently had malarial fever, but besides the fact that in some instances such attacks occurred in several members of the family of various ages there was nothing to mark the individual cases from those of ordinary severe colitis. A remarkable series of cases of what was symptomatically acute dysentery was admitted into Shadwell Hospital in the summer of 1890. These four patients varied in age from 1½ to 8 years, and their father was similarly affected. All the cases occurred between August 17 and September 18. The father and the youngest child died. In all the symptoms began with much abdominal pain, diarrhœa and tenesmus, followed by the passage of "blood and slime" and, in the later stage, of shreddy material or obvious sloughs with no fæces. In most there was vomiting at the outset, in the elder ones shivering, and in all rapid wasting. Convalescence was very slow in those who recovered. Inquiry revealed no dietetic cause, but the father discovered a much decomposed rat in the cistern. The mother who did not drink water, and three other children who did, were not affected. In the fatal case examined in hospital marked ulceration and sloughing patches were found along the whole of the large intestine.

In considering *melæna* as a symptom of intestinal disease we must remember, besides the above-mentioned causes, both tubercular ulceration presently to be mentioned, enteric fever and, though it is very rare, ulceration of the duodenum, one instance of which I have seen in the practice of my colleague Dr. Sturges.[1] The symptoms in this case were at first chiefly those of vomiting and diarrhœa, but some blood was passed per anum just before death. Two ulcers were found, one of which had perforated close to the pylorus, and there was a closely neighbouring circular patch of capillary hæmorrhage suggesting the first stage of these ulcers. Melæna is also a prominent symptom of great diagnostic import in intussusception, results often from a polypus in the rectum especially during or after defæcation and frequently with associated diarrhœa, and may accompany simple prolapse of the bowel.

The treatment of cases of enteritis is practically included in a great degree under that of severe gastro-intestinal disorder already dealt with. We must rely chiefly on the blandest diet, on the continued use of bismuth and opium, and on protracted cautiousness against possibly irritating articles of food, such as fruit and vegetables, even long after convalescence has been apparently established. In proportion as there is good evidence of marked intestinal inflammation or ulceration, or, in other words, the more confident our diagnosis is, the greater care we must exercise; for such lesions are slow to heal and quick to relapse.

Even in true dysentery and symptomatically allied cases, whether

[1] Reported by Dr. Hebb in vol. vii. (1891) of the Westminster Hospital Reports.

acute or chronic, I believe that bismuth in large doses and opium are, like the well-accredited ipecacuanha powder, among the best remedies. Starch and opium and other astringent enemata are sometimes very useful. Large injections of a solution of silver nitrate, half a grain to the ounce, have also been recommended in adult cases and may well be tried, though I have known relapse occur soon in two instances which had been regarded as cured by this method.

In treating melæna as such we must have due regard to its probable cause. If associated with hæmatemesis local treatment will probably be useless.

The blood which comes from local lesions may be bright or dark according to the seat and nature of the mischief, and mixed or unmixed with fæces. In intussusception and rectal polypus the blood is usually red, in ulceration higher up it is mostly dark. In every case of gastric or intestinal hæmorrhage a local source must be carefully searched for before making any definite diagnosis from inference alone. The rectum should always be examined with the finger. I once saw post-mortem in a case of enteric fever, where death occurred a few days after very severe hæmorrhage from the anus, two deep ulcers a little above the sphincter which were the undoubted source of the blood. Ulceration elsewhere was slight and quite superficial. The hæmorrhage here might possibly have been checked by local treatment had the rectum been examined.

In the severe forms where no local cause can be detected or reached an ice-bag should be applied to the abdomen and small quantities of iced milk given by the mouth. The child should at the same time have small doses of alcohol. It is useless in severe cases to waste time over the trial of styptics given by the mouth, their action being at the best highly uncertain ; but the subcutaneous injection of ergotine from $\frac{1}{4}$ to $\frac{1}{2}$ grain or more may be always practised. The same treatment is applicable to all cases with similar symptoms whatever their cause may be. Whenever examination detects local disorder local treatment is necessary, such as the removal of a polypus, or the injection of some astringent remedy as starch and opium or a solution of nitrate of silver into a rectum which is the subject of ulceration or severe catarrh.

Tubercular Disease of Intestines.

In all chronic and obstinate cases of diarrhœa in children, especially when accompanied by daily remittent pyrexia, tubercular ulceration must be suspected and careful search made by observation and inquiry for any corroborative evidence of this mode of causation. In infants under three months old tubercular disease of the intestines and indeed

E

of the abdomen generally is not frequent. After this age it becomes more common but most cases occur after the second year. Tubercular mesenteric glands usually coexist with tuberculosis of the bowel, but either condition may be found alone. In a large majority of cases there is clinical evidence of tubercle elsewhere, especially in the lungs. Out of 400 cases of tuberculosis I found but 12 per cent. clinically registered as "abdominal" and most of these had peritonitis either with or without intestinal ulceration. It must never be forgotten that tubercular ulceration of intestine is by no means always marked by diarrhœa. I have seen many cases with absolutely no diarrhœa where intestinal and other tubercle was proved to be very extensive. Tubercular disease of the intestine, but especially when there is diarrhœa, is apt to be accompanied by abdominal pain some time after feeding, and the motions are very offensive, brown and watery. Frequently there is more or less admixture of blood with the fæces. There is very often some tenderness and much peristalsis visible through the abdominal walls, due probably to the adhesion of coils of intestine ; and mostly, though by no means always, some rise of temperature. Tubercular ulceration is seen mainly in the ileum but may be much more extensive. It is recognised by its usual character of caseation often accompanied by gray granulations. The ulcers may be deep but seldom penetrate to the serous surface. Rarely they cause contraction of the gut by cicatrisation. In almost all cases tubercle elsewhere is found post-mortem whether detectable or inferable during life or no, and even if the lungs escape the bronchial glands are caseous.

The diagnosis of the tubercular nature of intestinal disorder pointing to ulceration depends mainly on the discovery of evidence of tubercular disease elsewhere, especially of the mesenteric glands, peritoneum or respiratory organs. Persistently remittent temperature is highly corroborative of the suspicion, but we must remember that in severe cases, while the disease is confined or mainly confined to the abdomen, the temperature may be normal. Although some examples of chronic tubercular ulceration may sometimes be at first unattended by marked wasting the co-existence of such wasting with other symptoms, especially when there is little diarrhœa, points strongly to tuberculosis. In many cases the suspicion of tuberculosis amounts almost to a certainty when anæmia and wasting progress and predominate over all local symptoms. It must however be recognised that in spite of all care tuberculosis is not seldom erroneously diagnosed on the ground of both intestinal, thoracic and even cerebral symptoms suggestive of tubercle, and that tuberculosis especially in the absence of diarrhœa is often found post-mortem when least suspected.

There would seem to be little doubt from clinical and post-mortem experience together that tuberculosis frequently supervenes on previous inflammatory conditions of the intestines, especially those of long standing. I have over and over again observed the late epiphenomena of pyrexia, of pulmonary trouble, and sometimes of peritonitis in cases long under observation for chronic diarrhœa with probable intestinal catarrh or simple ulceration, where the post-mortem examination has established intestinal and general tuberculosis.

All cases of suspected ulceration of the bowel should be sedulously *treated* according to the principles previously laid down—both local and general measures being taken—regardless of their possible or probable tubercular origin. It is perhaps only in those cases where rapid wasting, continued pyrexia and the positive evidence of tubercle elsewhere exist that we are justified in giving a hopeless forecast. I have too often been agreeably surprised at the recovery, after protracted care and nursing, of intestinal cases which I have regarded as almost certainly tubercular to feel able to give any hard and fast rules for diagnosis and prognosis, or any better advice than—Persevere in treatment.

CHAPTER VI.

CONSTIPATION.

By this term I would designate insufficient evacuation of the bowels as evidenced by one or more of the following symptoms :—pain or local discomfort during defæcation, prevailing abdominal distension or discomfort with small stools not necessarily either hard or very infrequent, or certain symptoms of ill-health concurring with markedly defective bowel action. Merely infrequent defæcation is physiological with many children, and when unattended by any local or general signs or symptoms of disorder is neither to be dreaded nor treated. Deficient muscular action of the intestines in weakly children contributes to infrequent defæcation, and completely digestible food, especially milk, will often occasion scanty and dry fæces.

The chief immediate causes of constipation are imperfect expulsive action of the intestines, either from inherent general weakness or defective stimulation, or some obstruction in their course. I shall speak here mainly of those cases which are unconnected with malformation or mechanical lesion, but would remark in passing that there are in all probability some cases which are symptomatically chronic constipation but really due to the congenital narrowing of the intestine at some point

—a condition which is not infrequently found post-mortem when it has never been suspected. I have met with a few cases which I believe to be of this description and it is probable that such cases may be more frequent than demonstrable. The following appears to be referable to this category. A girl of nine years old came under observation with a history of having suffered during most of her life from abdominal discomfort apparently connected with the fact that she had been for an equal time subject to periods of fæcal retention of two or three weeks' duration, separated by the occurrence of one and sometimes of two very large motions. This condition resisted all treatment, even frequently repeated large enemata seeming to be of use only about the times when the motions were in the habit of being spontaneously passed. The act of defæcation was rendered easier and much less painful but was very slightly anticipated by enemata, a certain amount of pressure from fæcal accumulation appearing to be always necessary for any passage to take place through the presumably narrowed part of the gut.

Some infants have deficient intestinal action from birth, even when breast-fed; the stools are lumpy and hard, passed with straining, often very pale or almost colourless like the stools of jaundice, and sometimes slightly streaked with blood from small erosions within the anus probably caused by the hardened fæces. Now constipation in infancy, a period when the bowels are as a rule frequently evacuated, should always be noted. Its cause is often not to be detected, for, although there may be sometimes signs of catarrhal derangement of the intestines, or of deficient action of the liver or possibly of the pancreas after such time as the pancreatic secretion is usually well-established, in very many cases the only symptom is the difficult or painful passage of hard dry fæces, and the dryness of the fæces is by no means always explicable by drain of fluid from skin or kidneys. Unquestionably a change in the diet may indicate the cause while working the cure of constipation, and such change should always be tried. Often the tendency remains until the child takes active exertion and a mixed diet, and is probably due to deficient nerve force causing intestinal atony. The frequency of constipation as a symptom of brain-disease, as in tubercular meningitis, must be remembered in this context, as also its very general concurrence with melancholia, hypochondriasis and other evidences of a disordered nervous system in adult life. In many apparently gastric attacks in nervous children, with or without fever, as elsewhere noted, constipation is often prominent. When local and dietetic causes can be excluded by careful examination and reflection over the case we must trust to time and general tonic measures, with such stimulation by food as may be consistent with the child's age, to aid in the cure. Frequent and prolonged kneading of the abdomen with the oiled hand is often very useful, and help is sometimes afforded by occasional small suppositories of soap or injections of cold water, or half a drachm of glycerine. In infants there is little advantage in the use of medicines which should be always avoided if possible. Sometimes however a purge will be found necessary when distension or discomfort is great. The sluggishness of the bowel may perhaps be lessened by the systematic administration of liquor strychniæ or tincture of belladonna. There is at least a considerable amount of authority in favour of this treatment, but I have seen little or no result either in children or adults with habitual constipation from the persevering use in many cases of such remedies, although in two instances, in an adult and a child respectively, of marked constipation with flatulent distension I have observed a rapid diminution of the abdominal swelling with expulsion of gas soon after the administration of one or two full doses of strychnia. Occasionally in infants constipation is due entirely to irritation of the anus caused by painful fissures and inducing contraction of the sphincter. Laxatives and local treatment by nitrate of silver or some other astringent, or the application of cocaine, or, failing these remedies, surgical treatment of the fissure must then be resorted to.

Apart from these instances of infantile affection which frequently lead to no further symptoms and do not necessarily interfere with health it may be said that constipation at all ages owns various causes, and that each case must be studied and treated by itself. The chief causes, outside general debility, nervous or otherwise, are want of exercise, too unstimulating or easily digestible or monotonous diet, and overfeeding with neglect of regular times for defæcation. A diet largely composed of milk greatly favours constipation, as also does prolonged rest in bed. This is seen markedly in most cases of convalescence from acute diseases, especially enteric fever. Sometimes constipation of considerable persistence is observed after severe attacks of diarrhœa or the frequent and unnecessary use of purgative drugs. In these latter cases time and a mixed diet are at once the most rational and successful medicines.

Habitual or marked constipation at any period of infancy or childhood may be accompanied by local symptoms of more or less severity or general ill-health, or a combination of these conditions. When constipation from any cause is marked or lasting there may be distension of the abdomen with tenderness, giving rise to a suspicion of peritonitis. Great loading of the lower bowel with fæces may be attended by the daily or more frequent passage of small liquid stools, sometimes amounting to what may be called diarrhœa. In their lesser degrees these cases are not rare. An extreme example, ending fatally, I once saw in a boy about three years old who had been medically treated without examination for some months for diarrhœa from which he had been suffering for over a year. He was admitted into hospital in a state of collapse with much pain and enormous distension of the abdomen, and died suddenly a few hours afterwards. The bowel was found post-mortem to be greatly stretched and full of a rock-like mass of hard gray crystalline fæces which nearly filled an ordinary bucket. In the mass there was a small sinuous channel through which the liquid evacuation, disastrously diagnosed and treated as "diarrhœa," had been constantly trickling. The diaphragm was pushed far up into the thorax, the boy apparently dying from cardiac paralysis. Eustace Smith relates a very similar case. The lesson is here learned of making a thorough physical examination of all cases of reported diarrhœa as well as of constipation before instituting a course of treatment. Constipation *per se* should never be fatal either directly or indirectly, nor give rise to serious symptoms. Complete occlusion of the intestine and typhlitis leading to perityphlitis or general peritonitis are said by various authorities to be the occasional or even frequent results of neglected constipation. These events are nevertheless exceedingly rare or, in my opinion, altogether apocryphal. Constipation preceding typhlitis is not, as often taught, the rule, but the exception, and their occasional connection is, I think, merely accidental. As a result however of typhlitis secondary to affection of the vermiform appendix constipation is often marked, and sometimes favours the mistaken diagnosis of occluded bowel.

The symptoms other than local which may accompany habitual constipation are numerous and often indefinite and require much diagnostic caution before being referred to any one cause. Many of the ailments attributed to constipation are doubtless part of the original malady out of which the constipation arises, as is exemplified in some neurotic cases, and in *rapidly developed* anæmia. Decided rise of temperature is a frequent accompaniment of a loaded bowel. Not only in enteric fever both during its course and in convalescence, but also in apparently simple cases, there is abundant clinical evidence of raised temperature resulting from constipation. From among numerous examples of this in my note-books I give the following extraordinary instances of two young children who were admitted at different times into hospital with great abdominal distension and pain, and temperatures of 107° and 108° respectively, in both of which all these symptoms rapidly and permanently disappeared after the evacuative action of simple enemas. I am inclined to refer such cases as this, with other instances I have seen of abdominal pain at all ages accompanied by high temperature of short duration, to a nervous origin through impressions made on the abdominal sympathetic, in view of the suddenness of the rise of temperature and the equal suddenness of its fall with coincident evacuation of the bowel. In more chronic cases of constipation some pyrexia may possibly be due to the retention of excrementitious matter or auto-infection. For the rest, as in adults, so in children, habitual constipation is often accompanied, be it causally or coincidentally, by a sallow or pale complexion, by languor, slow circulation, loss of appetite, foul breath, troubled sleep or headaches; and there are some cases where such symptoms disappear if the constipation be cured by diet or drugs. Recurrent sick-headache, or migraine— a very common disorder of childhood and almost always the outcome of a special or general neurosis—is not seldom attended by, and in my opinion most erroneously explained, both popularly and professionally, by constipation, the relief of which in no way lessens the frequency or duration of the other symptoms. I am aware that many authorities think that constipation in children is much more frequent and entails many more untoward consequences than I have above referred to. I have only spoken from my own experience. For detailed account of different views I would refer to an exhaustive article by Dr. W. Earle in Keating's *Cyclopædia*.

The general **treatment** of constipation, after a careful inquiry into its most probable causation, is rather hygienic and dietetic than medicinal. In all cases the abdomen must be examined, when scybala may often be felt, and the rectum, if found by the finger to be loaded, should be evacuated by an enema or other means. In recent acute cases accompanied by abdominal pain or other urgent symptoms rest, opium and expectance are the next steps, as will be further mentioned under the head of obstruction. We must always remember that there is no such thing as acute constipation requiring purgation for the purpose of avoiding a dangerous event. Constipation, with any other symptom of *recent* occurrence should never be treated *as such*, and occurring alone may be safely and wisely neglected for a while. In the absence of vomiting or other symptoms of obstruction and especially of intussusception, even when there is pyrexia, distension of the abdomen with a history of constipation sometimes indicates, as I have above shown, evacuation by enema. Even if there be mechanical obstruction one effort with this object will probably be harmless. General directions, however, as to the due diagnosis of such cases cannot be given.

In chronic constipation the diet must be first attended to. I have already indicated the general treatment of infantile constipation which rarely requires aperients. But, if there be reason to believe from the concurrence of gastric disorder or the appearance of the stools that a catarrhal condition is present, the infant's diet should be regulated accordingly, and milk and farina should for a while be given sparingly or not at all. Small doses of sodium bicarbonate and rhubarb will then often be found useful for a few days. If the supposed catarrhal condition of the intestine be due to chill either as a symptom *per se*, or be a part of more widespread evidence of cold, warm clothing to the body and especially to the abdomen is certainly indicated. At all events, however doubtful the exact diagnosis may be, chronic constipation is so often attended by sluggish circulation that due attention should always be given to this point. Whether or no deficient fluid be the cause of constipation a free supply of pure water may always be tried in cases where the stools are dry and hard ; it is at least a good stimulant to a torpid intestine. With older children a daily regular evacuation after breakfast or an attempt at it should be insisted on both as a preventive and remedial measure, many troublesome cases of constipation being initiated by neglect of this custom otherwise so important in civilised society. A mixed diet with plenty of vegetable and fruit and whole-meal bread is necessary, and exercise with gymnastics suited to the age is to be enforced. All hygienic measures should be taken before resorting to drugs, which are, however, in some cases required not only to remove the abdominal discomfort that sometimes results from a loaded bowel but also to prevent the possible ulterior symptoms of ill-health which have been mentioned. Merely purgative drugs, however necessary they sometimes may be, are but temporarily stimulant ; they should be used in the smallest doses which have the desired effect, and as seldom as possible. I believe that senna and sulphur, in the form of confection, or aloes are as good remedies as any, and cascara is often very effective. I have already said that, in spite of very many cases of habitual constipation in children being, in my opinion, primarily referable to nervous causes and not to local disorder or disease of the alimentary canal, I have usually been disappointed with the much-quoted action of strychnia and belladonna so widely used in this affection. It is of the highest importance to attend to all concomitant symptoms in cases of chronic constipation. I have had far greater success from the prescription of exercise, sunlight, fresh air, arsenic, cod-liver oil and iron, and the proscription of prolonged rest, either in bed or out of it, than from any systematic course of special drugs. There is in fact but little peculiar in the constipation of childhood ; it has much the same clinical aspect as that of adults. There are few affections where the therapeutist's temptation is so great to achieve an easy although but temporary success by symptomatic treatment while neglecting the causal conditions of the disorder.

CHAPTER VII.

INTESTINAL OBSTRUCTION.

THE subject of mechanical obstruction of the bowels in young children, apart from the fæcal accumulation dealt with in the last chapter, practically resolves itself into that of **Intussusception.** Occasionally we meet with other occlusions, such as partial or complete congenital stricture which may sometimes be due to fœtal peritonitis, hernias, strangulations from peritoneal bands or obstruction from malignant or other tumours, but these, with the exception of complete congenital atresia, present nothing special to childhood. It must never be forgotten that symptoms of *acute peritonitis* from whatever cause arising very often exactly simulate those of intestinal obstruction.

Intussusception is most common in infancy and early childhood and at this period generally consists in the invagination to a greater or less extent of the lower part of the ileum and ileo-cæcal valve within the colon, the cases of engagement of the small intestine alone being mostly confined to later childhood and adult life. The comparative frequency of the involvement of the cæcum in infancy has been referred to its looser connections in the iliac fossa at this age. The anatomy of intussusception is amply described in the general text-books. I need but mention here that small and multiple intussusceptions, easily reducible, are frequently found post-mortem, and doubtless take place during the act of dying. Those which cause symptoms are larger, at least two or three inches in length, and may involve several feet of gut and be felt in the rectum or even protrude from the anus. No exciting cause for the disordered peristalsis which must precede this affection can be demonstrated in most instances. Eustace Smith quotes cases occurring soon after a fall, and I have seen two instances of this possibly causal connection.

Intussusception leads to partial or complete obstruction of the canal and inflammatory strangulation of the two contained layers of bowel. Sometimes ulceration takes place with perforation through the outer layer into the peritoneal cavity, or there may be general peritonitis without perforation. The invaginated part may be so gripped as to become quickly gangrenous and may be discharged through the anus. In some cases with firm adhesions recovery takes place, the external layer of bowel forming a tube continuous with the unaffected part above the lesion.

Clinically, cases of intussusception may be divided into two classes, those where the diagnosis is almost certain, and those where it is only probable. The clearest cases are those where the child shows sudden evidence of abdominal pain, screaming and drawing up the legs. The pain comes on paroxysmally, vomiting follows and is repeated, and there is straining with the passage of fæces, if the bowel below the obstruction be not already empty, or with the passage of mucus or blood. There is generally no marked abdominal tenderness at the outset, and deep pressure can be made without finding any tumour. Later on a tumour, tender on pressure, may sometimes be felt externally, if the abdomen be not distended, and in the course of time a soft elastic swelling may be detected by the finger in the rectum. Whether or no there be one or more evacuations, liquid or otherwise, at the outset, constipation soon sets in. The child is very restless and may not sleep at all. If the symptoms do not soon remit with or without treatment they rapidly increase. Coils of dilated intestine are often seen through the walls, collapse takes place with small frequent pulse and cold extremities, and the child dies in from four to eight days from the outset of the attack. The temperature may remain normal, or may rise several degrees, especially when there is peritonitis, until it falls at the period of collapse. The more complete the strangulation of the bowel the more rapid is the course of the case.

Chronic intussusception of very various duration and symptomatic character may take place in childhood, but it is not common and perhaps never affects infants. It is marked by constipation either persistent or intermittent and by varying distension of the abdomen with peristalsis visible through the abdominal walls. In one case of this kind where I suspected imperfect bowel obstruction death with acute symptoms took place, as I heard, after several months. The peristaltic action was almost continuously visible for many weeks. Such cases are generally fatal in the long run and are as a rule beyond the scope of other than conjectural diagnosis.

Now, abdominal pain, vomiting, constipation and the passage of blood from the anus coming on in a previously healthy child are practically almost pathognomonic of intussusception, and the diagnosis is assured by the discovery by palpation of a tumour, especially in the right iliac region or in the rectum. As to tumours felt externally, their diagnostic importance varies greatly. In the absence of marked symptoms of obstruction it is sometimes exceedingly difficult to distinguish them from accumulations of fæces. I have known a case where there were symptoms, lasting some weeks, of constipation and occasional passage of bloody mucus with the scanty fæces. Complete constipation followed, with sickness for some days. An elongated tumour not indented by pressure was felt in the region of the descending colon but no blood was seen on the finger after rectal examination. The boy, aged 13, seemed very ill and in abdominal pain. Two copious injections of the bowel with warm water, not followed by any fæcal discharge, soon relieved the symptoms, which, however, returned and were relieved again by a simple enema. The tumour may be felt in the region of either the ascending, transverse, or descending colon. Authorities differ as to its most frequent position but the question is I think of but little clinical importance.

The following case which is typical of not a few is a good example of what I have called the group of probable cases. A boy of four years old, previously quite well, had a fall on his head on the evening of July 16, 1881, and seemed much collapsed soon afterwards. He was given some medicine, vomited, complained of much pain in the abdomen and passed a few loose stools. The vomiting continued, with great thirst, until admission into hospital the following midday. At 10 P.M. on the 17th vomiting was urgent with signs of collapse, the abdomen was soft, but a small hardish lump, tender on pressure, was felt near the umbilicus. There had been no motion since admission. Nothing was felt in the rectum. A subcutaneous injection of $\frac{1}{24}$th of a grain of morphia was given, and iced milk in small quantities. After a sleep of several hours the vomiting returned. A simple enema (one pint) brought away a little greenish fluid and a few small fæcal lumps. In the afternoon of the 18th two pints of soap and water were slowly injected into the rectum, the boy being held with upraised buttocks, and the abdomen simultaneously manipulated by pressure directed from the groin towards the umbilicus. Much of the fluid was retained. The child vomited but twice during the night and the next day, but the pulse remained very weak. On the evening of the 19th he had another morphia injection. After vomiting twice more on the 20th and retaining some iced milk given by the mouth he passed a solid but unformed stool in the evening, the lump was no longer felt, and he recovered quickly. This case may be put down by some to fæcal obstruction, but the suddenness of its onset and the unformed character of the first stool point strongly away from this explanation. The temperature was scarcely above normal throughout. In another instance very similar to this, where recovery made the diagnosis doubtful, the return of the child to the hospital in a few days and death with all the classical symptoms, followed by a post-mortem, proved the existence of intus-susception. I agree with Henoch that the absence of a palpable tumour is of but little diagnostic import, for the abdomen is frequently much distended when the case is first seen and it is but seldom that anything can be felt in the rectum during the early stage of the affection. I have been sure of this latter sign in only two cases both of which ended fatally, and on referring to a consecutive series of twelve cases diagnosed as intussusception by various members of the staff of the Hospital for Children I found that in none of the six cases which recovered after injection was intra-rectal tumour reported, while its absence was specially noted in four. ' The tumour caused by accumulated fæces can usually be indented by pressure, and the case is characterized neither by sudden onset nor discharge of blood. Peritonitis very often causes symptoms of obstruction but not of intussusception. The onset of vomiting may here be sudden but there is usually early distension of the abdomen with much tenderness and pyrexia.

Cases of sudden and repeated vomiting at the onset of acute febrile disease with constipation are occasionally diagnosed as obstruction. The two following instances which I have seen of pneumonia being taken and in one case treated for obstruction are of sufficient rarity and importance to be quoted. The first was a boy of six years old who was admitted to hospital with a history of repeated vomiting of yellow liquid for four days beginning after a meal of boiled pork. The bowels had not acted for six days. Three days before admission he complained of pain with abdominal tenderness, but there was no distension. On the day before admission his vomit was said to smell like motions. The next day the vomiting ceased. He had five enemata by medical advice with no result. On admission the temperature was over 102°; there were signs of bronchitis but no pulmonary dulness. The abdomen was distended and there was some tenderness, but nothing was felt by external or rectal examination. The bowel was ineffectually inflated. The next day the child vomited twice, but not fæcally, and the head was retracted. A gallon of tepid water was injected, retained for a while, and returned slightly discoloured. The child seemed to be dying. Abdominal section was then performed, everything was found perfectly normal, and four ounces of liquid fæces were passed from the anus during the operation. Two days afterwards the child died with peritonitic symptoms and the post-mortem revealed recent peritonitis with enlarged mesenteric glands but no stricture or intussusception. There was however double pneumonic consolidation, one lung being in the stage of gray hepatisation.

The second case, a girl of 4½ years, was sent by medical advice to Westminster Hospital on July 16, 1890, as intestinal obstruction demanding operation. She was said to have been well on the morning of the 14th. In the afternoon she looked pale and ill, went to sleep for an hour, and then woke up with abdominal pain and vomited. Very frequent vomiting with pain in head and abdomen continued until admission, following always on attempts to eat or drink. Between the attacks of vomiting the child was very drowsy. She had a slight cough which had lasted some months. Previous to this attack she had had some diarrhœa, but there had been no action of the bowels from the 14th to admission. When seen in hospital she was much exhausted, the temperature was 104.8, pulse 138, respirations 40, and the face was flushed. Nothing was detected on examination of the chest, and no pain was complained of. A normal action was induced by an enema which was easily administered. Some acute febrile disease, probably pneumonia, was suspected from the temperature and general appearance which was unlike that of abdominal mischief, and nothing was discovered by abdominal or rectal examination. The temperature and pulse remained high with no further action of the bowels, on the 18th signs of consolidation at the left apex were first discovered, and at midnight on the 19th there was a critical fall of temperature with sweating. The child was well and up on the 23rd. In this case the repeated vomiting, which is exceptional in pneumonia, and the suddenness of onset with abdominal pain fairly raised a suspicion at first of intestinal trouble, but the condition of the child on admission rendered this diagnosis at that time extremely unlikely.

A careful study of all the cases of intussusception which have come under my notice leads me to believe that the **prognosis** is generally grave when the diagnosis is certain and especially when the invagination is felt in the rectum. Yet many cases recover with prompt treatment. My colleague Mr. Parker[1] has made the observation from his cases that absence, or slight amount, of blood-discharge from the anus, and indeed a short period of pain or vomiting at the outset followed by a time of comparative comfort are by no means necessarily of good import, as these conditions are often concurrent with rapid gangrene of the intestine leading to death in collapse ; while a free discharge of blood signifies by itself incarceration rather than strangulation. On the other hand I would say that such a condition as causes a free discharge of blood must, if persistent, soon become one of strangulation and points to prompt remedial interference. It may I think be said that the cases which may recover from treatment are first, those where with initial pain and vomiting and obstinate constipation, with or without a tumour detectable through the walls, there is no discharge, or but slight, of blood from the anus and no considerable abdominal distension, and next those where there may be a notable amount of blood passed for a while, the abdomen being likewise undistended ; while the gravest of all and least likely to recover under any treatment are those where, with obstinate constipation, the abdomen is distended and the acute symptoms of pain and perhaps vomiting as well diminish early or cease altogether.

[1] *Clinical Society's Transactions*, vol. xxi.

Constipation therefore of sudden onset attended by pain or vomiting in a previously healthy child should always raise the suspicion of intussusception, and calls for early treatment without waiting for that abdominal distension which even in the absence of any palpable tumour will in many cases clinch the diagnosis. The discovery of a tumour or the discharge of blood would favour this decision and emphasise the need of immediate relief. It is thus clear that **treatment** must be begun in many cases before the diagnosis is settled; in waiting for certainty the chance of recovery may be lost. A definite case untreated and of three days' standing will generally die whatever is done, although in exceptional instances there may be recovery. I leave out of consideration here the rare cases of chronic intussusception which generally end unfavourably, and the occasional instances, also in later childhood, of recovery after sloughing away of the imprisoned gut.

When there is a reasonable suspicion of intussusception chloroform should be given or opium and morphia cautiously administered until drowsiness be produced. The bowel should then be copiously and slowly filled with tepid water, the buttocks being well elevated by placing the child on an inclined plane, and the abdomen simultaneously manipulated by pressure directed towards the umbilicus. Several cases with undoubted signs including the presence of a palpable abdominal tumour have been in this way successfully reduced at the Children's Hospital. When the gut is gangrenous and very thin the operation may cause perforation and rapidly ensuing death. This possible accident however in no way contra-indicates the invariable necessity of having recourse, with due care, to this procedure. If this method fail air may be very carefully injected; but after two unsuccessful trials at most with either liquid or gaseous injections abdominal section should at once be performed. At this stage, the treatment being surgical, I would but add that if reduction be impossible or the gut in a very bad state an artificial anus should be made and gangrenous parts removed. In cases where a tumour can be felt in the rectum gentle attempts may at first be made at reduction by means of a sponge-tipped probang or bougie, but little is to be expected from this procedure. In older children, as urged by Dr. Eustace Smith, when there are very severe symptoms with much prostration, operation by section is generally contra-indicated, as not only giving little hope of success, but also precluding what little chance there may be of cure by sloughing which occasionally takes place even in cases apparently desperate. I have known however but of one such instance, and of its after history I am ignorant.

Prolapse of Rectum.

This is a common affection in infancy and very early childhood and is generally characterized by the invagination of the middle in the lower part of the rectum and its protrusion through the anal ring in the form of a shiny red swelling which readily bleeds. A small prolapse returns by itself after defæcation while larger ones require replacement. Although some cases are doubtlessly referable to straining with constipation or to the lax and catarrhal condition of the middle part of the rectum in connection with severe diarrhœa I am convinced both from examination and inquiry in numerous cases of out-patients that this affection very frequently occurs with no discoverable intestinal disorder and is probably due to a feeble sphincter. It must be remembered, too, that the rectum is straighter in infancy than later on, and that the sigmoid flexure is in the middle line or sometimes even a little to the right. Straining in urination may excite prolapse in time, and it is generally recognised that stone in the bladder should be sought for as a possible determining cause of prolapse.

Some cases get well with only a few replacements, others may be very obstinate indeed, but most recover in the long run. The treatment is to return the tumour by firm pressure with one or two fingers on its central part and then to apply a thick pad to the anus, bandaging or strapping the buttocks tightly together. Henoch gives a qualified recommendation to the subcutaneous injection of ergotin in the neighbourhood of the anus, stating that it never does local harm, often seems to do good, but often fails. Nitrate of silver in strong solution may be applied, with frequent success, to the surface of the tumour. In some cases surgical treatment directed to the contraction of the sphincter ani is found necessary. Concurrent constipation or diarrhœa should be always appropriately treated.

CHAPTER VIII.

PERITYPHLITIS AND TYPHLITIS.

CLINICAL and post-mortem observation of numerous cases of the affection known by the above titles has for many years convinced me of the practical uselessness of endeavouring to make any distinction between the signs and symptoms which have been thus nominally differentiated. In every *fatal* case of so-called perityphlitis that I have examined, and I have seen many, I have found perforative ulceration of the vermiform

appendix and almost always a fæcal calculus which caused it. I therefore follow Dr. Wilks' teaching in this matter, insisting on the great
probability of the disease nearly always beginning in the appendix,
and believing that it is the varying intensity of the inflammation that
has brought the two terms, typhlitis and perityphlitis, into use. It is
now very generally allowed that all cases thus described, as well as
those called para-typhlitis, are really due to localised peritonitis in the
neighbourhood of the cæcum. There is, but rarely, apart from disease
of the appendix a true typhlitis which may lead to perforation and
peritonitis, but I know nothing of this affection in children.

Perityphlitis unquestionably occurs most often in early life, and, being
common in children, deserves notice although possessing no very special
clinical characters to differentiate it from the adult affection. My experience is quite in accord with that of many others that males of all
ages suffer much more often than females, that in infants the affection
is very rare, and that in children generally there is a greater proportion
of cases of short duration and apparently favourable termination than
in adults in whom the formation of abscess seems considerably more
frequent.

Besides fæcal calculi in the appendix, which are certainly by far the
most common causes of fatal, and probably of all, cases of typhlitis in
children, foreign bodies such as seeds, fruit-stones, &c. have occasionally
been found in the same position, and some of them may form the nucleus
of a subsequently harmful fæcal concretion. In infants the diameter of
the colon bears a much smaller proportion to that of the appendix than
in later life, and thus the entrance of fæces and foreign bodies may be
favoured. I have, as above stated, seen no instance myself of simple
catarrhal inflammation resulting in ulceration, or of a primary ulceration of the cæcum, leading to perforation ; but cases of this kind have
been reported, and it is possible, considering some instances of recovery
where subsequently gall-stones or other concretions or foreign bodies
have been discharged per anum, that an original typhlitis may have
occurred. It is stated by some authorities, with however but little
support from cases in point, that distension of the cæcum by fæcal
matter or as some put it, "constipation" is by itself the most frequent
determining cause of "typhlitis." I strongly oppose this view having
found neither history nor clinical evidence of chronic constipation in my
cases. The reported instances moreover of even fatal distension of the
intestines with hardened fæces do not point to ulceration as one of their
results. Only rarely does specific ulceration of the cæcum from enteric
fever or tuberculosis lead to signs of typhlitis and perforation, most cases
of perforation from these causes being, as is well known, sudden and not
preceded by local evidence of typhlitis.

The symptoms and course of typhlitis vary widely with the rapidity and other accidents of the process. The evidence of local inflammation without suppuration will often disappear in two or three weeks. If there be suppuration the case may be of much longer duration unless surgically treated, and the abscess may burst externally or, though less often, into the abdominal cavity. In some cases there is marked evidence of peritonitis which may become general, and will then argue certain perforation. In others, after apparent recovery, adhesions or chronic abscesses result which may lead to recurrence of the original symptoms. Thus on the one hand the signs may be merely local, such as pain and tenderness, fulness or swelling in the right iliac region, and this whether or not the inflammatory process may have spread from its original seat in the appendix or elsewhere and have involved, with or without per-foration, the surrounding tissues. In the probably frequent cases of perforation of the vermiform appendix with only local symptoms and favourable event adhesions doubtless shut off the morbid process from the general cavity of the peritoneum, a result which is favoured by the anatomy of the region. On the other hand the original ulcer may quickly proceed to perforation with scarcely any warning symptoms and thus communicate freely with the peritoneum, setting up at once the symp-toms of acute peritonitis or obstruction of the bowel the local cause of which can then be arrived at by inference alone. It may indeed be said that acute peritonitis coming on suddenly in a previously healthy child beyond infancy is in a large majority of cases due to perforation from appendicular ulceration. I have much evidence in my case-books to show that purulent peritonitis after a short illness in children is mostly due either to this cause or to external traumatism. Occasionally it is the result of perforation of tubercular ulcers or of suppurating mesenteric glands.

Common to most cases that I have seen are a more or less sudden origin with severe pain and tenderness in the cæcal region and, sooner or later, retching or vomiting. In some instances, however, there has been a previous history of the local symptoms for some time before their severity prevented the patient from walking. The temperature is generally though not always raised, and on examination a swelling may usually be found in the right iliac fossa over which the percussion note lacks resonance. Marked constipation mostly prevails, but sometimes there is even diarrhœa throughout. The right thigh is often flexed, and pain is caused by attempts at its extension. In numerous cases which end favourably the symptoms abate after some days, the bowel action returns, and convalescence is gradually established; but the tendency to recurrence must always be remembered. It is impossible in the cases which recover to say how far the typhlitic or appendicular inflammation

F

has progressed, or even whether perforation has taken place or not.
It is not improbable that there is some degree of localised peritonitis in
every case. In the more severe instances where the tenderness is very
acute and the fever high we have almost certainly to deal with a
localised peritonitis. A small perforation of the appendix may doubtless take place and the orifice be sealed up by adhesions in cases which
recover, evidence of which I have seen in a cadaver after death from
thoracic disease. On the other hand in the case of a boy of fourteen,
who was admitted into Westminster Hospital with acute symptoms of
bowel obstruction and rapidly following signs of peritonitis after a kick
on the abdomen at football, there was found post-mortem a congenital
and very narrow constriction of the small intestine about 5 feet from
the pylorus. The bowel at this point had become doubled on itself and
fixed by thick but quite fresh lymph which was a part of acute general
peritonitis set up by the perforation of a small sharp-pointed faecal
calculus found protruding from the wall of the vermiform appendix.
The perforation in this case was in all probability immediately caused
by the kick. When the perforation occurs posteriorly, and there is thus
a probability of the peritonitis being limited, the symptoms may be
chronic and an abscess may form after a considerable time. Here it is
important, as in all cases of insidious origin, to exclude the presence of
hip-disease by the ordinary tests. There may be nothing at this stage
to indicate involvement of the bowel, and there may or may not be the
usual constitutional symptoms, more or less severe, of a collection of
pus. Such cases are as a rule of very bad prognosis, although they may
endure for long. In those instances, then, where the first symptoms
observed are local pain and tenderness with constipation and retching
or vomiting and the diagnosis leading to appropriate treatment is made
early, the prognosis is as a rule good, with the exception of the comparatively few which in spite of all care go on to abscess. This class,
as has already been observed, may possibly include many of even perforation of the appendix. When, however, the early signs are those of acute
peritonitis, often accompanied by urgent symptoms of obstruction, the
prognosis is very grave. There does not seem to be any further ground
for distinction between the cases which die and those which recover.

As regards **treatment**, absolute rest in bed with the limbs supported
in the most comfortable position must be enforced in all cases suspected
or diagnosed as typhlitis. The event in many cases, especially in those
where only local signs exist, depends largely on proper treatment, no
small part of which is the avoidance of purgation which has been so
often resorted to, with or without medical advice, in cases admitted to
hospital. Opium should always be freely given, not only for allaying
pain and giving rest to the bowels but also for its probably antiphlogistic

effect. The diet should be liquid, and consist partly of milk and partly of meat-juices. Hot poultices to the abdomen give much relief, but they should be very light and therefore frequently renewed. If there be much pain and fulness leeching over the affected part is often of great service. When pus is indicated a careful examination should be made by aspiration, and indeed in all cases which are becoming rapidly worse an exploratory operation is advisable, even when no fluctuation can be demonstrated. When pus has been found it should be removed by incision. In all cases the more definite the swelling the less delay should there be in exploring for pus.

Careful dieting and rest should always be insisted on for a few weeks after recovery, for relapses may probably thus be prevented. If constipation persist, as it but rarely does, after all other symptoms have disappeared the bowels should be acted on by enemata or gentle purgatives. Otherwise no medicine is required.

The medico-chirurgical question of excision of the vermiform appendix for the purpose of preventing that return of the affection which clinical experience gives us cause to dread may be said to be still *sub judice*, but in properly selected cases the operation seems to be one of good promise. I do not think that relapsing typhlitis is common in children, but when it does occur and there remains any swelling in the iliac region to make diagnostic assurance doubly sure, surgical interference, and probably removal of the appendix, is indicated. Mr. Treves reports a case in a man with thrice-recurrent appendicular inflammation where the operation was successful and was undertaken at a time when there was neither sign nor symptom of active disease.

CHAPTER IX.

PERITONITIS AND ABDOMINAL TUBERCLE.

Acute Peritonitis.

As in the adult, acute peritonitis may be the essential affection, and is then generally due to perforation through intestine or traumatism ; or it may be secondary to more generalised disease. In the latter class of cases the symptoms of its onset may be but slightly marked or altogether absent. Abdominal pain and tenderness, fever with or without rigors, and vomiting are the main initial symptoms ; the belly soon becomes distended and almost motionless, the face pinched and pale, the nostrils distended, and there may be violent paroxysms of intestinal colic. There may or may not be evidence of ascites. In the majority of cases the distension is chiefly gaseous. Although constipation is the rule diarrhœa is sometimes present in the early stage or even throughout, and in some cases there is but little or no evident rise of temperature, the condition of nervous collapse so frequently incident on abdominal injury probably preventing this symptom.

The picture of acute peritonitis in children can best be drawn from the cases which result from perforative disease of the appendix as previously described ; for the rest, it is not common except in its secondary and often less marked forms. Generalised acute peritonitis is almost always fatal. The cases of purulent peritonitis which recover after discharge through the umbilicus or other parts of the abdominal wall, or after surgical incision, are probably more or less localized peritoneal abscesses. The best examples of this are supplied by some typhlitic events, and the results of falls, blows, or kicks on the abdomen. Of the latter class of cases Henoch gives several instances. The following case of my own is to the point.

A girl of six years old, previously healthy, complained of abdominal pain and vomited frequently a few days after falling on her belly over a ladder. The pain was continuous until admission to hospital a fortnight later. She was then very restless, and abdominal dulness and fluctuation were marked although there was resonance in the flanks. Two days after admission friction sounds were heard at both pulmonary bases, and especially at the right where there was some loss of resonance. After a fortnight, the abdominal distension being somewhat greater, I ordered tapping at a point near the umbilicus, and $2\frac{1}{2}$ pints of ordinary pus was withdrawn. General improvement followed, and the temperature rarely rose above normal ; but gradually there appeared signs of increased fluid, and in the course of three weeks Mr. Parker at my request made an incision in the middle line between the umbilicus and pubes and let out a pint and a half of pus. The intestines were found to be matted together and the liver, spleen and transverse colon were distinctly felt. The peritoneum was washed out antiseptically, a drainage tube inserted, and the wound nearly closed. After a discharge for a few days the patient quickly recovered and was quite well and about in a fortnight, the physical signs of the probably secondary pleurisy having disappeared.

Apart from typhlitis and external traumatism acute peritonitis in children may result from perforation of intestinal ulcers, as in enteric fever, and, but very rarely, from that of gastric ulcers or the suppuration of abdominal glands. It may also arise suddenly in connection with general tuberculosis either as an early or late symptom of the disease. In a secondary form it may occasionally occur as extension from purulent pleurisy, or from the breaking of an empyema through the diaphragm. Certain cases must be put down to more general causes, including septicæmia. Such are those which are seen in the fœtus or the newly-born, with or without jaundice, some of which are perhaps syphilitic; in the advanced stage of scarlatinal nephritis; and in occasional grouped or isolated instances reported by various observers where a purulent peritonitis seems to be the main characteristic of a generally fatal "blood-poisoning." Of such I have seen a few instances both in children and adults. I have no hesitation however in expressing the greatest doubt as to the existence of "idiopathic" peritonitis, or a peritonitis due to "chill" alone; and I can say nothing about rheumatic peritonitis, at least in children. A careful search and due inquiry or post-mortem dissection will as a rule negative this unsatisfactory diagnosis. Although I once saw a fatal case in a weakly child about four years old, where it was said that careful examination post-mortem failed to indicate the cause of the peritonitis, I would refer this in the light of general experience rather to some undiscovered lesion or to the possible cause of external traumatism than to a still more hypothetical idiopathy.

The **prognosis** of acute peritonitis is least unfavourable in cases where there is reason to believe it is more or less localised, and in those, probably of the same order, which are due to external traumatism. When this affection is due to a lesion which permits the escape of the contents of hollow organs into the cavity, or is secondary to other diseases, it is mostly of fatal augury. It must be remembered that secondary peritonitis, although generally causing distension of the abdomen, is not unfrequently, especially in exhausted patients, unmarked by clear symptoms and discovered only post-mortem.

As to **diagnosis**, the not infrequent mistake of taking peritonitis for mechanical bowel obstruction with the occasional maltreatment directed towards the latter affection is deserving of notice. The careful therapeutist will refrain from even accidentally clearing up the diagnosis between the constipation of peritonitis and that of obstruction by means of a purge or enema, and thus this difficulty at least must remain. Fæcal vomiting too, may take place in acute peritonitis, as I have seen in one unquestionable case with a post-mortem; and it must be remembered that both abdominal distension and tenderness without melæna may very occasionally occur even at the outset of a case of intussusception. Pleurisy, mainly localised at the diaphragm, may be mistaken for peritonitis, and there may be no thoracic signs in some cases, at least for some days, to establish the diagnosis. Possible mistakes, however, in diagnosis are not of much practical importance, provided we always treat even a suspected case of peritonitis by avoidance of purgatives, absolute rest, a scanty fluid diet, and sufficient opium to produce a slight degree of narcotism.

With ordinary caution we shall then never kill, though we may, if our provisional diagnosis be correct, but rarely cure. Attempts to relieve pain should be further made by the application of heat, without undue pressure, to the abdomen, and for this purpose a bran poultice is better than a linseed one. Leeches too are often of service.

Chronic Peritonitis and Abdominal Tubercle.

It is but very rarely that chronic peritonitis other than tubercular comes under our observation in children. Doubtless chronic peritonitis evidenced only or mainly by ascites may exist, as is shown by an important case of traumatic origin reported by Henoch ; and it is seen from time to time as the result of malignant growths, especially of sarcomatous tumour, arising generally in the post-peritoneal glands and sometimes, but rarely, in the peritoneum itself. The nature of a considerable number of cases which recover, especially in children, will perhaps always remain in some doubt, but the ample post-mortem evidence we now have of the obsolescence of both peritoneal and other tubercle favours, on the whole, the diagnosis of tubercular peritonitis in most recoveries where traumatism can be in all probability excluded.

Post-mortem examination proves that, while in a very large majority of cases peritoneal tubercle is coincident with tubercle elsewhere and notably in the mesenteric and bronchial glands and in the lungs, yet it may occur alone. I have seen one case where the abdominal organs were adherent to the walls, the peritoneal cavity being obliterated and the surface almost entirely invaded with nodules of yellow tubercle, while there was absolutely no other lesion than necrosis of the upper jaw. The chief symptoms of the disease are wasting, vomiting, abdominal pain and enlargement, diarrhœa, and some degree of fever, but these may occur in different combinations, and some may be absent altogether. Most cases begin insidiously with some pain, sickness, diarrhœa and fever, but the after-course varies much. In those which go on to enlargement from effusion of fluid all other symptoms besides a certain degree of wasting often cease ; the enlargement may sometimes be great, but in most cases the abdominal walls do not become very tense. In these cases the diagnosis from ascites of hepatic origin depends chiefly upon pathological experience which tells us that chronic peritoneal effusion in children is generally of tubercular origin. Not infrequently such cases recover, at least apparently, with good nourishment and hygienic surroundings, with or without repeated tapping. After drawing off the fluid we may often detect indurated masses in the region of the umbilicus or other parts. Other cases of peritoneal tubercle are characterized by absence of fluid, and the abdomen may be but slightly enlarged or even retracted. Hard masses of different size, shape and position may be felt through the walls, owing to a thickening of the omentum and its adherence to the transverse colon, to sacculated exudations or matting together of the intestines. There may be little or no pain or tenderness. Many cases are complicated with the signs and symptoms of intestinal or pulmonary tubercle, such as continuous diarrhœa, cough, fever and rapid wasting, and not a few which have run a very chronic course may suddenly sicken and rapidly die with evidence of tubercle of the pia mater. I have seen several times, as in adults, an inflammatory thickening with a red blush on the skin in the region of the umbilicus as described by Fagge. In two cases there was a discharge of sero-pus 'from the umbilicus. Often, however, this induration and redness after a while disappear altogether.

It has been well remarked by Dr. Osler that there is the closest analogy between tubercle of the peritoneum and of the lung, for we see in both the fresh miliary eruption, the caseous ulcerating masses, and the chronic fibroid and often pigmented nodules. The varying clinical course of this disease, from the acutest to the almost wholly latent form, is thus explained. *Pleurisy*, often of the dry form, is very frequently associated with this affection and is not always accompanied by actual tubercle of the pleura. Ascites, as we have seen, is not as a rule extensive, though most often present in some degree, but tympanites from gaseous distension of the bowels, sometimes extreme, is very common. In several of my cases, including those which were steadily becoming worse, I have noted the continuous absence of *pyrexia* and sometimes even a submormal temperature as described by others and dwelt on by Dr. Osler. The most chronic cases are those where the tubercle is limited to the peritoneum or slightly involves also the mucosa of the intestines and mesenteric glands. *Diarrhœa* of persistent character generally indicates tubercular ulceration of the intestines. In the majority of fatal cases there is found fatty enlargement of the liver.

The diagnosis of tubercular peritonitis presents far less difficulty in childhood than in adult life when the confusion of it with ovarian disease and also with malignant growths is frequent and sometimes unavoidable. In every case careful search must be made for the evidence of tubercle elsewhere, and due attention should be given to the previous and family history of the patient.

That tubercle of the peritoneum may spontaneously become inactive is now amply proved by clinical and anatomical research, and, although the cases in which recovery takes place are mainly those of mild onset and inextensive signs, even large effusions of undoubtedly tubercular nature have been known to disappear. I have no doubt of this fact from my own experience although I am unable to give a crucial instance with anatomical proof. We have therefore every encouragement to follow any line of **treatment** which experience may suggest to assist the arrest of the tubercular process. Cod-liver oil, iron and arsenic may be all useful either separately or combined. Possibly the hypophosphites, which are always harmless, may do good. I have tried them many times in deference to the authority of good observers but have not much reason to speak in their favour. Above all things fresh air with a nourishing diet should be strictly insisted on. In those not very frequent cases where there is distinct liquid effusion causing distension I believe tapping to be decidedly useful. I have not found any good to result from painting or anointing the abdomen with any counter-irritant or absorbent.

Of "laparotomy" I have had no experience in my own cases, and from the general clinical course of this disease in early life I am of opinion that treatment by this method, curative though it may possibly have been in some adult cases, will be but seldom applicable to children. According to the advocates of this operation those cases which begin suddenly with some fever and considerable ascites, without evidence of tubercle elsewhere, are the best subjects for this treatment, but I believe that this class of cases, rare altogether, is very rare in children. On the whole it would seem advisable in peritonitis deemed to be tubercular, apparently uncomplicated with tubercle elsewhere, and withal rapidly progressing with acute or sub-acute symptoms, that the abdomen should be opened and the cavity drained according to the approved methods of modern surgery. But in all cases of insidious onset and slow progress, whatever be the physical signs, the operative hand should be stayed for a while by the certain knowledge that the disease is sometimes naturally cured.

Chronic peritoneal mischief with ascites, due to sarcomatous disease of the abdominal glands, may generally be diagnosed or suspected by the presence of rapidly growing tumours which can usually be felt; but in those rare cases where the peritoneum itself is the main seat of the sarcomatous process nothing perchance may be established on examination besides ascites.

CHAPTER X.

ON ASCITES, JAUNDICE AND DISEASES OF THE LIVER.

Ascites.

PERITONEAL effusion in children whether it be of inflammatory origin or not requires but little special consideration. As in adults, so, although rarely, in quite young children it may be a part of the general circulatory stasis from heart disease, especially when there is tricuspid regurgitation, and it may also be seen in chronic renal disease. Abdominal tumours, mostly sarcomatous, beginning in the omentum or in the retroperitoneal glands and connective tissue, are the cause of some cases by obstructing the peritoneal circulation, and more or less ascites from acute or chronic peritonitis, mainly of the tubercular form, is of frequent occurrence, though in the majority of cases not excessive. Non-tubercular peritonitis, though rare as a cause of chronic ascites, must also be remembered. One such case at least I have seen in a child in whom the effusion appeared soon after a fall across a gate and nodular masses were felt in the abdomen after withdrawal of the fluid. Recovery with disappearance of all abnormal signs was ultimately complete.

Ascites arising from cirrhosis of liver, pylephlebitis, inflammation of the hepatic veins secondary to perihepatitis (Hillier), syphilitic disease either of the liver itself or causing obstruction of the trunk-vein in the portal fissure, or from glandular enlargement, tubercular or otherwise, in the fissure, requires no detailed description. In any case of ascites, otherwise unexplained, syphilis should always be thought of as a probable cause.

In children beyond infancy it is not very rare to meet with ascites even of great degree and long duration which is clinically referable neither to peritonitis nor to any discoverable cause other than, in some instances, *hepatic enlargement* which may be very patent to examination especially when the fluid has been removed by absorption or by tapping. I have seen three such cases which have begun gradually in children of previously apparent good health and have ultimately recovered with or without one or more tappings, the only permanent sign being an enlarged liver. Hillier refers to cases apparently of this kind, and suggests that they may be due to thrombosis in veins, or to pressure on them by external plastic exudation, enlarged glands or tumour. In one case of mine in a girl aged fourteen the abdomen was tapped four times at considerable intervals, 120 ounces being removed at the first operation and 12 at the last. Two years afterwards I saw her in apparently perfect health but with a much enlarged liver. On first admission there was a systolic murmur at the cardiac apex which soon disappeared. The urine was scanty throughout the course of the case which was under treatment for nearly a year, but there was never any albuminuria or abdominal pain. Cases of this kind are perhaps best referred provisionally, in our ignorance of their nature, to some chronic form of interstitial hepatitis.

In a case of a boy, aged 9½ when first seen, there were two attacks of ascites with scanty urine lasting about three months each with an interval of nine months. During the first attack there were auscultatory signs of some œdema of the lungs, with occasional slight albuminuria, but no other discoverable dropsy. There was no albuminuria or other symptom in the second attack from which he recovered with apparent completeness. The liver was much enlarged throughout. In this case, however, there might be some suspicion of other than hepatic origin.

It is often, of course, very difficult to eliminate *peritonitis* as a cause of some obscure cases of ascites. The following case would appear from the history to be of this nature. A girl aged 8, previously quite well, had severe pain followed by rapid enlargement of the abdomen. On admission soon afterwards there were found extensive ascites and double pleural effusion with friction sounds. During her stay of two months in hospital, after which she left perfectly well and with no abnormal physical signs, there was at first some intermittent pyrexia.

As a general rule a peritonitic cause may be inferred when there is or has been abdominal pain and tenderness, and especially when these symptoms follow on a febrile attack, with headache, vomiting, diarrhœa or wasting. If nodules can be felt in the abdomen the case is most probably, though not certainly, tubercular.

A large peritoneal effusion may perhaps cause secondary œdema of the feet and legs by pressure on the abdominal veins, and sometimes greatly contracts the thoracic cavity by raising the diaphragm and thus compressing the lungs with the result of producing much dyspnœa. The symmetrical loss of resonance often noted at both pulmonary bases is probably due to collapse of lung and must not be too lightly attributed to pleural effusion. It is unnecessary, owing to their rarity, to do more than barely mention ovarian cysts and hydronephrosis as possible sources of error in the diagnosis of ascites, even in young children. In the **treatment** of ascites we must take all pains to discover the cause and thereby direct our hygienic and medicinal remedies. It is well in all peritoneal effusions of such extent as to cause much inconvenience, and especially dyspnœa, to remove the fluid at once by *tapping*. In proportion as an ascites is symptomatically but little complicated and not probably referable to kidney- or heart-disease all methods, according to my experience, fail which are directed towards the promotion of diuresis, diaphoresis or intestinal flux. I believe that hot-air baths are here practically ineffectual, as are all the medicinal remedies in popular use. My case-books of past years contain many records of failure to reduce ascites by copaiba-resin, squills, digitalis, the alkaline diuretics and pilocarpin. Tapping therefore should be performed whenever the symptomatic distress is great, and should be repeated, but at intervals of some weeks, when without any distress the fluid is demonstrably re-increasing. I believe that repeated tapping makes for cure in some cases. The diet should always be nutritious, unless contra-indicated by other symptoms, and not lacking in fluid. Iron and other tonics with plenty of air and light are often of great importance. In such of my cases as ultimately recovered or seemed to recover much greater improvement was made at convalescent homes without active treatment than at any period of hospital residence, except so far as immediate relief by tapping was concerned in those patients who required it.

Jaundice and Diseases of the Liver.

Infants both at birth and soon afterwards may be the subjects of jaundice which is either transient, with no other symptoms, and of no serious import, or connected with bad nutrition with or without petechial hæmorrhages in the skin. Jaundice is also caused by congenital absence of the bile-ducts with sometimes no trace of a gall-bladder, by syphilitic disease affecting the portal fissure, or by a septic condition arising from umbilical phlebitis. The first-mentioned form known generally as *icterus neonatorum*, which is seen mostly, though not always, in weakly or prematurely born babies or in cases of delayed birth with impeded respiration, is due to true bile-staining, and, although its origin is doubtful, probably arises from hepatic causes connected with altered circulation and pressure on the bile-ducts in the liver. The conjunctivæ are stained as well as the skin and sometimes bile-pigment in small quantities is found in the urine. This condition, unaccompanied by any bad symptoms, generally passes off within a week or at most a fortnight from birth. There may or may not be varying degrees of staining of the urine or loss of fæcal colour. These cases are to be distinguished from a yellowish discolouration of the skin, without any conjunctival staining or other signs of jaundice, which sometimes supervenes on the great cutaneous congestion often seen at birth.

The next form, which is seen sometimes very soon after birth in connection with wasting, stomatitis, diarrhœa and vomiting, or occasionally, as mentioned by Henoch and as in a case seen by myself, with small cutaneous hæmorrhages, is probably due to *catarrh of the bile-ducts*. Although from their general conditions these cases are grave they may recover completely.

Jaundice from retention of bile owing to *congenital absence of the ducts* is always fatal. It may not appear until a few weeks after birth, and the child may live, without much wasting at first, for several months. Most however die much sooner. The liver in these cases is sometimes enlarged early, but in those that die after the longer course of the disease is usually found shrunken with destruction of the cells and is dark olive or black in hue. In one case of the kind where the jaundice was seen at birth and the child lived three weeks the liver was found by Dr. Hebb to be small and black and consisted practically of fibrous tissue in the meshes of which were granular pigmented and much degenerated cells. There was no trace of the bile-ducts and only a fibrous cord in place of the gall-bladder. In some cases of congenital absence of the ducts umbilical hæmorrhage is observed before death.

Syphilitic disease of the liver, involving or not the gall-bladder and large ducts, is a recognized cause of jaundice and is to be diagnosed by a large, hard and often uneven liver with other signs of syphilis. This form may be congenital and is always fatal. *Septic jaundice* connected with umbilical phlebitis is accompanied by fever, a sanguineo-purulent discharge from the navel and often by petechial eruption and vomiting. Convulsions are apt to precede death.

In the diagnosis and prognosis of infantile jaundice we are guided by increase of symptoms, by enlargement of the liver, by the concomitant signs of other disease and by the existence of wasting and general discomfort to form the graver opinion of hepatic disease or congenital malformation as its cause, while good health with no ingravescence of symptoms will usually justify a favourable forecast. It is a sign rather of serious disease than of the harmless "icterus neonatorum" when the symptoms do not set in till some time after birth.

Icterus neonatorum *per se* requires no treatment. Cases accompanied by disturbance of the alimentary canal or wasting must be duly dieted, and, for the rest, all possible causal indications for therapeusis must be followed. Constipation should be relieved by drugs in all cases; and for preference, but without any strong conviction, I always try for a while mercurial purgatives.

In children beyond infancy jaundice, though not common, may occur from almost all the causes which obtain in adult life. My notes include a few cases referable to gall-stones, syphilitic gummata of the liver, or chronic enlargement of probably cirrhotic character known as hypertrophic cirrhosis; two cases connected with acute yellow atrophy of the liver, and some due in all probability either to temporary hepatic congestion or to so-called catarrhal obstruction of the common bile-duct. I do not think, however, that the harmless and transient obstructive jaundice so common in adults and referable to one or other of the last named causes is of very frequent occurrence in children, and although jaundice due to gall-stones or to masses of inspissated bile in the ducts is said to be rarely demonstrated I have seen several cases which were symptomatically indicative of this affection, the existence of a calculus being proved in one.

Two cases of sisters aged $4\frac{1}{2}$ and $2\frac{1}{2}$ were of doubtful diagnosis. They both simultaneously had typically complete obstructive jaundice of a week's duration with severe paroxysmal attacks of pain lasting for three

or four days.   In the case of the eldest the attack began with acute pain
and vomiting as well.   Both children were perfectly healthy and their
history practically excluded any probable dietetic cause.   Painting was
going on in the house when the illness began, but recovery was rapid
and complete in unchanged conditions.

Acute yellow atrophy as a cause of jaundice, known also by the
clinical name of "malignant jaundice," is said to be very rare in childhood, but, as the jaundice may precede all other symptoms, I quote as a
warning the following case which, until alarming nervous symptoms set
in a day or two before I saw it a second time, completely misled me into
giving a good prognosis.   A girl of six years old was brought to me for
jaundice of about a week's duration with light-coloured fæces and dark
urine.   Her previous history was good, and beyond one or two rather
severe attacks of abdominal pain and a slight occasional irritability there
had been no other complaint.   Physical examination revealed no abnormality in the size or condition of the liver or in any other organ or
system, and the child looked in no wise ill.   Three weeks after the onset
of the jaundice the child one day became drowsy, having been previously
no worse than when I saw her, and had violent attacks of screaming in
apparently causeless passion.   She rapidly became worse with rising
temperature, headache and frequent irregular pulse, and died within
three days in general convulsions after several hours of coma.   Nystagmus was observed before coma set in, and she vomited "coffee-ground"
material twice during the convulsion.   The lower edge of the liver could
be felt the day before death but the vertical hepatic dulness was then
apparently diminished.   The liver only could be examined post-mortem.
It was of a dark gamboge colour, flabby and easily lacerable, and weighed
$14\frac{1}{2}$ oz.   Its section showed under the microscope that the cellular
hepatic structures had disappeared, only the portal-canal system being
discoverable.   There was also considerable fatty degeneration.   The
capsule was smooth and thin.   This case perhaps bears out the view of
some that the fatal toxæmia in this disease may be due to ptomaines
and leucomaines accumulating in the hepatic ducts, and that thus it
may be secondary to ordinary jaundice of congestive, catarrhal, or even
calculous origin.

In another case aged $2\frac{1}{2}$ years, of 11 days' duration, the symptoms
were very similar, there being nothing during the first week or more to
differentiate them from those of "simple" jaundice.   The liver was
enlarged and tender until a few days before death when its edge was no
longer felt.   The temperature rose and the child died with convulsions
and coma and a temperature of 107° F.   Just before death there was
vomiting of a coffee-ground material.   The liver was found after death to
be quite covered by the ribs, 15 oz. in weight (not therefore abnormally

light), of a deep orange colour and very flabby. The bile-ducts were
all patent. The stomach contained about 8 oz. of dark grumous fluid.
Leucin and tyrosin were found in the liver. Microscopically there was
no trace of normal liver-tissue although the portal system and hepatic
veins were recognizable. The rest of the lobules were made up of dis-
organized and granular remains of liver cells and connective tissue
débris. In a few places there were collections of round indifferent cells,
chiefly about the portal side of the lobules. In this case the correct
diagnosis was not made until shortly before death.

Diseases of the liver itself are not prominent in childhood and are
marked by little that is special to our subject.

Fatty liver, with or without enlargement demonstrable during life,
is frequent in tuberculosis and occurs also in connection with rickets
and chronic diarrhœa, and in grave cases of scarlet fever, diphtheria
and other diseases, especially when there has been prolonged pyrexia.
There are no special clinical symptoms of this condition in children any
more than in adults, and it may exist in its lesser degrees with fairly
good health.

Lardaceous liver, mostly in common with a similar affection of the
spleen and often of the kidneys and intestines, is perhaps the next most
frequent affection and occurs in tuberculosis with or without bone
disease, in suppurative bone-disease especially of the hip-joint and spine,
in syphilis, and sometimes after prolonged suppuration independent of
bone-mischief such as an uncured empyema. Of late years, however,
lardaceous disease from this last named cause has become, I believe,
very rare, owing probably to prompt and effective treatment by resection
of the rib in operating for empyema. Lardaceous disease causes far
greater enlargement of the liver than fatty or other changes, the organ
remaining about the normal shape and being hard and smooth to the
touch.

Interstitial hepatitis or cirrhosis is certainly rare in childhood,
especially in the form—ending in irregular contraction of the liver—
which is common in adults and mostly due to alcoholic excess. I have,
however, seen two cases in children of three and four years old where
both the typical clinical symptoms and signs and the history of either
much beer- or spirit-drinking rendered the diagnosis quite clear, and a
few cases have been reported with post-mortem corroboration. The large
cirrhotic liver with jaundice and with or without ascites is more common
in childhood, and this change is sometimes found after death, especially
from the acute exanthemata, when there have been no symptoms referable
to hepatic disorder during life. In syphilis and tuberculosis also cirrhotic
change in the liver often occurs to a greater or less degree with or without symptoms, and especially in syphilis the liver may be acutely tender

from perihepatitis. It may be said further in this context that both miliary and caseous **tubercle** is often found in the liver of children, sometimes with much enlargement of the organ which was evident during life with or without some jaundice, but more often with no abnormal physical signs, and that general **syphilitic** disease of the liver often with considerable enlargement and hardness and much cell-growth, or discrete gummata with varying degrees of fibrosis are not seldom seen in quite young infants. That infantile syphilis, other signs of which have passed away, may be the cause of some otherwise apparently idiopathic cases of cirrhotic liver in children seems, from what we know of pathology, to be at least probable though by no means certain.

There is a chronic form of enlargement of the liver with ascites of which I have seen several examples, mostly with symptomatic recovery and apparently permanent enlargement of the organ. I have already referred to these cases under the head of ascites. Of temporary congestion of the liver with some pain or tenderness and evidenced further by slight jaundice I have seen several probable examples, and some instances in childhood of undefined illness with pale coloured stools, digestive disturbance, and constipation may perhaps be referred, as Dr. West and others have said, to this condition. Certainly in many cases of this kind exposure to cold seems to be the exciting cause, and they are specially apt to occur in the early spring. I would however remark here, as I further urge in another context, that the recurrent and so-called bilious attacks, usually attended by headache and often by vomiting and some fever, which are often observed in children over eight or nine years of age are almost always instances of "migraine," and do not in any way yield to treatment by warmth or to medicines which may act on the gastro-intestinal or hepatic functions. Chronic enlargement of the liver due to passive congestion and the result of heart disease, especially mitral, is very common in children, but need only be mentioned here as always to be borne in mind in diagnosis.

Tumours of the liver are rare in childhood and have no special peculiarities. We should, however, always remember the not very uncommon occurrence of hydatid disease, and the possibility of medullary sarcoma and other malignant growths. **Abscess** of the liver, whether multiple from mischief in the region of the portal vein, or secondary to intestinal ulceration, and sometimes occurring in connexion with empyema or, though very rarely, with the irritation of lumbrici in the bile-ducts requires no detailed notice here.

In physically examining the region of the liver in cases of suspected hepatic mischief in childhood the much larger relative size of the liver in early life must be remembered, such physiological largeness being generally evidenced by the ready palpability of the edge of the liver up

to the end of the second year or even later, while in young infants the
enlargement is easily demonstrable both by the touch and by percussion.
Besides this the outspreading of the lower ribs, which is more or less
marked in infants and especially in those who are rickety, often exposes
the liver to such an extent as to appear like undue enlargement to the
unpractised observer.   Pleural effusion of the right side may also much
depress the normal liver, and an enlarged liver may escape observation,
without examination by percussion of its upper limit, by being hidden
under the ribs or by the abolition of its subhypochondrial area of dulness
through liquid or gaseous distension of the abdomen.

The question of *treatment* of hepatic affections in children resolves
itself mainly into that of the benign forms of jaundice and of other
symptoms referable with more or less probability to functional disorder
of the liver.   Of the treatment of ascites which may accompany cirrhosis
or other affections I have already spoken.   In all cases of jaundice where
there is constipation it is well to give laxatives to secure one or two daily
evacuations.   Saline or aloetic medicines or a combination of these may
be given for this purpose, or one or other of the mineral waters such as
those of Karlsbad or Vichy.   The best diet is that which consists mainly
or entirely of milk and farinaceous food, but in cases which do not soon
recover a more mixed diet, including some meat, will probably be necessary.   Our object is to avoid irritation of the stomach and duodenum,
and to unload the venous circulation as much as possible so as to allow
the hepatic functions to re-establish themselves whether they have been
arrested by catarrh of the ducts or engorgement of the blood vessels.   If
this method fail we may try repeated doses of calomel or grey powder, or
give alkalies or acids.   But in curable cases a simple diet with occasional
mild aperients is probably all that will be required.

I have but little doubt, from some well-marked cases I have seen
both in adults and children, that the liver, like that of the Strassburg
geese, may become congested and enlarged as the result of overfeeding
combined with overheating and want of fresh air.   With or without
jaundice there are symptoms of indigestion, languor and irritability.
The treatment of such patients is clearly indicated as soon as the
cause of their malady is suspected, and may be rewarded with a marked
degree of success.

CHAPTER XI.

ENLARGEMENT OF THE SPLEEN.

IN childhood the spleen is apt to become chronically enlarged in various general affections of which the most common are syphilis, tuberculosis and lardaceous disease. Leucocythæmia accounts for a few cases, as also does malaria of which I have seen several definite examples in children born and bred in the close neighbourhood of the London Thames. The spleen is enlarged in many cases of rickets, but from the great preponderance of normal spleens in pure rickets I do not regard this enlargement as an integral part of this disease.

I may also mention general lymphadenoma as an occasional condition in which marked splenic enlargement may arise. This affection, though not rarely seen in childhood, has no claim, in any difference from the well-known disease in adults, to separate consideration here. Enlarged spleen may also be seen in cases of chronic cirrhosis of the liver or other obstruction of the portal vein, as well as in some instances of long-standing heart-disease. But by far the most frequent examples of splenic enlargement have in my experience occurred in the ill-nourished children of the poor, and are marked by a *waxy pallor* of skin and mucous membranes, the blood being greatly deficient in red corpuscles but without the character of leucocythæmia as evidenced by the hæmocytometer. Doubtless very many of these cases are the subjects of rickets, but, just as we very often find pronounced rickets without splenic enlargement, so we meet with numerous examples of profound anæmia with enlarged spleen without any evidence whatever of rickets. Among my hospital cases syphilis occurs in many histories, but nevertheless its possible or probable absence is sufficiently frequent to throw considerable doubt on its claim to any essentially ætiological position. In this affection, which is often though by no means always fatal either from intercurrent lung- or other inflammation or from wasting with or without peritoneal dropsy, the spleen is found post-mortem to be large, hard, tough and simply hypertrophied, and, though sometimes the lymphatic glands and the liver may be more or less enlarged, no other morbid condition can be demonstrated in many cases. The enlargement of the spleen occurs in all degrees, varying from a size of bare palpability to a visible tumour which may occupy more than the left half of the abdomen. My experience of many of these cases is in

G

accord with Henoch's that their origin is mostly insidious and that the splenic enlargement and pallor may have advanced to a great degree before wasting, ascites or any marked digestive disturbance is observed. Œdema of the feet, hands or eyelids and sometimes of the body generally has occurred more or less in several of my cases, and sometimes there has been a small purpuric eruption or a tendency to epistaxis or other hæmorrhages ; but it is in the much rarer cases of true and always fatal *leucocythæmia* which may sometimes be demonstrated before the pallor becomes very marked, that such hæmorrhages take place. I have seen retinal hæmorrhages in two cases of this latter disease in children under three years old.

Children suffering from this "splenic anæmia" in a marked form are usually drowsy and very apathetic. Sometimes the liver is enlarged as well as the spleen.   In one case of mine aged four years, where the corpuscular richness was reduced to one half of the normal, both spleen and liver were large and hard.   There was absolutely no history or suspicion of other disease and the attack had begun three weeks before admission with shivering, vomiting, diarrhœa, headache and some delirium.   On admission there was a slight rise of temperature which soon subsided, and much sweating.   Recovery was far advanced on discharge.   Although the ultimate prognosis in profound anæmia with much splenic enlargement is, according to my own hospital experience, most often bad, yet several cases live on for some years, and some, as Henoch records and I have myself seen, seem to recover completely from all signs and symptoms of this affection.   An interesting and valuable account of a series of thirty cases from his practice at the Victoria Hospital for Children in Chelsea was brought by Dr. J. W. Carr before the Medical Society of London in February 1892.   Thirteen either greatly improved or recovered with apparent completeness, both the blood and the spleen regaining their normal character even in several instances where the spleen had been very large and the anæmia excessive ; while ten were known to have died either from increasing anæmia, exhaustion or intercurrent disease, and one after two years was *in statu quo*.   The rest could not be traced.   Most of those which recovered, Dr. Carr informs me, were seen as out-patients (and therefore probably more or less early in the affection) while the fatal cases were mainly in the wards.   My own noted cases being exclusively in-patients were comparatively severe and most of them either showed but little improvement or died.   Dr. Carr found either certain or probable evidence of syphilis in 14 cases, 7 of which died.   Most of them were more or less rickety but there was no connexion between the severity of the rickets and the size of the spleen or amount of anæmia. In some of the more severe cases hæmorrhages and irregular attacks of pyrexia had occurred. In seven necropsies the spleen was firm, dark and more or less hard, and the microscope showed only simple hypertrophy with some increase of fibrous tissue. There was no noteworthy change elsewhere.

I have **treated** nearly all my cases with arsenic and iron and many with quinine as well, duly insisting on warmth and attending to nutrition. In some few cases I have seen rapid improvement set in coincidently with the arsenic and iron treatment after long periods of deterioration or of at least no favourable progress. Dr. Carr gives the preference to iron over all other drugs. Whenever syphilis, malaria or lardaceous disease be evidenced or suspected appropriate treatment must be undertaken when possible; and in all cases the patients should be amply supplied with fresh air and light.

It seems probable that the splenic functions are exceedingly active in early life and that consequently affections of this organ are more frequent then than at later periods, but as we are in possession of no exact knowledge of the nature of the part played by the spleen either in the making or the modifying of the blood or in other physiological processes it would be unpractical in a clinical work to discuss the ultimate **ætiology** of the disorder we are considering. We must be content with the title of "splenic anæmia" without necessarily attributing the anæmia to primary enlargement of the spleen, and may provisionally regard the "symptom-complex" as an expression of bad nutrition perhaps rather especially favoured by syphilis. It must be remembered as bearing on the ætiological question that almost all the cases occur within the first three years of life, much the largest majority in the first year and a half, and many in the first six months. Dr. Carr's youngest case was two months old when first seen, and the oldest was $2\frac{1}{2}$ years.

Of more or less acute enlargements of the spleen we find examples in enteric fever as well as in some other febrile diseases. I have occasionally noticed in acute lobar pneumonia considerable splenic enlargement which receded on the patient's recovery. Enlargement may also occur in connexion with embolism in the spleen, and especially in ulcerative endocarditis.

I would state in conclusion that by "enlargement of the spleen" I mean an enlargement which can be readily demonstrated by careful palpation; for in children much more than in adults the results of percussion alone are seldom to be trusted as evidence. The resistance moreover of the abdominal walls and gastric or intestinal distension often make it difficult to detect an enlargement easily recognizable in other conditions. An enlarged spleen can usually be distinguished from other tumours either by feeling its notch, or by carefully observing its position and noting that it rises and falls with the moving diaphragm. It must always be remembered that thoracic deformity or left pleural effusion or other conditions that may displace the diaphragm may cause a healthy spleen to become palpable.

CHAPTER XII.

URINARY DISORDERS.

THE quantity and quality of the urine vary more within the bounds of health in early childhood than in later life. Heat and cold produce comparatively greater degrees of scantiness and profuseness of flow, and the imperfect stability of the nervo-muscular mechanism of urination is probably a not infrequent cause of marked quantitative variations. Urea is excreted in relatively larger amounts than in adults, and lithates are more often seen as a deposit owing not only to excess from a probably overdue amount of food, but also to diminution of the proportion of urinary water. Uric acid often appears in the urine without any symptoms of ill health, as the result not only of concentration but also of temporary hyperacidity of the urine, or from excess of nitrogenous food. When however marked quantitative or qualitative abnormalities are observed for any length of time search should be made for morbid causes. **Polyuria** may result from some dietetic errors or digestive disorder, barley-water being, according to Eustace Smith, a cause in some children. Ordinary *diabetes* is very rare. I have very little personal experience of this disease in childhood, and have never seen a case in my practice at the Hospital for Children. Authorities on the matter agree that diabetes is rapidly fatal in early life and often hereditary, and it is believed by some that it has a close connexion with hereditary phthisis and struma. A case I once saw, of about 10 years old, died with coma two months after the symptoms were first observed, the polyuria and glycosuria having much decreased under orthodox treatment. A similar case is reported in detail by Dr. G. B. Fowler in Keating's *Cyclopædia of the Diseases of Children.* The so-called "diabetes insipidus," where a flow of watery urine of very low specific gravity takes place out of all proportion even to the copious ingestion of fluids prompted by thirst, is more often met with. Such cases are sometimes, though rarely, known to follow on injuries or organic disease of the brain, but more are connected with functional nervous disturbance, and most which have come under my care with this diagnosis, suffering perhaps from some debility or other indefinite complaints, have shown little or no evidence whatever of the affection when placed under the ordinary routine of hospital treatment without special diet or medicine. In obstinate cases, however, valerian, or the valerianate of zinc recommended as useful by Trousseau and Sir William Roberts, may be tried. Before regarding any case as diabetes insipidus we must remember that polyuria may result from chronic renal disease of which I have seen two instances, at first sight simulating diabetes, where albuminuria was slight and frequently absent, and also from hydronephrosis of congenital, calculous, or other obstructive origin, in which it may be observed, as shown by one of my cases, before there is any abdominal distension or evidence of renal enlargement. In the case referred to both kidneys were greatly enlarged from pelvic distension, and the mucous membrane of the bladder much trabeculated, with numerous hæmorrhages. Towards the end there was some albuminuria and occasional hæmaturia. **Oliguria** or even temporary anuria is often seen with profuse diarrhœa, vomiting, great exhaustion from insufficient food, various febrile states, excess of uric acid in the urine, and in renal mischief with or without dropsy, and either primary, or secondary to cardiac disease.

Enuresis or incontinence of urine, although in children rarely due to local trouble, is from a practical point of view best treated in this context. In infancy this condition is physiological, the urinary reflex not being as yet under the control of the brain. In some children this imperfection of control lasts far beyond the normal period, and when not distinctly referable to deficient training and habit or marked local irritation is usually connected with other signs of nervous instability. Not only does it occur, sometimes in an obstinate form, in those who suffer from night terrors or epilepsy, but in greater or less degree it is frequently observed in stammerers, in those apt to start and tremble at slight causes and in emotional children generally. Often it may occur after some definite occasions of excitement. In many instances the affection begins from various causes in early childhood after normal control has been acquired. According to Goodhart there is a special liability to it in members of rheumatic families, affording one illustration out of many others of the neurotic relationships of acute rheumatism. The conventional reference of this phenomenon to spasm of the detrusor or atony of the sphincter muscles is little, if anything, more than a tautological statement neither explanatory nor therapeutically helpful. In the worst cases incontinence is diurnal as well as nocturnal, but all are usually curable alike by care and psychical treatment, except where there is local malformation and in some few where there may be a marked hereditary tendency to this special complaint. In the majority of instances incontinence occurs only during sleep, when any undue reflex excitability of the bladder is unantagonized by the extra contraction of the sphincter accompanying voluntary impulse in the waking state. When the incontinence is also diurnal we may infer some derangement of the nervo-muscular apparatus of urination which is frequently referable, in the absence of local trouble or detectable cause of peripheral irritation, to imperfect nervous control over the sphincter. My experience has taught me that, when local mischief can be excluded, almost all obstinate or really intractable cases of incontinence of urine by day as well as by night are marked by some degree of mental abnormality. Apart from cases due to bad training which are usually cured with ease the affection is perhaps most frequent in boys although its worst instances occur certainly much more often in girls. The difference is probably due on the one hand to more multiform sources of peripheral irritation in boys and on the other to a greater liability to nervous disturbance in girls.

The common nocturnal form of incontinence may occur every night or at longer and irregular intervals, and it is, according to many observers, especially apt to take place soon after going to bed or early in the morning—periods when although the higher control is in abeyance yet reflex irritability is less profoundly affected than in the deepest sleep. However this may be at least many cases are soon curable by being waked up and encouraged to urinate within a few hours after going to sleep, the time being fixed in each instance after careful observation, and by a second arousal early in the morning.

Causes of irritation leading to the necessity of frequent urination should always be searched for and removed when possible. Such are too much drinking, hyperacidity of urine, the presence of uric acid, probably late and improper meals, renal mischief both nephritic calculous and, though very seldom, hydronephrotic, and stone in the bladder. Diabetes insipidus must also be thought of. Threadworms in the lower bowel and other rectal troubles are often excitants, as also is vulvitis, or preputial irritation especially with adhesion or marked phimosis. I do not, however, think that a long and tight foreskin apart from a high degree of phimosis is so common a source as many seem to assume. Several protracted cases of enuresis previously circumcised with therapeutic design have come before me, and I have seen many children with very contracted preputial opening and no tendency to incontinence. Few cases last beyond the period of puberty even when neglected.

The proper *treatment* of enuresis is partially pointed out by the consideration of possible exciting causes. In nocturnal cases late meals and especially late drinking in any quantity should be avoided, alkaline medicines given if the urine be very acid, and all urinary and other indications duly attended to after careful examination of the case. The child should always be awaked at regular intervals during the night to pass water, and should be incited by any means other than punishment to make efforts to break the habit. Most cases are thus cured without drugs in a short time, as is amply evidenced in hospital practice. In many instances however and notably in those where the incontinence is also diurnal general tonic treatment both hygienic and medicinal is of indispensable value even when all psychical methods have more or less failed. I have no belief that strychnia cures by its alleged action on the sphincter of the bladder but with iron or arsenic it is probably a very helpful drug. I have so signally failed to produce any probably good effect with belladonna that I am constrained in the face of most authority to the contrary to pronounce at least against its frequent usefulness. I have never known improvement from it in cases with otherwise unaltered conditions, nor on the other hand in those to whose previously careful but unsuccessful treatment it formed the sole addition. I have pushed it in many simple cases, afterwards cured, as far as any of its advocates could wish, and have moreover incidentally found that symptoms of poisoning are by no means so hardly produced in young children as many modern authorities allege. In many cases, and especially when there is marked excitability, a temporary course of nervine sedative medicines such as the bromides, opium or other drugs, an extra dose being given at night, may be tried, and often meets with marked success. In such cases belladonna has probably gained some of its credit. The worst case that I have ever seen recover was that of a girl fourteen years old who had suffered all her life and whose mother and uncle had been similarly affected until beyond the period of childhood. I treated her for two weeks at home, giving careful instructions to the mother and rapidly pushing belladonna till throat symptoms and a rash appeared. She was then admitted unimproved into a private hospital, where with dry diet, direct instruction, routine living and neither medical visits nor drugs she became rapidly better and was permanently cured in one month.

Copious and continued **deposit of urates**, especially when accompanied by crystals of uric acid, is not seldom attended by symptoms both local and general and may indicate a tendency to calculous formation in the urinary tract. The vexed question of the ultimate pathology of those cases where there is an actual excess of uric acid excretion, being outside the domain of renal disorder and involving the consideration of hepatic and other functions, need not here be discussed, but it must be remembered that the symptoms of so-called "lithæmia" or "lithiasis" are any or all of them frequently met with in children and are often connected with marked deposits of urates and uric acid. Such are disturbed digestion with coated tongue, pains in the limbs and head, toothgrinding, irritability of temper and drowsiness followed by sleeplessness.

These attacks are often the sequelæ of excessive or indiscreet feeding and are, I think, frequently seen in children of distinct gouty heredity. They are also doubtless frequent in strumous and neurotic town children who are the victims of want of fresh air and exercise. It must be remembered that uric acid infarcts are sometimes found in the kidneys of recently born infants and that uric acid gravel is often seen as a deposit in the urine of quite young babies, becoming less frequent as age advances. Up to a certain point such a deposit is physiological, and it is further, as all observers know, quite common to find uric acid gravel from time to time in the urine of children before the age of puberty with few or no symptoms. The border line between the physiological and the pathological is here difficult to draw, but it may be said that frequency of this condition is usually accompanied by some symptoms, and is mostly confined to distinctly delicate children whose system, owing probably to deficient blood oxygenation, is not capable of sufficient urea formation. We also frequently meet with cases of somewhat suddenly occurring fretfulness or screaming in infants and young children which may last more or less for a few days and then disappear rapidly with a copious discharge of highly acid urine loaded with urates, and there are still more severe attacks with great pain where crystals of uric acid or of urate of soda are found in the urinary deposit. In these cases there may be frequent urination or occasional anuria, and the child may complain of pain in the lower part of the abdomen or the urethra. As far as the urinary phenomena are concerned in cases of so-called lithiasis we may find all grades of severity between those which from their triviality and evanescence can scarcely be regarded as pathological and others which are definitely recognisable as due to calculous formation in the urinary tract. It is well known that, although other deposits may be found, the calculi of children almost always consist either of uric acid or urate of soda. In spite, therefore, of occasional deposits of urates or uric acid being of little practical significance in young children their frequent occurrence should not be neglected. Symptoms of discomfort and pain, not always in immediate connexion with the act of urination, are common in cases of this kind, and hæmaturia is an important diagnostic sign of probable calculous irritation of the urinary passages. I have notes of many cases of paroxysmal abdominal pain without hæmaturia in children otherwise healthy which led to the discovery of crystalline deposits in the urine, relief or recovery being usually attained by careful dieting and alkaline medicine. In the numerous cases attended by hæmaturia we should always suspect renal, ureteric or vesical calculus, and even in the absence of hæmaturia the bladder should always be carefully and even repeatedly sounded when the case is doubtful.

Cases of frequent "gravel," especially when accompanied by symptoms, should be dieted at regular intervals by small meals consisting of very little meat, plenty of milk and vegetables and a moderate quantity of farinaceous food. Bacon may be given from time to time. Sweets should be avoided. Citrate or acetate of potash should be taken in such doses as to render the urine slightly alkaline, for which purpose about fifteen grains every four hours will probably be sufficient for a child about eight years old. In cases where renal calculus is suspected or diagnosed this treatment must be persevered with for several weeks. The child should always be made to drink copiously of water which must not be hard and should preferably be distilled, and due warmth of clothing and plenty of fresh air and exercise should be insisted on. A graphic exposition of the symptoms of "lithæmia" in children by the late Dr. M. Fothergill is to be found in vol. ii. of Keating's *Cyclopædia of the Diseases of Children.* Though somewhat high-coloured in my opinion for an accurate clinical picture the article is valuable owing to the too scanty attention usually paid to this subject by writers on the maladies of childhood.

With regard to **renal calculus,** frequent though it be, there is nothing peculiar to childhood. I would however emphasize the fact that in children as well as in adults the presence of calculus by no means necessitates hæmaturia nor even marked pain, and that crystalline deposits are often absent from the urine. Pyuria either continuous or intermittent with more or less frequent urination may be all that is observed.

Hæmaturia may occur, as in adults, from many causes, the most frequent of which by far are crystalline or calculous deposits in the urinary tract, usually demonstrable by microscopical examination of the urine. Most attacks of recurrent hæmaturia in children are preceded by pain in the umbilical, hypogastric or lumbar region, and are connected with more or less constant enuresis. The pain and enuresis may how-ever occur without hæmaturia. A careful examination and discovery in the urinary deposit of crystals of uric acid, urate of soda or oxa-late of lime will sometimes explain an otherwise mysterious case of hæmaturia attended by little or perhaps no pain, as I have more than once found in instances where patients had been long treated by styptics and other measures with the diagnosis of intermittent hæmaturia from hypothetical causes. In this context I may mention an interesting case, lately reported to me by Mr. Scott Battams, of a little boy who was brought to him suffering from profuse hæmaturia after eating for several days immoderate quantities of rhubarb supplied him by his mother in the belief that it was very good for him. The extent and suddenness of the bleeding, which was a first attack, prompted inquiry as to diet and thus at once revealed the true cause—the oxalate of lime in the rhubarb.

The hæmaturia rapidly and permanently disappeared on the discontinuance of this manner of feeding.   It must always be remembered that hæmaturia may result from small crystalline deposits in the kidney as well as from those which merit the title of calculus.   My notes of several cases due either certainly or with the greatest probability to crystalline or calculous deposits in the kidney show that the affection may recur for years with more or less protracted intermissions but often without increasing severity.   Sometimes, though rarely, they seem to cease altogether.

I do not fail, however, to recognise with Goodhart and others a class of cases, though I believe it to be but small, in which the absence of all signs and symptoms other than true hæmaturia and at least apparent recovery under ordinary care render a definite diagnosis impossible.

Hæmaturia occurs also, though not always and often but slightly, in tuberculous kidney and in sarcomatous kidney.   I have seen it in three cases of purpura and in two of diphtheria.   Its most common cause, however, apart from renal or vesical calculus, is acute nephritis, whether or not of scarlatinal origin, when the urine has usually a smoky tint.

Blood which comes from the bladder or ureter, in children almost always due to stone, is distinguished from that first shed in the kidney by its greater brightness, less intimate intermixture with the urine, and, above all tests, by the absence of renal blood-casts on microscopical examination.

Hæmoglobinuria or the presence of the colouring matter, without the corpuscles, of the blood, demands separate attention owing to its peculiar clinical relationships, although at least in Britain it is by no means a common affection.   This term is to be applied only to cases where absence of blood corpuscles is established by the microscopical examination of the urine immediately after it is passed, for under certain conditions, such as ammoniacal urine, the corpuscles may be very soon destroyed. True hæmoglobinuria implies hæmoglobinæmia or at least a partial destruction of the circulating blood, and may occur in cases of septic or medicinal poisoning, chlorate of potash in large doses having been especially credited with its production.   Of the salient, dangerous and mostly fatal example in new-born infants, sometimes of epidemic character, known as Infectious Hæmoglobinæmia or Winckel's disease and descriptively styled by that observer as "cyanosis infantilis icterica perniciosa cum hæmoglobinuriâ" I have no personal knowledge, nor, as far as I know, is there any important British literature on this disease.   A full account drawn from continental and American sources is given by Dr. Griffith in vol. iii. of Keating's *Cyclopædia of the Diseases of Children.* Its very probable origin in some microbic poisoning of the blood is important in studying the cases of paroxysmal hæmoglobinuria, elsewhere

alluded to in connexion with Raynaud's disease, which are from time
to time observed in children as well as in adults, especially when we
remember that probably the majority of adult cases are of malarial
origin. Paroxysmal hæmoglobinuria, however, may occur in children
without any traceable history of malaria or other definite antecedent
mischief. It is almost always associated with some degree of circulatory
stasis in the extremities or ears, known, according to its varying intensity,
as local syncope, asphyxia or gangrene, but, since all these latter phenomena may take place with no urinary abnormality, this form of hæmoglobinuria cannot be regarded as a substantive clinical affection. The
urine in these cases, devoid, as has been said, of blood disks, is of dark
port-wine or of porter colour and almost exclusively passed after exposure
to chill in some degree. My own clinical experience of this affection
has been hitherto confined to adults, but undoubted cases in children
have been reported from time to time. Dr. W. Pasteur has kindly shown
me the notes of two typical cases in girls aged 6 and 4 respectively,
occurring in his practice at the North-Eastern Hospital for Children.
Both were of undetected origin, and the blood drawn from affected parts
of the body at the time of the chills and the hæmoglobinuria was very
poor in corpuscles which were much altered in character, crenated or
otherwise, and showed little tendency to form rouleaux. The urine in
both cases contained crystals of oxalate of lime such as have been found
in other instances. Cases in children have been reported also by Drs. T.
Barlow, J. Abercrombie and others.

I have recently seen at Westminster Hospital two adult cases, similar
to the above, but both malarious, with signs of well-marked though
inextensive past gangrene of the ear-tips, and with marked pallor or
blueness of the extremities immediately preceding the appearance of
blood in the urine. Sometimes there is pain in the back or abdomen
during the attacks and sometimes there are periods of pyrexia. In both
of Dr. Pasteur's cases the temperature rose frequently to 103° when the
dark urine was passed, and was often considerably above normal at other
times. The *treatment* of hæmaturia and of hæmoglobinuria must of
course have respect when possible to the ascertained cause in each case,
and is elsewhere incidentally dealt with. The symptom itself, from
whatever cause arising, scarcely ever calls for attempts at styptic treatment. When the symptoms are associated with Raynaud's disease it is
advisable to give quinine, arsenic and iron, and in all cases the body-
warmth should be sedulously maintained.

Albuminuria as a symptom of kidney disease will be presently mentioned under that heading. It may occur temporarily in many febrile
and infectious disorders with greater frequency than in later life, and its
disappearance very soon after the patient's recovery is more often noted

in childhood. A salient example of this, among others, is the frequent albuminuria without dropsy or other sign of nephritis which so often accompanies or supervenes on scarlatina. Albumen may be also caused by the presence of pus in the urine derived from the urethra or from leucorrhœa and inflammatory conditions of the vulvo-vaginal passage, from the irritation of vesical and renal calculus, and from tuberculous or sometimes malignant disease of the kidneys; by hæmaturia and hæmoglobinuria; by chyluria; and by pressure on the renal vessels from tumours in the abdomen.

CHAPTER XIII.

ON ANASARCA AND KIDNEY DISEASE.

Anasarca more or less general but especially affecting the eyelids and extremities is the leading symptom of kidney disease in infants and children, the chronic nephritis of contracting kidney without dropsy which is so common in adults being but rarely seen in early life. Before speaking of kidney disease I shall dwell shortly on certain conditions apart from nephritis or at least without any evidence of its presence in which marked anasarca similar to that of kidney disease is observed.

We not rarely meet with general anasarca, without albumen or casts in the urine, which may be of considerable duration but is shown in fatal cases to be unconnected with renal disease. It must be insisted on however that many of these cases are completely indistinguishable from renal dropsy during life, that it is only a careful microscopical examination of the kidneys that can exclude disease, and that, as Henoch and others have shown, there are undoubted cases of readily recognized nephritis which have not been evidenced by any abnormality of the urine. It is especially anæmia in its various forms, splenic and otherwise, leucocythæmia, cardiac weakness, diarrhœa and exhausting pulmonary affections in which anasarca without albumen or casts in the urine is apt to occur. Many of these cases, of course, may recover perfectly. Most authors agree in stating that dropsy without albumen or casts in the urine may occasionally occur after scarlet fever; and the same sequela is seen after other acute diseases, mostly connected with considerable anæmia. I have seen also in weakly infants several cases of extensive anasarca generally fluctuating much in quantity where there had never been any albuminuria and where after death only slight renal congestion was found. Thus a child who had had convulsions from birth and jaundice for three weeks, and never seemed well, had general dropsy at the age of three months; the urine was very pale and became scanty shortly before death at the age of 4½ months. The child was wet-packed and treated with hot-air baths with no good effect. In another child, admitted at the age of 14 months with very variable anasarca and urinary flow, the attack had begun recently and suddenly with dark urine after four months of illness and wasting following varicella. There was great anæmia but no splenic enlargement and no qualitative change in the urine. In spite of care and treatment the child died two weeks after admission. Nothing was found post-mortem in either case but a little renal congestion and slight collapse of lungs.

General anasarca without albuminuria sometimes occurs after inflammatory skin affections such as erysipelas and urticaria. I have also seen two cases of extensive dropsy with albuminuria and scanty urine in immediate connexion with severe acute urticaria. In one, which was a single attack, all symptoms completely disappeared. In the other, possibly of a different nature, there had been several exactly similar attacks during one of which there were under my observation repeated uræmic convulsions with pericarditis and pleurisy. Slight albumen without casts in the urine remained after otherwise good recovery.

Acute nephritis in children is undoubtedly most often of scarlatinal origin, and may occur at periods varying from one to four weeks or more after the onset of the disease. It is mostly evidenced by dropsy, nearly always by albuminuria and the presence of epithelial or hyaline casts in the urine, and very often by more or less hæmaturia. Shortness of breath in some degree is mostly present and sometimes there is a little fever. Vomiting at the outset is frequent and may be repeated. Irregularity and slowing of the heart-beats, obscuration of the first sound or a systolic apex murmur, and accentuation of the second sound, especially at the aortic cartilage, are often to be observed. Great pallor of the skin almost always obtains, and in many cases, especially after scarlatina, this pallor, and œdema of the eyelids, hands and feet are the first symptoms complained of, and lead to the detection of scanty urine containing albumen or perhaps some blood. In some cases nephritis is distinctly evidenced by albuminuria, casts, leucocytes and epithelial cells when the urine is not scanty and there is little or no œdema. The specific gravity of the urine is usually low. Headache is a very common symptom, and there is generally constipation though occasionally diarrhœa. Fever is sometimes absent throughout the attack, and is scarcely ever high in nephritis without other complications. Uræmic symptoms, as in adults, may occur at any time, are not necessarily connected with very scanty urine or marked dropsy and may be followed by complete recovery. Pleurisy and, still more, pericardial effusion are usually of late occurrence and very serious significance. Œdema of lungs, fauces, and glottis are also signs of great danger. Ascites is mostly a somewhat grave expression of the disease but, nevertheless, often disappears. Acute pleuritis, pericarditis and peritonitis may each or all take place, but generally in fatal cases. Although in every case of acute nephritis in children safe practice requires us to inquire and examine for evidence of antecedent scarlatina, we must remember that the disease may occur with diphtheria, measles, some forms of pneumonia and other fevers, and may accompany the septic diarrhœa of infants which is so common in the hot weather. It is moreover sometimes found in children a few weeks old. I am convinced from experience of many cases that acute nephritis is not at all rare in young children, not only when there is absolutely no reason to suspect but also where it is possible to quite exclude a scarlatinous origin. Some of these cases have followed soon after a definite "chill" or exposure, others were in appearance idiopathic, and in still others there was suspicion or evidence of pre-existent chronic nephritis. On the whole I am of opinion that apparently idiopathic nephritis, including nephritis from "chill," is much more common in children than in adults. As regards severity and complications I know of no absolute difference between scarlatinous and other nephritis. In one of my cases with well-marked dropsy and nephritic urine of certainly non-scarlatinous origin there were pneumonia and slight pericarditis followed by recovery.

The *prognosis* of acute nephritis in children is on the whole good, the large majority of scarlatinous and apparently idiopathic cases recovering with seeming completeness. Nevertheless recurrence, after complete disappearance of nephritic symptoms, with slight or indiscoverable exciting causes is probably not very rare, several instances having come under my notice ; and the question may be asked though not definitely answered whether these are not really cases of chronic nephritis of almost stationary character or very slow progress. Chronic nephritis with its usual symptoms is however not a common sequela of the acute form in young children, but is seen more often in later years.

Of **chronic nephritis** of insidious origin it need only be said that it is of uncommon occurrence in childhood and has no special characteristics. Both the large and small forms of morbid kidney may occasionally be found. I have notes of three cases with dropsy, scanty urine, and plenteous albumen of which I could give no explanation, and of one ending in uræmia and death in which there had long been copious urine with little albumen and no dropsy. In all instances of chronic albuminuria the possibility of *lardaceous disease* should be remembered. Such cases are liable to attacks of acute nephritis, as was shown in a girl, aged 7, who had three years previously suffered from hip disease which had apparently been cured. She was attacked suddenly with hæmaturia, oliguria and dropsy and died after a month's illness. Lardaceous change was marked in kidney, liver and spleen. With respect to the prognosis of individual cases of nephritis, whether acute or chronic, there is nothing special in childhood. In chronic cases, as in adults, dilatation and uneasy working of the heart are of bad omen.

Without entering into any discussion regarding the possible existence of a persistent albuminuria without ill health which is not due to nephritis, or dwelling at all on the so-called "functional" albuminuria of adolescence, I would here only record my complete agreement with Goodhart and others who have given special attention to this subject that such conditions are at least extremely rare and should as a rule be ignored. Chronic albuminuria must for safe practice be regarded and treated as chronic nephritis.

The proper *treatment* of acute nephritis in childhood is in no way special. Besides carefully maintaining warmth in bed, and ordering a milk diet with no stint of water, of which the child may drink as much as he wants, we should endeavour to act on the skin, when symptoms indicate it, by hot-air baths daily repeated with due regard to their effects, or in default of this, by ordinary hot-water baths, with avoidance of subsequent chilling. Many order more or less continuous wet-packing instead of or in addition to the baths. From experience I strongly deprecate the use of pilocarpin, especially with children, but always try to increase the urinary secretion, when it is scanty, by the alkaline diuretics or caffeine. The benzoate of soda is also recommended for this purpose by Goodhart and others. As long as there is plenty of healthy renal structure we may often succeed in thus increasing diuresis, but all so-called diuretics are certainly impotent in proportion to the amount of renal mischief present. If no effect follow their administration the trial should be abandoned. Digitalis, I think, is only useful or not unadvisable in cases where the heart's action with undue frequency and irregularity otherwise indicates it. With a slowly acting heart, in spite of scanty urine and dropsy, it should not be given.

It is probably on the whole well to endeavour to promote daily evacuation of the bowels, but much purgation is to be avoided, as never diminishing otherwise recalcitrant dropsy but often causing harmful or even dangerous depression of the patient. Marked diarrhœa indeed may occur in cases of nephritic œdema, the dropsy remaining unaltered. In many cases only occasional aperients are advisable.

I omit all detailed treatment of urgent symptoms and complications, as common to the affection at all ages, and as regards chronic nephritis I would only say that in children or in adults each case should be

treated on its own merits and according to general indications, and
that a continuously monotonous and non-nitrogenous diet theoretically
prescribed will often bring about multiform harm with no corresponding
advantage.

Tubercular disease of the kidneys, both in the miliary and caseous
or strumous forms, is common in children. Miliary tuberculosis, occurring usually as a part of the general disease and having no special
symptoms, can only be approximately guessed at. Often there are no
clinical signs whatever. The strumous form frequently affects only one
kidney, at least for a long while, may be unaccompanied by any signs of
disease elsewhere, and is generally characterized by pyuria, hæmaturia
and, sometimes, by the presence of a tumour in the renal region. It can
frequently, though not always, be diagnosed from calculous kidney by
concomitant symptoms and by the fact of pyuria predominating over
hæmaturia, the converse being mostly the case in calculous mischief.
In a case of mine, however, which began three years before admission
with nocturnal enuresis soon followed by hæmaturia, the hæmaturia
ceased for two years, but the enuresis continued and there was much
wasting. Hæmaturia then recurred for two months. On admission
there was plenteous urine, with no albumen but a considerable
quantity of mucus; the right kidney was felt to be enlarged, and
there were some slight signs of disease at the right pulmonary apex.
Two ounces of thick pus were evacuated by Mr. Parker from the
right kidney and the very large cavity was plugged antiseptically.
At the necropsy both kidneys were markedly "strumous" and there
was some slight caseous change at both pulmonary apices. In all
ingravescent cases, where the diagnosis is made of strumous kidney,
nephrotomy and drainage should be practised, with instant or subsequent nephrectomy if the disease prove to be very extensive or
the symptoms increase. In one case of this kind in a young woman,
where nephrectomy was performed for me at Westminster Hospital by
Mr. Stonham, good health and freedom from pain, which had been
severe and prolonged, has been secured for at least two years, although
continuous slight pyuria gives evidence that the remaining kidney, like
the excised one, is tuberculous.

Of **malignant growth** involving the kidney I need only mention round-
celled *sarcoma* which is not uncommon in children under three years
old and may be found at a later age. It is of rapid growth and usually
forms an enormous abdominal tumour before death which is rarely later
than a year from the onset and most often much earlier. Like tubercular disease of one kidney sarcoma may be set up by calculous or other
irritants, but more often its origin is inexplicable and it may be congenital. Most of the symptoms besides wasting are due to interference

with other organs; hæmaturia, however, is frequent and, unlike that due
to renal calculus, of constant occurrence in spite of rest, but there is
often no other symptom of kidney disease. In a well-marked case,
aged $2\frac{1}{2}$ years, the first complaint was right-sided swelling and hardness of the abdomen, with fretfulness and occasional pain and vomiting
three months before admission. A very large tumour was found in
the right renal region, overlaid by the colon, but there was no other
abdominal abnormality and the child did not look ill. She, however,
rapidly became worse, and the tumour grew; the urine was loaded with
urates and had a slight trace of albumen, and the liver increased in size.
Œdema of the legs followed, and with epistaxis and bleeding from the
ears the child died in general convulsions a fortnight after admission.
Should a case of this kind be suspected or diagnosed early from signs of
progressive renal tumour the question of operation to prolong life may
be entertained, and an exploratory incision made. If the diagnosis be
confirmed nephrectomy should be performed.

CHAPTER XIV.

WORMS.

Of the various parasitic worms which infest the human body only the
two more or less common forms known as the thread- and the round-
worm need for practical purposes be dwelt upon. The tape-worm of
either variety is not common in young children, and hydatid disease,
though met with from time to time, claims no clinical distinction from
the affection as seen in adults. Out of the several subjects of *tape-worm*
which I have seen in young children none presented any marked clinical
symptoms, and some were markedly robust and healthy. I cannot therefore agree with the view that emaciation is more often to be attributed
to tape-worm in children than in adults. Both are frequently known
to be harmlessly affected thus for years. As regards treatment I have
found that the oil of male-fern in a large dose, preceded by very scanty
diet for a day and a purge over night and followed after a few hours by
another purge, is very often successful in removing the head of the worm.
The frequent failures to bring away the head are, I believe, mainly due
to want of thoroughness in carrying out this well-known method.

Concerning the clinical import of the presence of **round-worms** in the
body my experience teaches me that they are often expelled from the
anus, and sometimes from the stomach by vomiting, with no previous

H

symptoms or signs of ill health, and that alimentary disorder is no necessary condition of their existence. The nervous symptoms which have
been attributed to worms, such as strabismus, convulsions, chorea, night-
terrors, cough, headache and the like, have but slight, if any, claim to
consideration in this connexion, and markedly nervous and rickety
children may frequently void round-worms with no concurrent nervous
display. Abdominal pain and nausea, however, may undoubtedly precede the vomiting of round-worms which have found their way from
their habitat in the small intestine into the stomach, and in such cases I
have occasionally observed convulsions which, from their isolated occurrence in this association, may probably be regarded as due to irritation
caused by the worms. Diarrhœa, moreover, seems sometimes to be set
up by the presence of this parasite, ceasing with its expulsion. Among
the rarer symptoms are sudden dyspnœa, due to migration of a round
worm into the larynx, jaundice, from a similar occupation of the common
bile-duct, and, perhaps, obstruction of the bowel from great masses
formed by very numerous worms which are occasionally present. A
practical point to remember is that children who eventually drift into
tubercular peritonitis are often treated in an earlier stage for round-
worms. In such cases, where worms happen to be passed, the true cause
of the symptoms is likely to be ignored. Mr. Scott Battams has seen
several instances in point, and informs me of one child he saw suffering
from attacks of vague abdominal symptoms who had previously been
treated with worm-medicines followed by the expulsion of many lumbrici. The symptoms were said to be similar to those of the first attack,
but the case proved to be tubercular peritonitis.

It is not to be denied, in spite of what has been said, that many
children who pass round-worms are the subjects of alimentary disorder,
and that it is probable that an unhealthy condition of the intestinal
mucosa favours the life and multiplication of these parasites. When
therefore the presence of round-worms in the intestines is suspected or
proved, careful attention to diet and to the state of the alimentary canal
is advisable ; and we should always remember that round-worms are more
often multiple than single. The best medicinal treatment consists in
giving santonine with a little calomel for three or four nights in succession, followed by a morning dose of castor-oil. Oil of turpentine with
castor-oil is also useful and may be given when santonine is not well
borne. In all cases plenty of salt should be taken with meals.

Thread-worms infest the large bowel, especially inhabiting the cæcum
according to some authorities; but they are usually numerous in the
rectum and may often be seen moving about the anus and vulva. Their
chief and most troublesome symptom is itching in these localities, which
at night is generally excessive. Often there is evidence of concomitant

catarrh of the lower bowel, such as abundant and often sanious mucus, loose motions and tenesmus, variously regarded as cause or effect of the presence of the parasites.   Children who have abundant thread-worms often suffer from incontinence of urine, and some are said to find difficulty in emptying the bladder and to retain their urine for several hours.

It may be admitted, as in the case of round-worms, that in a large number of instances considerable alimentary disorder accompanies and precedes the local effects and detection of thread-worms, and that probably such a condition is greatly contributory, if not necessary, to their extensive propagation.   I have nevertheless seen a number of cases, which were brought on account of the local trouble alone, where thread-worms were abundant and the general health and state of nutrition thoroughly good.   On the other hand among the very many cases brought with the maternal diagnosis of "worms," based on the popularly-believed symptoms of ravenous appetite and nose-picking, I have almost always failed to obtain evidence of the existence of either thread-worms or round-worms even after repeated doses of vermicide and purgative medicines.   It is probable that whenever thread-worms exist in any quantity local irritation is always complained of.

I have found nothing better for the treatment of thread-worms than copious and high injections of salt and water in the proportion of half an ounce to a pint, coupled with saline or aloetic purgatives.   The local irritation is remedied by mercurial ointments which at the same time serve to kill some of the parasites.   Whenever there is any indication of gastro-intestinal disorder appropriate treatment, both as to diet and medicines, should of course be instituted.

CONCLUDING REMARKS CONCERNING THE DIAGNOSIS OF ABDOMINAL DISEASE.

There is doubtless a certain *facies* which suggests painful abdominal disease, characterized, as has often been pointed out, by more or less deep lines extending from the alæ nasi to the corners of the mouth, by some dilatation of the nostrils, and by a generally distressful expression.   An exaggerated form of such expression is seen, as it were diagrammatically, in any case of severe colic.   Grave and chronic abdominal mischief may, however, occur without these distinguishing marks.

Enlargement of the abdomen is of various import.   It is often temporarily marked in babies, and due then to distension of stomach or intestine with gas from imperfect digestion or sometimes from swallowing air.   The diagnosis of this condition rests on the absence of marked tenderness and of other abnormal signs on percussion.   Rickety children have frequently large abdomens from flatulent distension, flabby muscles and thoracic deformity, and this sign is often seen in chronic diarrhœa so often popularly diagnosed as "consumptive bowels." In some cases persevering deep palpation will detect the glandular masses of various size which point to tuberculosis. In older children there occur from time to time cases of extreme distension, most marked in the epigastrium, which are due to dilatation of the stomach, as described by Henoch. I have seen a few instances both in boys and girls. This condition may be temporarily relieved or soon disappear entirely, but may persist for long. It is undoubtedly of the nature of hysterical maladies, and is possibly connected with spasm of the gastric orifices. Enlargement with marked pain and tenderness suggests peritonitis or a latish stage of bowel obstruction. Although intussusception is not usually thus characterized at the outset I have seen more than one case where sudden enlargement with tenderness was the first observed symptom. In enteric fever, even without peritonitis, there is sometimes very notable abdominal enlargement. Enlargement with localised or general dulness points to disease of one or more of the abdominal organs, especially of the liver, spleen, kidney or the retro-peritoneal or mesenteric glands; to inflammatory or other growths of the peritoneum or of the pelvic organs; or to effusion of fluid. Sarcomatous disease of the abdomen, especially in the omentum, forms large and hard growths, and very distinct and considerable tumours may be formed by fæcal accumulations which are sometimes so hard as not to be indented by pressure.

Flattening of the abdomen may be seen in some cases of simple atrophy, and occasionally in peritonitis with extensive adhesions; but when much marked in any but very chronic cases of wasting it is often indicative of cerebral disorder, and especially of tubercular meningitis in association with constipation.

Want of normal movement of the abdominal walls during respiration should be noted as suggesting peritonitis or paralysis of the diaphragm.

In palpating the abdomen the flat of the hand, which must not be cold, should be used at first, and the enlargement of any organ or a morbid growth must be defined by the fingers with as little force and movement as possible. Prodding the abdomen with the points of the fingers is useless for defining the spleen, liver, enlarged glands or growths, and causes the child to hold its breath after full inspiration or to cry. The time when the child is quietest, be it early or late in the examination, should be seized for manipulating the abdomen, for restlessness and screaming interfere more with this than with auscultation or even percussion of the chest.

SECTION II.

GENERAL DISEASES.

UNDER this somewhat vague heading I shall treat of a heterogeneous group of diseases which have the common marks of profoundly affected nutrition and, for the most part, of more or less chronicity. It is not possible, nor would it be useful from the clinical point of view, to make any rigid classification on ætiological grounds of the affections to be considered here, or to separate them strictly from the group of general febrile disorders described in the subsequent section. In each of these groups, which are characterized by generality of symptoms, there are instances of both specificity of origin and of febrility, and in each are found examples of what is commonly known as "constitutional" disease, a term in my opinion too equivocal to be maintained. For clinical purposes alone the following subjects are grouped together.

CHAPTER I.

RICKETS.

FAMILIAR and for the most part easily recognised as this disease is, it nevertheless hardly lends itself to accurate definition. The multiform conditions out of which rickets seems to arise and the complex character of its morbid processes, which cannot be arranged in strict pathogenic order, forbid us to dogmatize concerning its essential nature. With a practical end, therefore, in view we must pass over several questions of morbid anatomy and pathology, confining ourselves mainly to what seem the chief conditions out of which the disease springs and to the most important points in its symptomatology and treatment.

The most salient marks of rickets are enlargement of the growing ends and borders of bones, especially of the ribs and limbs, more or less sweating, particularly of the head, much muscular weakness and a great liability to convulsions with or without laryngismus. In many cases, though by no means in all, there are other signs of widespread bad nutrition with digestive disturbance, and, although very often all traces of the disease may disappear, many cases die directly or indirectly from the affection itself or from abdominal or pulmonary disorder, and in many others deformities and stunting of the skeleton persist as results of that affection of the bones which may be regarded as the central fact in the morbid anatomy, though not in the pathogeny, of the disorder. So liable indeed are rickety children to pulmonary and alimentary affections, and so frequently fatal, owing to weakness both of bone and muscle, are the effects of the former upon them, that this disease must be regarded as the true cause of death in large numbers of cases usually registered under these headings. It is among rickety subjects that whooping-cough, as one instance out of many, is especially grave in its course and event.

The **symptoms and signs** of rickets are, generally speaking, as follow, but are liable to much variation especially according to their earlier or later onset. I have seen some well-marked cases where there could scarcely be even question of the affection described as "acute" or "scurvy-rickets" which, beginning more or less acutely with *pyrexia*, head-sweating and some tenderness of body, soon showed the pathognomonic epiphysial enlargement and ran an ordinary course to recovery. I believe that such an onset in its milder forms is probably not seldom overlooked, and that the pyrexia is perhaps symptomatic of the quasi-inflammatory process of perverted bone-formation which may be exceptionally rapid in these instances. *Sweating*, especially about the head, is seen in nearly all cases, being often the first observed symptom ; and *digestive disturbance*, marked by vomiting or diarrhœa or both and by a large belly, is very frequent. *Wasting* plainly appears in most instances while the rickety process is active, but the disease may attack apparently well-nourished children, and many retain their fat, though rarely or never their firmness of flesh, throughout the course of the affection. *Convulsion* is very often an early symptom and, even when unattended by other indications, its frequent recurrence, especially after the first few months of life, is mostly followed by demonstrable rickets. Too much clinical stress can scarcely be laid on the close association of the rickety process and the convulsive habit. Laryngismus is almost exclusively met with in connexion with rickets, and tetany, an expression of the convulsive tendency, is largely referable to the same origin. Infantile convulsions are the result of the very unstable equilibrium of the undeveloped and, at the same time, rapidly developing nerve centres, and this condition is markedly enhanced by the imperfect nutrition which is the probable cause of rickets. Convulsions may occur throughout the course of rickets and are not seldom coincident with death. Out of one set of 33 cases dying from various causes with severe rickets in my wards 5 were convulsively moribund. Laryngismus is sometimes immediately fatal.

The characteristic *osteal* changes are of the most prominent diagnostic

importance, the earliest to appear being the enlargements at the epiphysial ends of the long bones and at the borders of the skull bones ; while
the deformities which often result from the bony softening during the
rickety process are usually of later date and mainly due to movement
and pressure.   The earlier the disease sets in the more marked are all
the osteal changes.   Epiphysial enlargement is observed first and almost
always at the junction of the ribs with the costal cartilages, and more
markedly at the lower and more movable ribs, such as the fifth, sixth,
and seventh, than higher up.   When marked this beaded condition of
the ribs is named the "rickety rosary."   In some cases there may be no
other definite signs of rickets than slight and almost invisible beading
of the ribs, to be detected with certainty on careful palpation only, and
a tendency to head-sweating.   From the great softness of the bones in
many cases the thorax is deformed by marked depressions outside the
rows of "beads," and often by another, known as Harrison's sulcus,
running transversely from about the root of the ensiform cartilage
towards the axillæ on either side.   These grooves are probably caused
by atmospheric pressure on the morbidly soft ribs and are deepened
with the indrawn breath.   The lower part of the breast-bone is proportionately thrust forward, often giving rise to a special shape of thorax
the transverse section of which in its lower part has been well likened
to the shape of the body of a fiddle with its shoulders foremost.   Great
thoracic deformity is seen mostly in badly nourished children who
have suffered from pulmonary affections that put much stress on the
softened ribs already yielding to ordinary pressure.

Scarcely less frequent than the "beaded" ribs is enlargement at the
ends of the limb-bones especially at the wrists and ankles—the most
mobile and hard-worked joints in infancy—both before and after the
child begins to crawl and fall about.   In marked cases this epiphysial
enlargement is often called "double-jointedness."   The bony shafts of
the legs and arms may become bent in various directions and degrees,
mainly from the pressure of use, or of position while the child is carried,
these deformities being comparatively seldom seen in the youngest subjects of the disease.   The upper extremities are usually most distorted in
children who crawl, the lower are very often bowed outwards in those who
have walked, and other flexures may occur as well.   What is known as
"green-stick fracture" of the clavicles and other bones is frequently
seen at this stage, the time of its occurrence being unknown.   Complete fracture but very rarely takes place, and only, it is generally
believed, in cases which date from fœtal life.   Bending of the bones
is sometimes the first symptom noted when the onset of the disease has
been either too late or too mild to prevent the child's walking.

The vertebral column may be bent backwards in the dorsal region, and there is frequently lateral curvature as well; such deformities, however, disappearing when the child is lying prone or is suspended by the arm-pits, except in cases where the sitting posture has been long maintained. These phenomena are referable to muscular weakness. Neglected rickets is doubtless a frequent cause of round backs and drooping shoulders, which are often and deplorably maltreated by orthopædic artifice.

The skull often shows marked signs of affection at a very early period owing to the perverted bony growth presently to be noticed. " Bossy " swellings appear on the bones, especially on the frontals and parietals, and there are often elevations along the sutures. The forehead may appear large and square, the top of the head flattened, and a ridge may be seen at each side of the median groove. In many cases the head is long and often broad posteriorly, and there is occasionally a prominent vertical ridge in the middle of the forehead. Sometimes there is marked cranial asymmetry, or the head may be abnormally large. Hydrocephalus, be it remembered, is not seldom coincident with rickets, and in some cases there is reason to believe that the brain itself is hypertrophied.

Yielding areas of imperfect ossification are often felt close to the lambdoidal and other sutures, and localised spots of thinned bone may appear at a distance from the sutures on the occipitals and parietals during the first year. This latter condition has been long known as "cranio-tabes" and considered as a special mark of rickets. Leading modern authorities have affirmed or denied this position. For myself, after examining very numerous rickety cases for this phenomenon, I incline to agree with the original opinion of Dr. T. Barlow that cranio-tabes is but seldom seen in rickets where syphilis is not markedly present, and hardly ever where it can be with any great probability excluded; and I am convinced that many reported cases of rickety cranio-tabes are referable to the imperfect ossification in the neighbour-hood of the sutures which I have mentioned above. Further, a condi-tion closely allied to this, though less well localised, is to be found in badly nourished children with thin skulls, apart from any evidence of other disease.

The sutures are generally late in ossifying, and the anterior fontanelle, closed as a rule in health at the age of $1\frac{1}{2}$ years, may be open until the fourth year or later. The lower jaw, as justly described by Fleischmann, is sometimes much altered in shape, losing its curve and turning sharply backwards beyond the lateral incisors. The teeth are of late and irregular eruption, and soon decay. It must be remembered that both the cranial and dental abnormalities are most prominent in early rickets, and that in cases where the disease is delayed till the 18th month or later they may be altogether absent. At whatever period, however, rickets begins, dentition and ossification are arrested or modified.

I would add here that, in spite of all clinical and anatomical research, it is an almost insuperable practical difficulty to the physician in some cases to differentiate between rickety and syphilitic bony overgrowth, whether it be at the epiphyses or elsewhere, and we must ever bear in mind that rickets attacks considerable numbers of syphilitic children.

Restlessness and *irritability* are often present, the child frequently throwing off the bedclothes and lying completely doubled up, head foremost. My experience bears out the old teaching that tenderness of the body and especially of the limbs on movement or pressure is a very significant symptom. Excessive tenderness, however, according to many authorities, is confined to a class of cases presently to be noticed, which is variously described as "acute rickets" or "infantile scurvy."

Rickets is so often associated or complicated with profound digestive disturbance and pulmonary affection, and is so indefinitely prolonged by the continuance of its causal conditions, that there is little of practical value to say concerning its *duration.* Of the above-mentioned 33 deaths 20 were due to bronchitis and broncho-pneumonia, 2 to general tuberculosis with meningitis and 6 to gastro-intestinal disorder with profound atrophy. There was positive evidence of congenital syphilis in 14 out of 100 in-patient cases which I have recently tabulated, and a previous syphilitic history in 10 others. I would remark here, however, with reference to the once burning but now extinct question of the exclusively syphilitic causation of rickets, that in a majority of these 100 cases there is recorded an absence of all evidence of syphilis, and can state moreover, from a prolonged experience of both in- and out-patient children in a district of London where both syphilis and rickets are truly endemic, that, while very large numbers of markedly syphilitic children which survive for six months become rickety, a very great proportion of rickety infants are free from all past or present evidence of syphilis. I have further, outside hospital practice, seen rickets develop several times where syphilis could be positively excluded.

In cases which do not succumb to some of the many accidents of the disease there is a marked tendency, with the suspension of the causal conditions, for the essential symptoms to diminish gradually or disappear. In many instances, both slight and otherwise, the epiphysial enlargement may be altogether obscured in the process of healthy growth, and the crooked limbs become straight with a frequency surprising to the inexperienced. Even in the severer cases, without complications, early treatment will often give pause to the activity of the disease, leaving the resulting deformities alone to signify its occurrence. Sometimes, and especially when neglected, the symptoms of the affection are continuous or recurrent till the fourth year or later. Of the permanent results of rickets there are abundant examples in stunted growth,

deformed limbs, distorted pelves and mis-shapen heads. Thoracic deformities in adults are not very numerous, owing to the prevalent and early
fatality of cases thus characterized. The most typical rickety form of
pigeon-breast is not often seen after early childhood.

Enlargement of the liver and spleen is frequently dwelt on as part of
the indications of rickets, but my experience does not permit me to adopt
this teaching. As regards clinical evidence of this, a large majority of
my cases are noted as normal in these respects. In 9 out of the 100
above mentioned both organs were palpably enlarged, but not excessively; in 5 the liver alone; and in 1 the spleen alone. In early rickets,
however, with active symptoms and much wasting the liver is very often
enlarged. With regard to marked enlargement of the spleen I am quite
in accord with Henoch who, with an enormous experience, regards this
phenomenon as only fortuitously connected with rickets. Such enlargement is frequent enough with profound anæmia without leucocythæmia,
as my own experience abundantly testifies, and is certainly very often
connected with syphilis. Enlarged spleens and livers in rickety cases
examined post-mortem show nothing really distinctive either macroscopically or microscopically.

The **morbid anatomy** of rickets is a subject of great extent and difficulty, and I can but touch on it in the most general terms. As the most
prominent symptoms of the disease are those of the bone-affection, so
the only approximately pathognomonic changes found post-mortem are
confined to the bones and bone-forming tissues. The exact nature and
starting-point of the process is not agreed upon, and I must content
myself with quoting, with one addition, the words of Dr. Barlow that it
is "an irritating overgrowth of the osteogenetic tissues, and that this
and not the deprivation of lime is the primary (anatomical) fact in
the disease." There is great vascularity and swelling both at the cartilaginous and periosteal seats of the change, formation of spongy though
chemically normal bone, absorption of normal bone, or the deposition of
soft bone containing little or no lime.

In some cases there is swelling of the connective tissue around the
bones and among the neighbouring ligaments and muscles, which contributes to muscular and articular weakness. The resulting deformities
depend on the predominance of one or other of the above changes, the
epiphysial enlargement being due mainly to the cartilaginous swelling,
while the skeletal deformities are largely referable to destructive changes
or deficient formation of the bones. In many cases of advanced rickets
the proportion of lime salts is very much diminished, but this condition
is said to be by no means so general as was formerly believed. The
morbid process in the bone-forming tissues may be hypothetically attributed to irritation induced by the results of that perverted nutrition to

which I shall presently refer, and may perhaps be regarded as causing in its turn the complex of symptoms which is clinically known as rickets. We are at present, however, ignorant of what factors of faulty nutrition form the connecting links with the bone-affection. We must look for ætiological material, in default of sufficient demonstration, to the peculiarities of the subjects of the disease and to the conditions in which it seems to arise.

Rickets makes its clinical appearance in a large majority of cases between the *ages* of six months and two years, but may be occasionally observed at birth in unquestionable form quite apart from those cretinoid cases inaccurately entitled "fœtal rickets," for an account of which I must refer the reader to the text-books. The age on admission of 100 cases taken into my wards, either in the active stage or showing marked deformities, varied from three months to four years. Six were under six months, eleven between six months and a year, forty-eight between one and two years, seventeen between two and three years and twelve between three and four years. From the frequently insidious and gradual nature of the onset of rickets it is not possible to fix statistically the earliest date of its appearance, but, seeing that microscopical examination may detect the special bone changes where there is little or no clinical evidence of the disease, it may be assumed that it often exists before its recognition is possible. I am sure, however, as far as most careful clinical inquiry and, in several instances, positive knowledge of infants observed from birth can be taken as proof, that rickets often sets in very soon after weaning-time in cases which, according to all available evidence, were previously healthy. I have moreover known a few children in whom the disease apparently began after the second year, but I can say nothing from personal experience of what has been described as "late rickets."

In the East London Hospital practice at least, marked rickets is seen nearly twice as often in boys as in girls. This is shown not only by my detailed cases admitted as rickets but also by a vastly larger number of rickety in-patients, registered in other categories, and still more numerous out-patients. If this remarkable difference in the sexual incidence of the disease be universal its explanation seems far to seek.

Rickets is rare in tropical climates, common in cold and damp regions, comparatively rare in the country, and commonest of all in the children of the poor in crowded towns. Although the existence of rickets in fat children and even in those of otherwise fairly healthy appearance is not to be denied, I cannot but strongly dissent from those who teach that obviously bad nutrition and pallor are absent in the majority of cases. The question of the rôle of improper *diet* in causing rickets is one of great difficulty. Certainly all special dietetic and chemical theories of causation have been found erroneous when tested by close observation and logical reasoning; and, although there is good ground to believe that a diet defective in fat is a very important factor in many or perhaps most cases, the prevailing creed that rickets is wholly a diet disease cannot be unreservedly accepted. It is true that any deviation from sucking only at the breast increases the chance of rickets; that the children of the poor, although suckled for long, have usually the most inappropriate food as well; and, further, that improvement often follows a discontinuance of unsuitable food. Yet it is equally true that town children are far more often rickety than country children, however they may be fed, and we not infrequently find rickety subjects who have been suckled by apparently healthy mothers of the well-to-do classes. I can further testify that even in the East of London there are large numbers of children who have remained free from rickets in spite of most unphysiological diet. We must therefore recognise other factors of great importance conducive to rickets, and these we shall find in the badly-nourished mother who gives her infant the worst start in life, the deprivation of sunlight and fresh air, the inhalation of air-borne poisons and all the other evils of crowded and insanitary dwellings; and we must remember that it is in these circumstances that most of the badly-fed children are found. The profound disturbance of nutrition, moreover, induced by syphilis, even when all attention is paid to diet, must be borne in mind as a factor in the production of many cases of rickets.

We must therefore at present be content to use the term "perverted nutrition" to express the earliest link in the morbid chain at the end of which is rickets, in our ignorance of whatever essential factor there may be in the production of the essential changes which mark the disease. It may be reserved for future study to show that the whole series of nutritive disturbances in rickets are due to some original vice of the nervous system, or perchance to the operation of organic germs, but such suggestions are at present neither very luminous nor helpful in practice. This much at least in my opinion is certain, that among children of healthy parentage who enjoy plenty of sunlight and fresh air and are in generally good surroundings rickets is decidedly rare, and that such children are not readily made rickety even by inappropriate methods of feeding. But, on the other hand, I am equally convinced that in the large majority of children who have not these advantages bad feeding is one of the leading excitants of rickets, and that sometimes the disease arises without any other.

Of the class of cases described as *acute-* or *scurvy-rickets* and now generally referred, according to the teaching of Drs. Cheadle and Barlow, to a scorbutic origin I shall say but little, my own experience of this affection being not very extensive. It seems clear, from the literature of the subject and from several inquiries I have myself made of practitioners abroad, that this complex of symptoms, though not of very frequent occurrence, is much commoner in England than elsewhere. Very marked tenderness of limbs, often with apparent swelling of the bones and especially of the femora, wasting, anæmia, purpura, and bleeding from spongy gums and elsewhere are the chief points to be noticed in the clinical picture of this disorder, which has been fully described, from his original researches, by Dr. Barlow. The limb-swelling has been shown to be due to sub-periosteal hæmorrhage or extravasations between the muscles. Evidence has been brought to show that in many cases of this affection there has been a notable absence of fresh food from the diet, the infants being brought up exclusively on condensed milk and on starchy or other patent foods. Several cases which I have seen myself were undoubtedly rachitic; a few but slightly so, to all appearance; some had had a diet much deficient in fresh milk, potatoes, and antiscorbutic material generally; and one at least had been throughout fed in a manner quite antagonistic to the production of scurvy. In one case with advanced rickets, which was examined post-mortem, sub-periosteal hæmorrhage was found, but no other bleeding was evidenced either then or during life. I shall again allude to scorbutus in infants, which in some degree is perhaps not very rare, and will only say here that I think there is reason to doubt the constantly scorbutic causation of sub-periosteal hæmorrhages occurring in connection with rickets. What may be called "acute rickets" quite independent of scorbutus, without marked swelling but with tenderness of limbs and some fever, has, as I have already said, a claim in my opinion to clinical recognition.

Treatment.—Bearing in mind the view above advocated that rickets is due to defective nutrition in the widest sense, and may even be developed in intra-uterine life, it is clear that the prophylaxis must include the hygienic treatment of the mother as well as of the child. It is unquestionable that poverty, with its frequent accompaniments of repeated child-bearing and prolonged suckling, strongly predisposes to rickets, seeing that the disease is rare in the children of those who are at once healthy, well-to-do and well instructed. Although I cannot dwell here on the matter of maternal hygiene, and can but refer the reader to what has been already said on the subject of the normal feeding of infants and the treatment of simple wasting, I must emphasize the contention that on due attention to these points depends the most important part of the prophylactic treatment of rickets. Rickets must be suspected when a child is late in walking or cutting its teeth and has an unduly open fontanelle, and in all convulsive or "croupy" babies; and such children, as well as those with more pronounced symptoms, must be placed in conditions excluding as far as possible all the hygienic and dietetic errors which we regard as conducive to the malady.

For the rest, the patient must be kept lying down as much as possible, and not allowed to walk or crawl until there is reason to believe that the disease is arrested. Every opportunity must, however, be seized for supplying the child with fresh air and sunlight. Scrupulous cleanliness should be observed, for there is generally much sweating, and frequent tepid baths are to be given. A bandage round the abdomen in cases where the thorax is affected is often useful, and when there is much restlessness with head-sweating the advice of Dr. Charles West, that the child should be provided with a horse-hair pillow with a central hole, may be followed with much benefit. In most cases I give, and I believe with much advantage, cod-liver oil; and iron preparations as well as arsenic are of considerable use. Especial stress must be laid on the paramount importance of protecting a rickety child, as far as may be, from all sources of catching cold and from all conditions, infectious or otherwise, out of which pulmonary affections may be suspected to arise, and on the extreme care that is necessary in the conduct of such cases as may be attacked by bronchitis or broncho-pneumonia. It is in rickety children pre-eminently that bronchitis passes on to broncho-pneumonia, and that fatal collapse of lung takes place owing to the feeble inspiratory power which is due to softened ribs, weakened muscles and generally impaired nutrition. I deprecate strongly the employment of rigid mechanical appliances to any part of the body of the rickety child, with the exception of well-padded splints projecting below the feet to prevent walking in cases where constant watching is out of the question; and I postpone all surgical interference with permanently distorted limbs until the disease is no longer active and the child has attained to some degree of vigour. From the artificial sequelæ of ill-developed limbs, impeded bodily activity and impaired health, which sometimes supervene among the moneyed classes on the unnecessary use of mechanical appliances, the children of the poor are as a rule, with a sort of poetical justice, fortunately exempt.

CHAPTER II.

SYPHILIS.

ALTHOUGH syphilis may infect children after birth by inoculation from primary or secondary sores, I shall treat here only of that form which is generally known as "congenital" and is of ante-natal date. Omitting discussion of the modes of syphilitic inheritance I shall simply state my present views on the matter, which are the outcome of the study of authorities and of experience at a hospital where there is no minimum limit of age for admission. Apart from the large numbers of out-patients that I have seen, nearly 200 cases of unmistakable syphilis appear in my in-patient case-books, besides very numerous instances of children, admitted for other affections, in whom there was either present evidence or past history of this disease. In many of these a fairly definite family history was obtained. The worst cases are those where both parents are syphilitic, and both gravity and fatality are generally in proportion to the activity and recent date of the parental affection. That maternal syphilis alone, especially when active, is a fertile cause of the infantile disease seems as certain as it is difficult to prove by mere inquiry; but I am sure that whether or no the mothers of infected infants are themselves syphilitic by fœtal contamination or otherwise, and whether or no they are, as is usually taught, incapable of infection from their sucklings, they are in a very large number of instances, both during pregnancy and for an indefinite after-time, perfectly free from all symptoms of the disease. Both hospital and other cases give ample evidence of this very frequent maternal immunity from all apparent symptoms, while, both from personal interviews with the fathers and from frequent though unwitting accounts given by the mothers in answer to my inquiries, I have been able to establish a clear evidence of paternal syphilis in very many of the cases alluded to. As a matter of practice, seeing that congenital syphilis is eminently contagious, and not being quite convinced of the received doctrine of maternal protection, I advise an apparently healthy mother not to suckle her syphilitic child, and positively refuse to sanction the substitution of a wet-nurse whether the child has visible oral mischief or not. Further, without deciding on the possible infectiveness of the milk of a syphilitic woman, I would not allow an apparently healthy infant to be suckled by any one affected with even a suspicion of syphilis.

I

Syphilis very frequently causes abortion, and repeated abortions of macerated fœtuses are practically always syphilitic. Still-birth, at or about term, often takes place, the infant being wasted, the skin dull, dry, inelastic, and extensively discoloured or desquamating, and sometimes there are bullæ on the palms and soles. Lastly there is a large class of cases, born alive, more or less affected or soon to be affected with some of those signs of syphilis presently to be noted, among which obstinate **wasting** is pre-eminent and very often of fatal import.

The more extensive the skin-affection, eruptive or not, the worse is the prognosis. I have seen many cases marked at birth with an almost universal scaly red rash, or less often with a bullous eruption most prominent on the extremities, among other signs of syphilis. These babies are small, old-looking, and monkey-like, and nearly always die within a few days or, at most, weeks of birth in spite of the best care and most orthodox medication. They usually lie perfectly still, occasionally uttering the feeblest whine, and all vital signs are often so slight as to obscure the moment of death. Some die with the symptoms and post-mortem evidence of sudden and extensive collapse of lung, and others with acute pulmonary affections. From this class are furnished most of the specimens of visceral syphilis and also of the now well-known bone-affections which many believe to be equally characteristic of the disease.

While life lasts these infants, as well as others who with less marked skin-disease may live for a longer period, often feed with apparent voracity, and may be affected neither by vomiting nor diarrhœa. The fæces in such cases, where assimilation of food is at a standstill, are usually copious and pasty or lumpy, changes of diet and drugs are alike vain, the alimentary canal seems little other than an inorganic tube, and for a while, except by intervals of sucking, life may be evidenced alone by the feeble pulse and almost imperceptible breathing. In several of these cases I have seen post-mortem no macroscopical evidence of visceral disease, but only more or less atrophy of the stomach and bowels and general dryness of the mucosa.

Various grades of this syphilitic malnutrition may exist, and are usually in proportion to the early appearance of specific manifestations. I have often thought that many infants of apparently healthy parentage which are merely wasted from birth, and are either largely or wholly irresponsive to dietetic treatment, are really syphilitic ; and this suspicion has been frequently justified or rendered very probable by the subsequent appearance in such cases of specific lesions, or by the confession of parental syphilis. In some instances, doubtless, this malnutrition is largely contributed to by the many evils incident on poverty, and then may soon recede with proper treatment, even without specific medication.

Slight cases of wasting, indeed, even when attended by unquestionably syphilitic lesions, often recover well and quickly under general hygienic and nutritive care alone.

Of the numerous syphilitic infants which show no definite signs of the disease at birth many are apparently healthy, while others are puny with discoloured and inelastic skin. Within the first, not very often later than the second, and rarely later than the third month, special symptoms and signs arise. There is **wakefulness at night,** often with apparent pain probably referable to the bones, and **nasal snuffling** with visible discharge. This last symptom is almost invariably present and is, further, strongly indicative, though scarcely quite pathognomonic, of syphilis. Simple nasal catarrh without cough is very rare at this age, and diphtheritic coryza, from its rapid increase and concomitant symptoms, seldom causes diagnostic difficulty. The breathing is noisy and the nasal mucosa swells and often ulcerates so that the discharge, at first glairy, becomes sero-purulent or bloody, and crusts form which may cause the child to stop sucking in order to breathe. The upper lip is often much excoriated by the nasal discharge, and in some cases there are said to be mucous patches or "condylomata" on the nasal membrane. The ulceration may spread to the bone, leading to necrosis. The depression, however, at the root of the nose, so frequent in syphilitic infants and often observable in later life, is not necessarily due to this cause, for it may occur when the snuffles have been slight. The cry is often hoarse, feeble, or silent from varying degrees of laryngeal involvement. Most cases are marked by anæmia and great fretfulness, and there is sometimes a little fever.

Skin-affections may be absent, but are usually seen in some degree, appearing somewhat later than the coryza. The inner surfaces of the nates and of the thighs and, next, the neighbourhood of the eyebrows, nose and lips are most often occupied, small patches of erythema, which may quickly coalesce and soon take on a shiny or a coppery appearance, being the most frequent form. The skin of the palms and soles is often red and freely desquamates, and sometimes there is a diffused scaling over large areas, most marked on the extremities. As in adults, the flexures and other parts exposed to any kind of irritation are most likely to suffer from syphilitic dermatitis in any of its forms.

Papular rashes are common and may be very extensive. At the corners of the mouth and nose, round the anus, or in other places subject to friction and moisture, they are apt to develop into raised patches known as "mucous tubercles" or "condylomata," and often become fissured and ulcerated. Such patches are certainly not seldom seen in very early cases, appearing, however, most generally after a few months. Besides this affection at the corners of the mouth there are

often vertical fissures along the lips, extending sometimes from the neighbouring skin which may be ulcerated and crusted.   Ulceration in the
mouth and fauces is frequent, the gums, tongue and tonsils being especially affected.   Slight enlargement with hardening of the lymphatic
glands in the neck, arm-pits and groins is very common after the lapse of
some time, and the nails may suffer from suppuration of the matrix or
from fissuring without ulceration.   Pustular eruptions, taking on sometimes a bullous form, are not rare, and are generally, according to my
experience, the mark of grave or fatal cases.

Roughly speaking, the syphilitic rashes in childhood correspond to
those in adults and very often require concomitant symptoms of the
disease to establish their true character.   It is especially difficult to
distinguish some of the slighter forms of syphilitic erythema on the
buttocks and pudenda from the results of irritation and excremental
soakage, aggravated often by neglect and friction with hard napkins.
When marked erythema spreads far down the lower extremities its
syphilitic origin is very probable, and the more papular in character it
is, whether extensive or not, the stronger this probability becomes.

In concluding these brief allusions to cutaneous syphilis I must record
a warning to exercise the utmost care in differentiating between syphilitic
rashes and those which mark the exanthematic fevers.   Mistakes are
sometimes made in this matter by the expert as well as by the igno-
ramus in dermatology, as the officers of fever hospitals well know.   One
important lesson to be learned is not to trust too much to the appearance
of the rash for the necessary differentiation, but to rely mainly on a
thorough examination and careful consideration of the case in all its
aspects, not forgetting to give due weight to its history.

Late dentition or the rapid decay of normally evolved teeth may be
among the results of infantile syphilis.   I have observed this in cases
where there was no appearance of rickets, but it must nevertheless be
borne in mind that syphilitic malnutrition frequently prepares the way
for rickety developments.   Iritis is but rarely seen until after infancy,
but slight cases may be easily overlooked.   The **bones** are often affected
in the cases which are still-born or soon die, and sometimes in those
which survive.   It is chiefly the long bones which suffer, at the junction
of the epiphysis with the shaft.   The inflammatory process may cause
enlargement, and sometimes suppuration with complete separation of the
epiphysis.   Loss of movement of the extremities, known as syphilitic
pseudo-paralysis, chiefly affecting the arms is occasionally seen, with or
without epiphysial enlargement, but always with evidence of tenderness,
which may indeed be the first observed symptom.   The skull may be
bossed with periosteal or bony overgrowth, especially in the neighbourhood of the anterior fontanelle, giving rise to the appearance known as

"natiform." I know, however, of no condition of the infantile skull, recognisable during life, which may not occur in cases of rickets quite unmarked by any sign of syphilis. As to the well-known *cranio-tabes* or localised thinning of the skull, especially in the occipital and posterior parietal regions, I can state from numerous observations, made after the publication of the researches of Drs. Barlow and Lees on this point, that it is not often found in cases where syphilis can be excluded, and I therefore agree with those who regard this phenomenon as a much more probable mark of syphilis than of rickets. In differentiating between the bone-deformities of rickets and syphilis it must be borne in mind that the syphilitic affection shows itself almost always long before the sixth month, the rickety but rarely before the eighth or ninth; while with the former there is, as a rule, other marked evidence of syphilis.

A peculiar form of affection of one or more phalanges of the hand or foot, especially of the proximal, is not uncommon, and is known as "dactylitis syphilitica." The skin tends to become involved and there may be suppuration. Sometimes the metacarpal bones are diseased. These appearances are, however, not easy to distinguish from those occurring without evidence of syphilis and known as strumous dactylitis.

Definite local symptoms of disease in the **nerve centres** are not of very great frequency in syphilitic infants. The sleeplessness, which is common, and the headache, which to all appearance not seldom exists, may probably be accounted for by pain in the bones. Chronic hydrocephalus is often preceded by definite signs of syphilis, and I believe, though I cannot prove, that this affection is largely of syphilitic origin, in spite of the popular objection that it does not yield to "specific" treatment.

There are enough recorded cases both of hemiplegia and one-sided convulsions in markedly syphilitic infants to establish a reasonable belief in a syphilitic causation, the lesion being probably a localised arteritis. I have seen several cases of this kind, including one of four years old with typical aphasia, where the symptoms have either completely or partially disappeared concomitantly with treatment by iodide of potassium, and others which have persisted, with rigid contractures, under the same régime. I have never treated such cases without "specific" drugs, but in two instances in adults who were definitely known to me as syphilitic, both from history and subsequent symptoms, typical hemiplegia soon passed away without any "specific" treatment, the origin of the attacks being unsuspected by those in charge at the time. Other paralyses of separate nerves, one or more, may occur in childhood as in adults, and several cases have been recorded of mental deficiency, sometimes congenital but more often of later appearance, in connexion with infantile syphilis. Epilepsy seems sometimes to be referable to this disease.

Generally speaking, irregularly distributed and multiple nervous symptoms should always excite a suspicion of syphilis.

In this context and in illustration of some other points I subjoin a short account of a remarkable case of late wasting with convulsions in a deeply syphilitic child who ultimately recovered good health, though with much cerebral impairment.

A boy, aged three, was admitted suffering from great emaciation of four months' duration and from frequent fits, headache, pains in the limbs, occasional vomiting, and some diarrhœa of much more recent date. He was an eight-months child, born with a "blistery" eruption on hands and feet followed by bad snuffles and a red rash on his buttocks. He was persistently treated at a metropolitan hospital with grey powders from the first week up to the eighteenth month of his life. He never had any other illness. He cut his first teeth at fourteen months, first walked at 2½ years, and never talked much. His father had suffered very frequently from sore throat and ulcers on his tongue. His mother, aged 28, had never been ill and looked healthy. She had had two miscarriages, one still-birth, two children who had lived a few hours and one who had lived ten weeks. One besides the patient was alive, aged nine months, suffering from snuffles, eruption and anal condylomata. On admission the boy was found to be unable to utter sentences, but appeared to understand most that was said to him. The temperature was normal. He had recently had sixty-one fits in forty-eight hours. Soon after admission he had a general convulsion, the right side being chiefly affected, and, a few days afterwards, eight similar attacks in quick succession. In the intervals he lay on his back apparently unconscious, with flexion of arms, thighs and legs, and rigidity of the limbs much more marked on the right. After a few days he became sensible, but the already excessive wasting increased, and the rigidity of limbs, though less, continued. In this condition he remained for three months when, with some very slight improvement in nutrition, he was taken home. A year afterwards he was brought again. He was then a bright pleasant-looking well-nourished boy, but could not talk and seemed very deficient in intelligence. He could move all his limbs freely in bed, though the left leg was always kept partially flexed, and there was no wasting or rigidity in any part. But he was entirely unable to stand or sit, and always passed urine and fæces under him. After six weeks he went home in the same condition.

Early demonstrable **enlargement of the spleen and liver,** or, I think, more often of the spleen alone, is common, especially in otherwise grave cases. The liver is hard from interstitial inflammation or may contain isolated growths. I have seen a case with jaundice where the liver was studded with softening gummata, the portal fissure being involved.

SYPHILIS. 135

It is possible that many of the cases, familiar to us among the infants of the poor, and otherwise ill understood, of fatal anæmia with much enlarged spleen but without leucocythæmia, are of syphilitic origin ; and it is certain that the spleen is enlarged, and sometimes excessively, in a considerable proportion of syphilitic infants.

The lungs, heart, kidneys, and testicles are occasionally found to be the seat of fibroid disease of probably syphilitic origin ; and the involvement of the pancreas seen, according to some authorities, in some of the worst cases, may help to explain the gravity of their course, and indicate the possibility of the digestive disorders in the less severe and more chronic cases of malnutrition being connected with impaired function of this organ.

As an illustration of congenital syphilis with renal disease I may quote the following case. A boy of nine months old, who had had snuffles soon after birth and whose mother had had three miscarriages, was admitted with a history of wasting and œdema of all extremities of six weeks', and a rash on the buttocks of three weeks', duration. He had general anasarca and a dark red scaly eruption on the buttocks and the inner side of the thighs, but no albuminuria. After a week the child had a convulsive fit, affecting the right side, with some rigidity, and became unconscious. No urine had been passed for several hours. He was put in a hot bath, but very soon died. Recent collapse was found at both pulmonary bases, and the kidneys were extremely tough with wasted cortex and somewhat adherent capsules, containing also a few very minute calculi. The brain appeared quite normal.

In all probability some cases of *purpura* may be explained by syphilis. Among others I have seen a case of chronic malnutrition eighteen months old, with extensive external and internal hæmorrhages, enlarged spleen, and bossy skull, in which there was a history of marked snuffles and characteristic rash soon after birth. Syphilis, it must be remembered, is a profound disturber of tissue-nutrition generally. Many children are seen after all definitely recognisable specific symptoms have disappeared, and I have noticed in them, more often than in adults, how great the disproportion may be between the marked wasting and cachectic appearance presented and the slight degree of malaise or debility evinced or complained of. Equal wasting due to deficient food or to gastro-intestinal disturbance would have inevitably brought down the cases I speak of to a much lower level of vitality. When inquiry reveals the cause of this condition appropriate treatment may often lead to great improvement or complete recovery, but in some cases that I have seen at various ages there has been permanent stunting of growth, children of seven or eight years old being emaciated and of the stature of half their age, without any trace of rickets or other deformity.

I have known a considerable number of syphilitic and wasted infants which, without much or any skin-affection or any pyrexia, steadily deteriorated, in spite of all treatment, and were found after death to be the subjects of advanced but unexpected tuberculosis of the peritoneum or intestines. A few others, with but indistinctive physical signs, had extensive tuberculosis of the lungs and pleura. With few exceptions, however, tuberculosis of the lungs is accompanied by more or less fever.

Relapse of infantile syphilis is said by most authorities to be rare, but this statement is vague, of little meaning, and incapable of proof. I believe on the contrary that, in one form or other, symptoms of infantile syphilis are very apt to recur. It is true that the majority of cases either die early or apparently recover while under observation and treatment, many of the latter class showing, perhaps for long, no further symptom, and many in all probability remaining perfectly well. But in hospital practice, on which most large statistics and important records must be based, the cases are not long enough under observation to justify the prevalent dogmatism on this question. Among other cases I have seen one of marked syphilitic snuffles, with laryngitis and skin-affection, become rapidly well while under treatment with mercury which was continued for many weeks before discharge. It was re-admitted three months later with fresh eruption, and marked cranio-tabes which was certainly absent before. The apparent frequency of unrelapsing syphilis in infants accounts largely for the oft-repeated statement of the much greater success of mercurial treatment at this age than in later years, the true explanation of the element of fact in this clinical observation being rather, I apprehend, that, when the syphilitic poison is not enough to kill, it is more readily thrown off during the active metabolism of early life. In some cases syphilitic symptoms continue to recur from infancy to later age, however treated, of which I have seen a few examples with grave and multiform effects. In others, conventionally known as "late hereditary syphilis," there is an interval of several years before fresh manifestations are noticed, or symptoms may be seen for the first time as late as from after the second dentition to puberty, with no obtainable history of an infantile attack.

In **late hereditary syphilis**, of which I found 38 cases recorded as such in my note-books, whether preceded or not by a definite infantile attack, stunted growth and deficient mental development are not seldom seen and are probably far more frequent than reported. Such cases seem never to have been able to regain the ground lost during the tissue-starvation of their earliest years. The skull may be bossed on the forehead or the sides or round the anterior fontanelle, or there may be a marked prominence along the line of the frontal suture; the head may be square and flat at the top; or there may be chronic hydrocephalus.

The bridge of the nose is usually depressed, being either destroyed by old ulceration or congenitally flat without bone-disease. Bony enlargements, especially of the arms and legs, are frequently seen, the tibiæ being often apparently bowed. The peg-top and notched teeth and the interstitial keratitis described by Hutchinson are frequently seen, and are among the most valuable indications of syphilis. Iritis and choroiditis are less common. Sometimes, though rarely, complete blindness is caused by atrophy of the discs. Necrosis of bones, nodes, ulcerating gummata and other affections similar to what is known as tertiary syphilis in the adult may occur, and suppuration of the middle ear is frequent. There is also a form of deafness which may appear almost suddenly and become complete with no detectable lesion. It is said indeed by some authorities that syphilis is a common cause of congenital deaf-mutism. Severe headaches, especially vertical and temporal, are very frequent, and pain in the long bones is often complained of. In a girl of twelve, with distinct Hutchinsonian teeth and a definite history of infantile syphilis, who suffered from epilepsy and choroiditis with atrophy of discs going on to complete blindness, there was well-marked spastic paraplegia; and in two other cases very similar to this there was a high grade of imbecility in addition. It must be admitted that it is sometimes difficult to make a diagnosis between late syphilis and so-called " scrofula," especially in bone- or joint-diseases without a history of infantile syphilis, and some of the bony deformities attributed to syphilis are indistinguishable from those due to rickets. Those who are content to infer the cause of mischief from the effects of specific treatment are not much aided here, for the later manifestations of syphilis of whatever kind and at all ages are notoriously often recalcitrant to all medication.

The **prognosis** in infantile syphilis is really good only in those cases where the general nutrition is not greatly impaired and the skin-affection not extensive. In a vast number of cases death follows sooner or later on malnutrition with marked dulness and inelasticity of skin, even when other specific signs, cutaneous or otherwise, are slight or almost absent. In cases where a child wastes almost from birth, with an appearance of skin at all suggestive of syphilis, the prognosis must be very guarded, and, when more definite specific signs supervene, mostly unfavourable. Pustular eruptions have a bad significance even when not extensive, being usually associated with obstinate wasting, and gastro-intestinal and pulmonary attacks much increase the gravity of all cases. The very numerous and slighter instances which are seen in out-patient practice, and at least apparently recover, form the main basis of most clinical accounts of infantile syphilis, and the fatality of the disease as a whole is thus perhaps scarcely realised by many readers of text-books, or by those who have not had experience at hospitals where the youngest infants of the poorest classes in large towns are freely admitted. Out of a series of 286 syphilitic infants admitted as such into the wards of Shadwell Hospital and treated in nearly all instances by mercury from the outset, certain cases having potassium iodide as well, 173 died. A large majority of these were under six months old.

The proper **treatment** of syphilis consists in endeavouring to improve nutrition, and to arrest symptoms whether clearly specific or not. The diet should, therefore, be most carefully arranged on general principles and to suit each individual case, and, likewise, all gastro-intestinal symptoms, such as vomiting and diarrhœa, should be treated as they arise. Cod-liver oil, iron and arsenic are often quite invaluable aids. The patient should be kept perfectly clean and warm and should have the benefit, with due precautions, of all possible fresh air and sunlight. In all bad cases, where rapid improvement does not set in within a few days of their coming under care, I have been in the habit of giving mercury in some form—usually the Pulvis Hydrargyri cum cretâ in daily quantities of from one to three grains, increased according to age—as also in all cases, however good their nutrition may be, where any specific lesions persist. The Liquor Hydrargyri perchloridi may similarly be given in quantities varying from half a drachm to a drachm a day. There is no doubt that many syphilitic events can be lessened or abolished by mercury or iodide of potassium, the latter drug being especially useful when there is active ulceration of mucous membranes or evidence of bone or periosteal mischief. I must however record my strong opinion that the question of the power of mercury in eradicating or "curing" the disease appears to me, after long experience and reflection, to be no nearer solution from the results of treatment of infantile syphilis by this drug. Syphilis, however treated, is a very malignant and fatal disease in infancy. In the less severe cases, characterized mainly by snuffles and slight eruptions chiefly confined to the nates and pudenda, recovery, which is frequent, is usually put down to the mercurials almost always prescribed. I know, however, from experience that many such cases recover equally well without mercurial treatment; and also that definite symptoms may make their appearance while mercury is being given in full doses on mere suspicion of syphilis, and may mark a severe or even fatal course of the disease. I would particularly remark here that it is only in comparatively few (though numerically many) of my cases, and such only as I have above alluded to, that I have omitted the orthodox treatment, almost all the fatal ones having had mercury systematically from the beginning, and many having been already under such treatment as out-patients. I believe that the effects of mercury and potassium iodide on syphilitic manifestations is no greater in infants than in adults ; that, as proved by the common mortality of the infantile disease however treated, it is certainly not more curable but rather much more serious than at a later age, owing to the profound malnutrition it engenders ; and that the much greater number of apparently or really complete recoveries in infancy is due to the more ready throwing off of a less than lethal quantity of poison during the active metabolism of early life.

I agree with the somewhat inconsistent practice of some of the most thorough-going advocates of mercury in discontinuing the drug very soon after the subsidence of specific symptoms, differing herein from other believers in the antidotal powers of the drug who logically urge a more continued course of treatment. Such a course, in my opinion, is even more often decidedly harmful to infants than to adults, and I have seen several cases rapidly improve when a long mercurial course was discontinued.

I substitute mercurial inunction for internal treatment only when the latter seems to disagree, although I have little reason to believe that inunction in such cases is much more appropriate. I find potassium iodide as markedly useful in checking symptoms as it is with adults, even in cases which ultimately waste and die ; and both in adults and infants I frequently prescribe this drug with success when symptoms are active and increasing, regardless of all doctrinal rules as to the so-called secondary or tertiary stage of the disease. While somewhat narrowing, then, the therapeutic field of mercury I would widen that of the iodide, as defined in each case by most syphilologists.

For severe mucous ulcers and skin-eruptions which do not tend to heal quickly the best applications are calomel powder, the ammoniated mercury ointment, or various dilutions of the nitrate of mercury ointment, the efficacy of which is often very manifest in lesions other than syphilitic.

I have written thus on the specific treatment of syphilis in order to emphasise my strong conviction that, important and often indispensable as mercurials may be, we must never forget the paramount duty of endeavouring to improve the general nutrition in syphilis by all available means. Simply to prescribe mercurials for syphilitic babies without insisting strongly on general treatment and improved hygiene is, in my belief, little more than trifling at the best.

CHAPTER III.

SCROFULOSIS OR STRUMA.

In speaking of this subject at the present day, when precise conceptions and definite statements in pathology are justly demanded, we are met by many and diverse difficulties. Not the least of these are the demonstrated facts of the identity or similarity of the anatomical processes in affections formerly known as "scrofulous" and "tubercular" respectively, and the ascertained presence of the bacillus of tubercle in many morbid products, such as for instance in those of joint- and bone-disease, which were previously described as "strumous" and regarded as a class apart. In a practical work, however, confined to disease in childhood where scrofulosis and tuberculosis or, according to the now prevalent nomenclature, tuberculosis alone figures so pre-eminently, it is perhaps less necessary to express a dogmatic creed on the difficult pathological questions involved in this subject, and more excusable to refrain from any detailed discussion thereon. I shall therefore only say that I cannot regard the presence of the tubercle bacillus as demonstrating the ætiological identity of all morbid processes with which it may be found associated. It is clear that the causal rôle of this bacillus is of a different kind from that of the assumed organic germs of such affections as the acute infectious fevers, seeing that both "predisposition" and well-recognised favouring conditions for its reception and growth are so prominent in the matter of tuberculous disease ; and, for all practical purposes, it can scarcely be denied that an isolated and chronic affection of a single joint, in a body proved by post-mortem examination to be entirely free from tubercle or any disease elsewhere, is not to be classed clinically with the familiar affection known as general tuberculosis. The definite limitation of some cases of local disease in which tubercle is found seems to be in itself a more important pathological fact in the ætiological study of the case than the mere presence of the tubercle bacillus ; while, on the other hand, the development of tubercle in cases clinically described as general tuberculosis seems to dominate or supplant most other causal considerations.

Without therefore discussing the still bewildering question of how tubercle is related to the various "inflammatory" and "caseous" processes with which it is frequently associated, or whether all caseation is of ultimately tubercular origin or not, I must incur the condemnation of many modern exponents of a perhaps premature finality in this pathological matter by simply stating that I still recognise a marked and clinically useful difference between scrofulosis and "tuberculosis" as presently to be described, although I am fully convinced of the long-known and excessively frequent association of the scrofulous habit with what is now regarded as the infection of tubercle. I cannot see my way to dispense altogether with the old conception of "diathesis" or "tendency," which is nevertheless open to a perhaps hypercritical charge of want of precision.

By the clinical term scrofulosis I would indicate a largely hereditary tendency to congestion and inflammation of various parts and organs, which is especially marked in the lymphatic structures and glands. Imperfect nutrition and circulation underlie these morbid expressions, which are seen chiefly in children born and reared in bad hygienic conditions with, frequently, a vicious heredity as well. Parental syphilis and phthisis underlie struma in numerous cases, while bad food and the deprivation of pure air and light are probably guilty of many more. Tonsillar enlargement, with recurring inflammation, is a common mark of this condition, mostly accompanied by a similar affection of the glands of the neck. These glands may only swell and after a while subside, but frequently suppurate, and then nearly always become more or less caseous. In close connection with this glandular affection we find disease of bone, swelling of joints, and various affections of the skin and mucous membranes. Now, although we know that general tuberculosis and acute tuberculosis of certain organs are frequent events in the "scrofulosis" above described, and that we can draw no important histological or other pathological line between the cases we call scrofulous, where there is caseation, and those we call tubercular, both the microscopical and bacillary evidence of tubercle being mostly found in all, yet there is a vast number of "scrofulous" children who never become the subjects of general or acute tuberculosis, and seem therefore to be proof against the special infective action of tubercle which is so marked in other cases. It is believed by some that caseation is a mark of the local entry of tubercle. Whether this be so or not it is certainly indicated by many inoculatory experiments that many of the non-caseous processes in the skin and mucous membrane of the so-called scrofulous have no power of tubercular infectiveness. It may probably, then, be held that caseation is a result of more morbid processes than one ; that there is a class of cases denoted by the term "scrofula" which is marked by a great liability to inflammatory enlargement of glands and lymphatic structures generally, arising from conditions of malnutrition both hereditary and acquired ; that the inflammatory process in many of these cases is exceedingly apt to become caseous ; and that in caseation, while there is an evident risk of the occurrence of acute or general tuberculosis, the process very frequently remains absolutely localised and its products become entirely obsolete.

It is especially, I think, in the glands of the neck that well-marked and long-standing caseation is so often seen to become ultimately obsolete, with no subsequent outbreaks of tuberculosis or lung-disease during even a long life; and the same may be said of many of the numerous cases of bone- and joint-mischief now so generally classed as tubercular by surgeons. It must happen to many hospital physicians to be frequently asked by their surgical colleagues to report medically on such cases, and almost as frequently to find, to the apparent surprise of the surgeon, no evidence whatever of tuberculosis elsewhere.

The chief mark of so-called scrofulous inflammation, as distinguished from that occurring in healthy children as the result of irritation, is chronicity. This is well seen in the common affection of the glands of the neck in children. Added to this there is a dominant tendency to catarrh of mucous membranes. Intestinal and perhaps gastric catarrh is common, often accompanied by fever, and intestinal ulceration may ensue. Nasal catarrh may be obstinate, as also ozœna with chronic bone-disease.

There may be tarsal or general ophthalmia, with much suppuration, keratitis, and all grades of affection of the pharynx and of the external and middle ears. Bronchial and pulmonary catarrh are prominent, ekzema is frequent, and slight injuries to the skin may cause obstinate dermatitis. Hard painless subcutaneous lumps, gradually softening and ultimately discharging cheesy matter, are common in strumous infants; they may be very numerous and run a lengthened course before finally cicatrising. Lastly, disease of the bones and joints must be mentioned, and especially the familiar caries of the vertebræ, which in its earlier stage so often comes under the notice of the physician. The pain in this disease is often for long referred to other parts of the body than the spine itself, according to the distribution of the involved nerves; and the early diagnosis is to be made, as urged by Eustace Smith and others, much more from the observation of increased pain on movement and its abolition by rest, and from careful examination of the attitudes of the child and the mobility of his spinal column, than from percussion over the vertebræ in order to elicit evidence of tenderness, which is often absent even in advanced cases of caries.

The glandular enlargements in the thorax or abdomen which may occur in scrofulous children will be mentioned elsewhere. Just as the neck-glands enlarge mostly in association with pharyngeal inflammation, so the thoracic and abdominal glands may swell and caseate owing to catarrhal inflammation in the pulmonary and intestinal tracts.

The old description of the general appearance of strumous children as to face and build of body is very often justified; but in many cases it is by no means applicable, pronounced strumous affection being seen in children of very different appearance. I however quite agree with those who recognise the extreme frequency, in cases which from other reasons may be regarded as strumous, of a redundancy of hair on scalp and body, of a rough and often scaly skin, and of rapid growth of nails.

The greatest danger to strumous children is that of ensuing tuberculosis, whether in the form of pulmonary phthisis, generally after the age of six or seven years, or of abdominal or cerebral mischief at any age. All inflammatory affections and diseases of infective nature have, moreover, their special dangers for those of the scrofulous diathesis, and we have seen that much enlarged glands in the chest and abdomen, inducing their own special troubles, may result from pulmonary or intestinal catarrh. At the same time large numbers of markedly strumous children, and especially those whose main affection is pharyngeal with enlarged cervical glands, make a good recovery and enjoy long and healthy lives.

The general **treatment** of struma requires much attention to hygiene in the widest sense. The child should have abundance of sunlight and fresh air, an ample diet, regular exercise, and daily bathing in salt water. Innutritious ingesta, such as sweets &c., between meals should be strictly forbidden, and all care taken to avoid the establishment of depraved appetite. Dryness of climate should be secured when possible, and the body-warmth carefully kept up by efficient clothing. I know no better home-resorts for strumous children than some of the East Coast watering-places such as Whitby, Cromer, Felixstowe, Margate or Westgate, in the summer, and, in the winter and early spring, Aberystwyth, Bournemouth, and Freshwater or Sandown in the Isle of Wight. The continuous use of cod-liver oil should, I think, always be advised, and especially insisted on, whenever it is well assimilated, in cases with active glandular enlargement or signs of catarrh. Iron with small doses of iodide of potassium, or the syrup of the iodide of iron, should also be given. Local treatment of the pharyngitis and tonsillitis, which generally precede enlargement of the cervical glands, should be instituted early, and may lessen or prevent the glandular trouble. If the glands be already much enlarged, and especially if they appear to be caseous, little can be expected from local treatment by "absorbent" unguents containing either iodine or mercury. From sufficient experience, and not only in view of the frequently tubercular relationships or sequelæ of these glandular affections, I am entirely in accord with the modern surgical practice of enucleating strumous glands in their early stages. When suppuration sets in the abscess should certainly be opened at once, and never allowed to burst. All breaches of surface, spontaneous or surgical, in scrofulous children should be carefully cleansed and drained antiseptically; and all discharges from mucous surfaces, especially from the auditory meatus, should be treated as soon as possible with astringent applications.

CHAPTER IV.

TUBERCULOSIS.

MUCH of the clinical subject-matter of tubercular disease is set forth, for practical purposes, under the special headings of cerebral, thoracic and abdominal disorders. It is nevertheless to be borne in mind that the younger the subjects the wider is the distribution of tubercle, and, moreover, that in a large number of cases proved at death to be general tuberculosis there have been few and sometimes no local signs or symptoms definitely demonstrative of this disease. In such instances the diagnosis, depending mainly on such general symptoms as wasting, remittent fever and the like, is often doubtful and may be altogether missed; while in a scarcely less number of instances of failing nutrition, with or without special signs or symptoms, tuberculosis is diagnosed where it does not exist. There are also some cases of general tuberculosis which, owing to the prominence of special clinical symptoms, are apt to be regarded under the aspect of local disease. I could quote numerous examples, where abundant and widely-distributed tubercle was found in all the three great cavities, each of which during life wore the semblance of either pulmonary, abdominal or, though rarely, cerebral disease alone. In illustration of the proportionate frequency of the forms of tubercular disease, as regards their chief clinical characteristics, I have referred to a series of 400 in-patients at Shadwell Hospital registered as tubercular or phthisical either on clinical or post-mortem evidence. Of these about 25 per cent. were found in the category of phthisis or pulmonary tuberculosis; 25 per cent. in that of general tuberculosis; 12 per cent. in that of abdominal tuberculosis, mainly peritonitis; and the remainder, 38 per cent., in that of tubercular meningitis or cerebral tuberculosis. Tuberculosis thus described forms rather over 5 per cent. of all admissions to this hospital.

By tuberculosis we understand generally a specific infectious disease characterized by the presence of those bodies, consisting either of miliary nodules or of more or less aggregated masses, which are usually described as tubercle and contain the "tubercle" bacillus. Although it is true on the one hand that there is no absolute histological test of tubercular products, the presence of Koch's bacillus being the only positive criterion, and on the other that the bacillus is not always found in some cases which are in every other respect, both clinical and anatomical, identical with those of the typical tuberculosis, yet the post-mortem recognition of tubercular disease is, as a rule, of no great difficulty ; and, where both the clinical facts and the post-mortem appearances of any given case coincide with what is generally known as tuberculosis, the non-discovery of the bacillus should be referred rather to an imperfect knowledge of all its conditions than to a mistaken diagnosis.

General or acute tuberculosis is an exceedingly common disease in young children. It may occur at any age, though but seldom in infants under three months old. While adhering strongly to the old belief that, whether tuberculosis is directly transmissible by inheritance or not, there is a remarkable proclivity in certain families to this disease in at least its pulmonary form after early childhood, even when the members of such families have lived in very diverse conditions ; and while recognising, further,· that there is a very frequent history of phthisis or other tubercular disease in the families of young children dying from acute or general tuberculosis of various forms, I must yet emphatically state that in a large number of cases, either diagnosed or proved to be tubercular in infants and young children, which have come under my observation there was no reason, as far as the history went, to suspect the occurrence of any tubercular disease in the patients' immediate families. Indeed from the clinical side only the features of a specific disease are more saliently marked in *general* tuberculosis as occurring in young children than in almost any other of its forms or subjects. Concerning this and other points the observations and comments of Dr. Sturges are of great value.[1] He took consecutively from the post-mortem records of the Hospital for Sick Children 1420 cases of all kinds, of which more than 30 per cent. were deaths from tubercle. Analysis of these showed that it is from birth up to five years old that "tubercular development has its chief activity and widest range," and further, with due allowance made for the imperfection or absence of family histories in many, that "inheritance itself fails to account for the enormous frequency of infantile tuberculosis." With respect to 204 cases of tubercular deaths trustworthy family records showed that both parents were healthy in 107 instances, one parent phthisical in 44, and both parents phthisical in but one.

Besides the many well-known unhygienic and individual conditions favourable to tubercular development, which largely account for the ravages of this disease among the children of the working classes of our large towns, the frequent and close association of measles with sequent tuberculosis must strike all who are widely acquainted with disease in children. On this point again I would refer to the writings of Dr. Sturges, with whose experience my own is in full accord, my note-books containing numerous instances of tuberculosis following in a few weeks

[1] See Westminster Hospital Reports, vol. iv., Churchill, 1888.

K

or a few months on measles, in children who were previously in perfect health. We may well believe that not only measles but also other diseases which are not seldom followed by tuberculosis, such as whooping cough, broncho-pneumonia, and, though to a less degree, enteric fever, may prepare the soil for the reception and growth of the tubercle bacillus by injuring the structure of the mucous membranes.

Into the general question of the many ways by which the tubercle bacillus may find its way into the body, or into the matter of prophylaxis, I cannot enter here, but would insist, in passing from this all-important subject, on the advisability of thoroughly boiling for the use of young children all milk from doubtful sources (which is tantamount to recommending this practice universally to the poor), and of spreading as widely as possible the knowledge, acquired of late, concerning the infective nature of the sputum of consumptive persons. Doubtless, as we have seen when treating of scrofulosis, there is an intimate connexion between the caseous products of this affection and both localised and general tuberculosis. Caseous lymphatic glands, whether tubercular or not from the outset, are frequently the starting-point of widely distributed tubercle ; and we can draw no hard and fast line between those cases of "strumous" or "tubercular" disease of bones and joints which remain localised, and those which are followed by abdominal, thoracic, or cerebral tuberculosis. We may certainly assume without controversy that caseation of any part or organ, from whatever cause arising, is in itself a favourable nidus for tubercular development; and, though we from time to time meet with undoubted cases of acute miliary tuberculosis in bodies absolutely free from any caseous focus, it is certain that in a large majority of instances such a focus is found, and especially often in the bronchial or abdominal glands, from which further disease has clearly been disseminated in the lung or other organs.

Chronic tuberculosis in children is usually, in its clinical aspect at least, of either the predominantly abdominal or pulmonary form, and is dealt with under these headings, although there is a class of cases which cannot be called acute, lasting as they do for many weeks or months, where general symptoms, such as wasting and anorexia, often endure for a considerable time with but little and sometimes no fever, and where, after death, caseous and not miliary tubercle is found widely disseminated through the body. For the most part, however, miliary tubercle occurs, with varying degrees of caseous degeneration, in cases known as general tuberculosis, and these are described as acute or sub-acute tuberculosis. Miliary tuberculosis runs the acutest course, and is thus liable to be mistaken for other diseases. When there are few local signs or symptoms, or when these mainly point to the abdomen, enteric fever must always be thought of, and often cannot be excluded until the lapse of time brings differentiating facts to light. Should the main stress of the disease be pulmonary the case may closely simulate broncho-pneumonia, and should it be cerebral we may have to hesitate for some time before definitely deciding on the nature of the attack. In every case of suspected tuberculosis we should carefully and repeatedly examine the chest for pulmonary or pleuritic signs, explore the abdomen for enlarged glands, and, when possible, search with the ophthalmoscope for choroidal tubercle in order to facilitate our diagnosis. In a great majority of cases there is a history of more or less marked ill-health and wasting of some duration before the clinical outbreak of acute tuberculosis, whether such outbreak be general or apparently localised.

So insidious, at least very often, and slightly characteristic are the early symptoms of acute tuberculosis that little can be gained by an attempt at detailed enumeration. Loss of appetite, wasting, anæmia, some œdema of the feet, remittent pyrexia, slight though it be at first, and general irritability are a group of phenomena at once frequent and full of warning import, but unless some localising signs appear, as indeed they usually do sooner or later, either in chest, abdomen or head, our diagnosis must often be doubtful until near the end or until after the necropsy. There may be much disease of the lung with but slight signs or only those of localised pleuritis, and very little or no cough ; there may be abundant intestinal disease without diarrhœa ; and multiple caseous tumours in the brain, or even miliary tubercle, with no local indications in paralysis or spasm, and no clinical or post-mortem evidence of meningitis. I have seen several cases of general tuberculosis beginning apparently suddenly with severe pain in the abdomen, and both with and without vomiting, this last symptom, when present, being doubtless due to the acute peritonitis found post-mortem.

I have often seen a small *purpuric eruption* develop, more especially in the feet and legs, shortly before death in all kinds of tuberculosis, and considerable *œdema* of the feet at this period is very frequent.

With all these facts in view it is clear that we can lay down no diagnostic rules for all cases, and that consequently, though **prognosis** must always be grave, it should never be hopeless in the early stages of suspected tuberculosis. It is not very uncommon to meet with cases where much wasting, fever, cerebral symptoms, vomiting, diarrhœa, or even a prolonged condition of deeply defective consciousness may, any or all of them, be present and endure for weeks, to be followed by gradual or rapid restoration to perfect health. Whether such cases be tubercular or not I am inclined to lay it down as a general rule, that where *acute* tubercular disease is deliberately diagnosed by an experienced physician, and the patient ultimately recovers, the case is one which points to or simulates the cerebral form of tuberculosis. That tubercle may become obsolete in any of its haunts has been proved by post-mortem observation, and I have myself seen cases which show that even a localised meningitis of tubercular origin may recover. Tubercular peritonitis, whether ultimately followed by renewed or further tubercular development or not, unquestionably recovers with no great rarity, but in these cases its onset is mainly insidious and its course chronic. When acute tuberculosis, giving rise to marked signs of bronchitis, attacks the lungs, its event is always fatal; and the same may, I think, be said of tubercular disease, whether ulcerative or not, of the intestinal mucosa.

Of the **treatment** of acute tuberculosis generally but little can be said in a practical clinical work, for the question is mainly one of preventive medicine. In the early stage of all cases of suspected tuberculosis of whatever form I am in the habit, following Sturges, of giving the hypophosphite of soda or lime in full doses,—10 grains three times a day to an infant of a year old. This practice is wholly empirical, but being sometimes coincident with recovery is perhaps something more than harmless. In the later stages, or when the disease is apparently established, little can be expected, according to my own experience, but the partial relief, in various ways, of some of the suffering. An interesting and important series of cases, however, carefully detailed by Dr. Sturges in vol. i. of the Westminster Hospital Reports 1885, showing recovery in several instances of what was to all clinical appearance advanced tuberculosis, both meningeal and peritoneal, coincidently with the administration of the hypophosphite of soda, deserves the closest attention. Dr. Sturges' comments are marked by the strictest logical acumen, and his advice to try large doses of the drug in all such cases, as well as in the early periods, may be followed always with safety and sometimes with hope.

The treatment of the chronic expressions of tuberculosis is dealt with in connexion with their special forms.

CHAPTER V.

ON ANÆMIA, PURPURA AND SCURVY.

Anæmia.

PALLOR of the skin and mucous membranes in association with deficiency of hæmoglobin, with or without numerical diminution of the red blood-corpuscles, is very prominent in many of the nutritive disorders of childhood both acute and chronic, and is brought about with great readiness.

Familiar examples of profound anæmia are seen in cases of chronic diarrhœa, nephritis, valvular heart-disease, tuberculosis, rickets, syphilis, rheumatism and malaria, and the seemingly special form of so-called "splenic anæmia," which I have already described in connexion with enlargement of the spleen, is almost exclusively an affection of childhood. Of *Hodgkin's disease* too, described as lymphadenoma, malignant lymphoma and otherwise, which is characterized by a widespread enlargement of the lymphatic glands and by lymphatic growth in other tissues with marked and progressive anæmia and wasting, many instances are found in later childhood and youth. I have seen several under the age of puberty, and some have been recorded in the first year of life. Anæmia resulting from metallic poisons, such as lead or mercury, and from the presence of intestinal or other parasites such as Anchylostoma, Filaria, or Bilharzia, need only be mentioned as necessary to be borne in mind, though mostly of very exceptional occurrence in English practice.

The variety of anæmia known as *leucocythæmia* or *leukæmia*, marked by great positive increase of the number of white blood-corpuscles, diminution of that of the red, and overgrowth of the spleen and often of the bone-marrow, has been previously alluded to. Although undoubtedly occurring in childhood and occasionally in early infancy, either in connexion with various and general disorders of nutrition or in apparent independence of such affections, this disease is, in my experience, rare before adult or middle age, and, important though it is from its prevailing fatality and its hæmorrhagic incidents, needs no special consideration here. It is probable that several cases reported as leucocythæmia in infants and young children should be classed rather with the "splenic" anæmias already described.

Besides the above-mentioned secondary or more or less specialised forms of this blood affection we meet with instances where simple anæmia obtains in various degrees in all the same circumstances as in adults, and sometimes with no ascertainable exciting cause and unmarked by any concomitant sign of organic disease. There are examples, in fact, as well in childhood as in later life, both of the apparently idiopathic anæmia or even chlorosis which recovers and, though this is certainly rare, of the pernicious anæmia which tends to death. Anæmia of more or less sudden onset, following on fright, or accompanying other nervous disorders, must be recognised in children as well as in adults. I have seen several instances of this in various degrees, and some pronounced examples have been recorded by Gull and others.

The subjects of marked anæmia are usually languid and fretful, with bad appetite, imperfect digestion, frequent headache and a tendency to constipation often alternating with attacks of diarrhœa. Œdema of the extremities is common, short breathing is sometimes marked, and the murmurs over the præcordia and in the neck, so familiar to us in adult cases of anæmia, are often heard. It is unnecessary to enter here into any discussion on the differentiation of anæmias by estimating the varying proportions of red corpuscles and hæmoglobin, for such cases as we are considering are, although less common, quite the same as those we meet with in adults; but it may be said that, as a rule, the less diminution there is in the number of the red corpuscles, however deficient the hæmoglobin may be (as appears to be usually the case in so-called chlorosis) the more favourable is the forecast, and the greater the hope from appropriate treatment.

It should be remembered that the normal blood in early childhood is richer in leucocytes, poorer in hæmoglobin, and of lower specific gravity than in later life. In it are found also in some degree the nucleated red corpuscles which in adults are only seen in anæmic states.[1] These and other differences, pointing to a less stable condition and readier depravation of the blood in young children, may perhaps explain in some degree the often rapid onset and progress of infantile anæmia in many various circumstances, and its frequent occurrence as a sequel of apparently slight or indefinite causes.

On the whole it may be said that the nutritive failure in infancy and early childhood so commonly resulting from numerous diseases and disorders, or from hygienic and dietetic errors, is especially often productive of prominent anæmia, which in adults has a more limited clinical ætiology.

The **treatment** of anæmia of course depends largely on whatever ascertainable condition may be found to underlie the symptoms, and all therapeutical indications offered by the history or evidence of antecedent or more widespread disease must be sought for and duly attended to. In all it is imperative to take every trouble to establish an ample and easily digestible dietary and to avoid monotony in the meals, and it will be well in some to seek to improve the appetite by the occasional administration of a drop or two of dilute hydrochloric acid, with or without tincture of nux vomica, in some aromatic water. Equally important are the preservation of the body warmth, and an ample supply of fresh air and sunlight. Close confinement and darkness will soon blight young children and mark them with an anæmia which admits of even less doubt as to its causation than the examples so familiar to us among girls of the working classes. We must therefore give due care to the ventilation of rooms by night as well as by day. In bad cases there is cardiac weakness and sometimes cardiac dilatation as a part of the general condition. Rest, therefore, should always be enjoined when exercise causes fatigue,

---

[1] See Dr. C. Griffith's article on "Diseases of the Blood" in vol. iii. of Keating's *Cyclopædia of Diseases of Children.*

especially if there be any tendency to breathlessness or œdema. In such cases children should not be allowed to walk when taken out of doors.

Constipation should be relieved by occasional purgatives such as aloes, senna, or sulphur when dietetic and other measures fail. Daily tepid baths, however, with subsequent rubbing of the chest and body, and kneading of the abdomen will often relieve this symptom while improving the general condition.

As regards medicines I believe none are better, when not otherwise contra-indicated, than cod-liver oil, arsenic and iron. In some cases all of these may be given with advantage. Rapid improvement has more often seemed to me to follow on the use of the two former, either singly or in combination, than on that of the latter alone. Cod-liver oil is especially useful in the case of thin and puny children with capricious appetites, who frequently take it better than anything else. The value of arsenic in anæmia generally appears to be as unquestionable as that of any drug in any disease. From half a minim to two minims of Fowler's solution three times a day appears to be a sufficient dose for children from one year old and upwards. I have several times seen untoward symptoms, such as sore eyes and sickness, arise when larger doses have been given, and am of opinion after a large experience of this drug in patients of all ages that, although adults can well take considerably more than the official doses for an indefinite period, there is no disproportionate tolerance of the drug in early life.

The preparation of iron that I usually give to anæmic children, whether alone or in combination with the liquor arsenicalis, is either the tartrate or ammonio-citrate in doses of from one to five grains according to age.

Purpura.

In considering the clinical aspects of purpura as an affection of child-hood I must confine myself mainly to those cases where extravasations of blood into the skin, mucous membranes, internal organs, or joints occur with more or less apparent independence of other distinct morbid pro-cesses. In children as well as in adults hæmorrhages of this kind, but especially those into the skin, are apt to take place in various blood con-ditions such as the exanthemata, septic endocarditis, syphilis, kidney-disease, heart-disease, anæmia, leucocythæmia, scurvy (especially in the complex of symptoms often described as "scurvy-rickets"), and many other cachexiæ, among which may be mentioned the state of low vitality caused by prolonged acute illness, as, for instance, enteric fever, after which a purpuric eruption, chiefly on the legs, is apt to appear when the patient begins to walk about. Purpura may also be the result of the administra-

tion of certain drugs, among which iodide of potassium is prominent. In
a not uncommon class of cases, probably peculiar to childhood and elsewhere referred to, where there is profound anæmia with enlarged spleen
often unconnected with either rickets or syphilis, I have several times
observed both cutaneous and mucous hæmorrhages, and in this as in other
purpuric conditions there may be extravasation into the structures of the
fundus of the eye.

Of the ultimate causation of purpura and the nature of the part played
by changes in the blood or vessels or by the vaso-motor or trophic nervous
system, or of the question of septic origin, it is unnecessary to speak, for
our knowledge here is very uncertain and therefore but little helpful at
present to practical clinical study.

In children the hereditary affection known as *hæmophilia*, or the
"hæmorrhagic diathesis" or "bleeder's disease," is a recognised cause of
hæmorrhages into skin, mucous membranes and joints, whether appearing
spontaneously or as the result of injuries. This morbid tendency is
usually manifested early, though rarely in infancy, lasts through life, and
generally leads to death in childhood or before mature age. In its typical
form the disease is rare, and for its detailed description I would refer to
the text-books. There may be bleeding from the nose or mouth, hæmatemesis, melæna or hæmaturia; excitement, headache, feverishness, or
other symptoms of illness may be present in the attacks; and occasionally
there are convulsions.

In considering purpura as an apparently independent affection we are
at once struck with the existence of a class of cases which, but from their
non-hereditary character and the absence in many instances of a tendency
to recurrence, are indistinguishable from so-called true hæmophilia. In
some which are recurrent the likeness approaches to identity. Some of
these cases begin suddenly in the midst of seemingly good health, and
may or may not be ushered in or subsequently attended by some pyrexia,
headache, irritability or other symptoms. In some, too, besides the spontaneous hæmorrhages, there is a tendency to bleed or bruise on any slight
injury. When we reflect on the close similarity of these two sets of cases,
and, further, on the numerous general causes of secondary purpura above
referred to, and remember the tendency to prolonged hæmorrhage in leucocythæmia after the extraction of a tooth or the application of leeches,
it is difficult to draw a hard and fast line between primary and secondary
purpura, or to reject the conclusion that in some at least of the non-
hereditary cases the pathology is similar to that of "hæmophilia." My
experience, moreover, forbids me to make a material distinction between
" purpura simplex," a term usually applied to hæmorrhages limited to the
skin, and " purpura hæmorrhagica " which includes bleeding from mucous
membranes. The first of these forms often passes into or accompanies

the second, and both may be transient, recurrent, or fatal, apparently idiopathic, or clearly secondary.

Purpura of the skin only, as far as can be observed in life, consisting in small circular or more extensive extravasations soon taking on a bruise-like appearance, is not uncommon, and often passes off in a few weeks or less, without recurrence, after rest in bed and appropriate nutritive and tonic treatment. Such cases are known as *simple purpura*. In my experience this condition, though most often justifying a good prognosis, rarely occurs in children even apparently healthy, but rather is symptomatic of bad nutrition or previous disease. It is unaccompanied, as a rule, by any febrile disturbance. Fresh crops of spots or patches may occur for an indefinite time if the patient be neglected or allowed to get up. But what seems to be this form of purpura in cases which die from underlying conditions is sometimes found post-mortem to be "hæmorrhagic" as well, and to extend to internal organs. In the case of an anæmic and probably syphilitic child, with a small and not very extensive purpuric eruption during life, I found post-mortem numerous hæmorrhages in the subcutaneous tissue, the pleura and the kidneys; and in another, which died with signs of broncho-pneumonia, diarrhœa and vomiting, many extravasations were discovered in the liver, the lungs, and the ileum besides the cutaneous purpura which alone was seen before death. There is, further, a class of cases described by Henoch and others under the name of "*purpura fulminans*" which are marked by large and rapidly spreading cutaneous extravasations only, without fever, and end in death after a few days' course. One such case I saw in a boy of nine, beginning six weeks after the onset of scarlatina. The post-mortem examination revealed nothing but general anæmia of internal organs. Of the few cases of this latter kind hitherto reported some followed acute disease, such as pneumonia, but others were to all appearance independent.

So-called simple purpura is sometimes accompanied by pain and tenderness in the limbs and joints, and occasionally by articular swelling or œdema of the feet, hands or eyelids. Erythema nodosum may also concur, as also more or less bullous eruption. Such cases are often called "purpura rheumatica," but are rarely marked by the clinical signs of true rheumatism. The joint swellings are certainly sometimes, and perhaps most often, hæmorrhagic in nature.

Among the cases known as "purpura hæmorrhagica," which are marked by mucous hæmorrhages or joint-affection as well as by frequently extensive cutaneous purpura, and are seemingly idiopathic or at least not seldom occur in children of apparently good health, some fever and constitutional symptoms are not rare either as heralding or accompanying the special symptoms. In some instances there is abdominal pain and tenderness with vomiting, profuse melæna and hæmatemesis, and frequent swelling of joints. Some are gradual in onset and frequently recurrent, with healthy intervals of months or even more than a year, and may occasionally end fatally from hæmorrhage either into the brain or from mucous surfaces, or with gradually increasing anæmia. Other, and perhaps the more numerous, cases begin suddenly, and, though marked by profuse epistaxis and bleeding from the gums and other parts of the mouth, with sometimes considerable melæna, hæmatemesis, hæmoptysis or hæmaturia, recover completely without fever in a week or two from the onset. In many cases of purpura, and especially in those connected with mucous hæmorrhages, the large bruise-like patches or ecchymoses which are seen appear first as such, and are the result of bleeding into the deeper tissues. They are often characterized by more or less swelling.

I select the following cases from my note-books in illustration of my remarks :—

(1.) A boy of 10, with a very good personal and family history and previously well dieted in every particular, had one week before admission a painful swelling of the right elbow, followed the same day by a purpuric eruption on arms and legs. Two days afterwards he vomited blood, was very drowsy, and three days later had great pain in his abdomen and passed much blood from the rectum. These bleedings recurred a few times, but the boy quite recovered in a fortnight.

(2.) A boy of 3½ was admitted with extensive cutaneous purpura, epistaxis, bleeding from the mouth and melæna, with a history of a similar attack, about a year and a half before, during whooping cough, and a subsequent tendency to bruise severely after any slight knock or injury. After a few days there was much bleeding from under the thumb-nail, which lasted several days and was difficult to arrest; the nail separated, and a slough came away from the dorsum of the thumb. But for the absence of any hereditary history this case might be regarded as one of hæmophilia.

(3.) A boy of 10, with good personal and family history, was admitted with profuse purpura, consisting of small spots and large bruise-like patches all over the body except the face, and with sanguineous bullæ on the lips of a few days' duration. He had an exactly similar attack one year previously. Both attacks lasted three months, and both began with oral hæmorrhage and purpura on the legs. During the second attack he was kept in bed and had several slight rises of temperature. There was never any epistaxis or undue tendency to bleed or bruise from traumatism.

(4.) A girl of 12, with rheumatic parents, who was very anæmic, always delicate and had signs of past rickets, suffered from a purpuric eruption, chiefly confined to the legs, and pain, but no swelling, in her ankles with each successive crop of spots. During her stay of three months she had several attacks of hæmaturia. Œdema of the legs had preceded this affection for two months. She had no fever nor any other sign of rheumatism. Recovery was complete, and the general health and appearance much improved, when she left hospital.

(5.) A girl of 7, with good family history, was admitted after suffering for ten days with epigastric pain, headache, and green vomiting. Melæna soon set in, and two days after admission she passed blood thirteen times and had a severe general convulsive attack for half an hour, followed by coma for six hours. The temperature rose to 104°, large blood extravasations appeared on the buttocks and eyelids, and there was some bleeding from the lips. She improved after a week and went out well in a month. Two quite similar attacks without convulsions had occurred at the age of 4 and 5½ respectively. In connexion with this case I would mention another in which there was right-sided hemiplegia with rigidity, afterwards recovering, during an attack of what seemed to be exclusively " simple " purpura.

As regards **prognosis** it must be acknowledged that, while we remain in ignorance of the pathological changes and events underlying the frequent cases of purpura of all kinds which are seemingly neither hereditary nor connected with any other demonstrable disorder, it is well to be very cautious in pronouncing our opinion. Doubtless a very large majority of all cases recover, excluding those known as typical " hæmophilia " and the rare instances of " purpura fulminans " which are marked, as we have seen, by sudden and large subcutaneous hæmorrhages ; but we have no definite ground for prophesying non-recurrence in any. It would appear that the apparently simple purpura following on remediable cachexiæ, and ordinary hæmorrhagic purpura, especially when mainly evidenced by epistaxis and bleeding from the mouth, and beginning acutely in a state of seeming health, eminently justify a good prognosis ; and my own experience agrees with that of Henoch in showing that recovery may almost always be expected, though it should not be foretold prematurely, in those apparently grave cases which are marked by abdominal pain, hæmatemesis and melæna even of great extent. Purpuric eruption with renal dropsy is probably almost always of the gravest augury.

In the **treatment** of purpura the first necessities are to keep the patient in bed however slight the attack may be, and to use appropriate remedies for any discoverable or hypothetical causal conditions, such as rickets, scurvy, anæmia, heart-disease, or septicæmia in any of its numerous forms, whether nosologically specified or not. I should be inclined in most instances to give arsenic and some preparation of iron, and quinine and sulphuric acid are sometimes useful. In cases where there is constipation, and others which do not soon improve by rest in bed, purgative treatment may be tried for a short time, as strongly recommended by Eustace Smith. In definite hæmophilia, however, and other cases beginning acutely without precedent anæmia or other general disturbance, complete rest and symptomatic treatment to check hæmorrhage are all that is indicated. The liquid extract of ergot may be tried for a while but is probably of little use; nor does hamamelis, in my opinion, deserve the reputation it possesses in some quarters. Profuse epistaxis must be arrested by plugging the nares, external hæmorrhage by the application of the perchloride of iron or some other astringent with pressure, hæmatemesis by ice and perhaps small doses of opium, and melæna by injections of starch and opium or of perchloride or pernitrate of iron of the strength of one drachm or half a drachm to the ounce of water respectively.

The diet in all cases should be regulated according to individual indications, chronic cases with anæmia requiring the best and varied nourishment, and acute cases being often advantageously fed on milk alone for awhile.

Scurvy.

I have already incidentally alluded to the subject of scorbutus in infants under the headings of infantile wasting and of rickets. From the reports of many cases by Drs. Cheadle, Barlow and others it can scarcely be doubted that there is an association between the occurrence of hæmorrhagic symptoms in infants, shown especially by sub-periosteal hæmorrhages in the thighs, bleeding gums, and sometimes by cutaneous and other extravasations, and a diet markedly deficient in fresh animal and vegetable food. These cases, as we have seen, are usually rickety and, before the theory of scorbutus was put forward, were often described as acute rickets; while those where the most prominent or only symptoms were swelling of the limbs with tenderness, owing to sub-periosteal bleeding, were previously known as "hæmorrhagic periostitis." Whatever be the true pathology of this class of cases, and whether or no they be rightly regarded as all of one class, it seems clear that sub-periosteal hæmorrhages occurring for the most part in rickety children with or without bleeding gums or extravasation elsewhere, and with or without fever, are more common in England than in most other countries where they have been looked for; and that in many instances, where a marked absence of fresh food from the dietary has been noted, the symptoms have rapidly improved when fresh food and lemon-juice have been given. It must, however, be borne in mind that in a considerable number of cases of so-called acute- or scurvy-rickets there has been nothing else to mark the case than much tenderness and swelling of the limbs presumably or demonstrably due to sub-periosteal hæmorrhage, and that at least in some of them, as I can myself testify, the diet has certainly been in no case lacking in antiscorbutic elements. On the other hand spongy and bleeding gums or, in young infants, ulcerative stomatitis, sometimes associated with purpura, are from time to time observed in cases where the diet has been singularly deficient in these elements. While, therefore, the contention that so-called "acute rickets" always implies scorbutus seems not to be established, and there is a probability that much tenderness and swelling of the limbs in rickets, and even sub-periosteal hæmorrhage, may be referable to other causes and perhaps to the malnutrition of rickets alone, we may on the other hand regard certain cases of bleeding gums, with or without extravasations into the skin or elsewhere, as in all likelihood of scorbutic origin when the nature of the diet is favourable to such disease. No one need question the self-evident proposition that in all cases of infantile scurvy the diet is deficient in some necessary material, but the diagnostic difficulty lies in the question as to what cases we are to call scurvy.

I have already alluded to the contention of some that boiled or otherwise sterilised milk is deficient in antiscorbutic qualities, and have brought clinical evidence to show that this statement, at least without qualification, is incorrect. Whether or not milk which has been subjected to prolonged or repeated boiling or steaming, and preserved for an indefinite time before use, is antiscorbutic may perhaps be doubted; and, in view both of the very scanty clinical evidence available on this point and of our yet imperfect knowledge of the possible changes induced by these processes, it would seem on the whole advisable to boil or sterilise the daily supply of milk and use it at once, according to a practice which has been amply proved to be unobjectionable.

In all cases of suspected or established scorbutus in infants appropriate **treatment** should be at once instituted. All preserved foods should be abolished, and the child fed on fresh milk thickened or not with potato-flour, or with raw meat-juice, according to age. A few drachms of lemon-juice should also be given daily in bad cases.

SECTION III.

ACUTE FEBRILE DISEASES.

SECTION III.—ACUTE FEBRILE DISEASES.

In this section I include diseases marked prominently by fever and general symptoms whether of demonstrably specific origin or not. In deference to long-established custom pneumonia, more properly ranking here, is dealt with in the section of diseases of the respiratory system.

CHAPTER I.

PYREXIA.

RISE of temperature in infants and young children is of much more various significance than in later years, and occasions more frequent diagnostic difficulty. At the same time it is very often of far slighter import. The cause of these peculiarities doubtless lies mainly in the greater instability of the higher nervous apparatus which numbers amongst its functions the regulation of the body temperature. The nervous factor in pyrexia is thus especially well exemplified in children, and we find both increased production and increased loss of heat as the result of comparatively slight causes. The normal temperature of a child in the first few months of life is always subject to more marked fluctuations than in the adult, frequently falling some degrees below 98° F. without apparent cause. In cases of markedly bad nutrition, and especially of diarrhœa and vomiting, the drop may amount even to ten degrees from time to time, and there may be for several days a temperature of three or four degrees below the normal. A persistently or remittently subnormal temperature is almost the rule in cases of rapid convalescence from enteric fever and other less definite febrile conditions, especially when there has been much wasting, but this is to some extent common in adults also. It may be due, as I have thought, in the light of a theory propounded long ago by Dr. Ord, to a lessened production of heat as such during the process of tissue-building. The upper limit of a healthy infant's temperature, except just after birth when it is sometimes lower, is probably about half a degree higher than in the adult. With regard to this point, however, as indeed to the whole subject of

161   L

temperature in children I would refer the reader to a valuable paper by
Dr. Sturges in vol. ii. of the Westminster Hospital Reports (1886). It
has been a common observation at Shadwell that the temperature of
young children is very frequently raised some degrees not only on
admission to hospital, but also on visiting days, the chart in many
instances giving a fairly accurate record of those occasions. Fits of
crying, moreover, convulsions, or other excitements, are often accompanied by a marked rise of temperature. Even in the course of enteric
fever the irregularity of the pyrexia of childhood may be exemplified,
the temperature being often remittent and daily touching the normal
line even at the height of the illness. This fact adds obscurity to
some cases of difficult diagnosis between tuberculosis and enteric fever.
Notable instances of high temperature may frequently be observed in
cases of affections of the nervous centres in childhood, such as tumours
or sometimes hæmorrhages, which in adults would be unaccompanied by
pyrexia. Again, in a case at Westminster Hospital of what proved post-
mortem to be extensive hydrocephalus in a child of two and a half years
old whose fontanelles were quite closed, the temperature had almost daily
risen to 103° or 104° during some months, the child being apparently
well in the apyretic intervals and showing no signs of local lesion until
some weeks before death, when clear evidence of headache and recurrent
convulsions led to the erroneous diagnosis of cerebral abscess. The
tympanic membranes had been punctured late in the case on the suspicion of an undiscoverable otitis, but the ear structures were found to
be quite healthy post-mortem.

In the light of these considerations we should not wonder if we are
from time to time quite unable to make any definite diagnosis of the
cause of both ephemeral and persistent pyrexia in children ; but we must
none the less make searching and repeated examinations of the whole
body before pronouncing on the nervous origin of any given case, bearing
in mind that very slight organic trouble, which in adults would be marked
by but little disturbance, may cause a considerable amount of fever in a
child. Passing over most local inflammations and the exanthemata, which
must of course always be thought of and are dealt with elsewhere, I
would emphatically mention here the frequency with which a pneumonia,
with but slight or sometimes with no discoverable signs of lung-consolidation, may escape notice in young children ; and also the highly important
fact that suppurative otitis may be the cause of long-enduring fever,
either of a remittent or less often of a continued type, which, in the
frequent absence of evidence of any local pain, especially in infants,
may remain undiagnosed for long unless the tympanic membranes be
examined or punctured, and may be regarded as enteric fever, tuberculosis, or some other less definite febrile affection under the vague

name of septicæmia. The fauces should of course be examined in any
case of doubtful pyrexia, for, although there may be no symptoms, a
tonsillitis may exist; and I need hardly add, except for the frequency
with which small empyemata are missed, that we need not hesitate to
puncture the pleura when we have a suspicion from physical signs of a
collection of pus.

A common form of febrile attack may be most appropriately mentioned here, as it is with difficulty placed without dispute in any definite
category. I allude to cases which are very frequently recurrent in some
children, and characterized by a sudden and considerable rise of temperature, rapid breathing with perhaps some rhonchi or scattered râles
discoverable on auscultation, headache, drowsiness and sometimes vomiting. These symptoms may last for some days and gradually subside.
By many these attacks are regarded as due to gastric catarrh, even when
unaccompanied by vomiting or intestinal disturbance and not preceded
by any dietetic error; and by others they are looked upon as "gastro-
pulmonary" fever. Dr. Goodhart, using the latter term, insists upon the
frequency of marked signs of acute bronchitis in these cases, and believes
that they are the analogues of asthma in older children. In my own
experience it has been by no means the rule to find any physical signs
in the chest at all, but usually only a much increased rate of breathing
in proportion to the height of the fever; and the extremely acute cases,
simulating bronchitis but rapidly convalescing, described as so frequent
by Dr. Goodhart I have not as yet recognised. I have, however, long
been impressed with the fact that the attacks I mention, which seem to
be of the same nature, are in no demonstrable way connected with gastric
disturbance, although almost always both medically and popularly regarded
and treated as such. They are much more often the immediate sequel
of marked excitement, and are especially apt to occur in nervous children
with other evidence of neurotic disturbance. I therefore strongly incline
to Dr. Goodhart's interpretation of their meaning and pathology.

Having seen several typical cases of ague in children who have always
lived near the London Thames, both in the neighbourhood of the docks
and in Westminster, I cannot but recognise the possibility of some of the
numerous obscure cases we meet with in children of intermittent pyrexia
(without rigors or a marked sweating stage or splenic enlargement or even
much anæmia), being due to the malarial poison. I have notes of not a
few cases of this kind in both young and older children, where the fever
has yielded to quinine or recurred on its omission, and has disappeared
entirely after a more or less prolonged use of this remedy.

Febrile attacks in infancy and early childhood, apart from the specific
exanthemata, are often accompanied by a more or less extensive cutaneous
blush, especially on the face and trunk, which, although usually evanescent

and shifting in position, may last for a few days. This may be seen in tonsillitis, in pneumonia, in acute gastric catarrh with definite cause and symptoms, in the presumably nervous fevers above mentioned and in other febrile conditions. It is often mistaken for and sometimes with difficulty distinguished from a scarlatinal rash ; but usually it occupies the whole of the face without leaving the white margin round the nose and lips which is almost always seen in scarlatina.

In all febrile conditions of children, especially in infants, wasting both of fat and muscle is much more prominent than in adults, and there is often marked weakness of limbs with considerable pain. Sweating is on the whole of rarer occurrence in the fever of childhood, and it is exceedingly difficult, as, for example, in the case of pneumonia when the skin is often dry and pungently hot, to bring about diaphoresis by even more than the adult doses of such a medicine as the liquor ammoniæ acetatis. Herein lies a partial explanation of the higher range of temperature so often maintained in children, and an encouragement to allow the feverish child to be as slightly covered as possible and to have plenty of fresh air. Delirium, as a rule, is but slightly marked in early childhood.

It must be confessed, as it will be inferred from the above remarks, that, owing probably to the nervous peculiarity of childhood, there remains, even after the expenditure of all diagnostic care and acumen, a certain number of cases of pyrexia with no local or otherwise definite explanation. We must, however, never forget the frequency of tuberculosis, in which remittent fever and some wasting may be for long the only discoverable signs.

CHAPTER II.

DIPHTHERIA.

By this term we understand a contagious febrile disease of both sporadic and epidemic occurrence, marked for the most part by a more or less tenacious pellicle or membrane in the fauces and naso-pharynx which often involves the larynx and trachea as well, and in many cases occupies the smaller bronchi, frequently setting up varying degrees of broncho-pneumonia. Albuminuria is observed in a considerable proportion of cases, and certain paralyses are apt to appear either in the course of the disease or after general convalescence is established. Both the frequency and the fatality of this affection are by far the greatest in children under two years of age. The cause of diphtheria is in all probability, a specific bacillus which, under certain conditions, invades the mucous membrane of the pharynx and nose, and generates there a rapidly absorbable poison with frequently wide-spreading constitutional effects. Nearly half of those attacked under ten years of age die, the immediate causes of death being mostly either profound blood-poisoning; obstruction in some part of the respiratory tract; a combination of these conditions; cardiac paralysis, which in most, although not all, of its instances takes place after the subsidence of the acute symptoms of the disease; or suppression of urine.

The question whether a membranous deposit is always and everywhere of diphtheritic origin must in my opinion, with our present knowledge, be answered at least provisionally in the negative. There are not only many cases of scarlatinous and other affections of the fauces with a membranous appearance indistinguishable from that of diphtheria, but also, and especially in quite young children, frequent instances of laryngo-tracheitis which have nothing in common with diphtheria, either in their course, conditions or sequelæ, other than the presence of membrane in the respiratory tract strictly below the epiglottis, the fauces and nares being found free from involvement both during life and after death. Of this latter fact I have quite convinced myself by necroscopical observations, bearing ever in mind that diphtheritic deposit in the nares, with little or no faucial involvement, may be undetectable at the bedside and overlooked in the post-mortem room. I have elsewhere stated some grounds for my belief in a non-contagious membranous laryngitis apart from diphtheria, and space forbids me to enter again at any length into this much-debated question. I would however point out here that, besides the generally admitted occurrence of membranous laryngitis from a traumatic cause, such as scalds from boiling water, there is the frequent and unquestionable fact of membrane being coughed up for the first time, some days after tracheotomy for laryngeal obstruction, in cases where there had been no sign of membrane anywhere either before or during the operation. I have seen several instances of severe laryngitis with normal fauces and no nasal obstruction or discharge where, at tracheotomy, after careful search and feathering-out of the larynx and trachea, no trace of membrane was discovered; but where, some days after the tube had been in position, coherent membrane was frequently expelled, unaccompanied by any febrile or other symptoms of advancing diphtheria, and sometimes disappearing rapidly after the removal of the silver tube or its substitution by a soft one. Such cases as these, which usually recover without any sequelæ, point strongly to the probability of the membranous deposit being due to the traumatism of the tube, and, in default of any bacteriological or other definite

evidence of the universally specific character of membranous deposit
anywhere, I cannot avoid the conclusion, based at least on ample
clinical grounds, that many apparently idiopathic and uncomplicated
cases of membranous laryngitis in quite young children are simply
inflammatory or at least non-diphtheritic in origin. I would add here,
moreover, that such laryngitis is sometimes concomitant with tonsillar ulceration or general pharyngitis, of which measles supplies some
examples ; and would therefore endorse the teaching of the late Dr. Fagge
that not all pharyngeal accompaniments of membranous laryngitis are to
be regarded as evidence of diphtheria.

The main practical lesson I draw from this belief in a non-diphtheritic
membranous laryngitis is that, without good evidence of diphtheria other
than the suspected or proved existence of laryngo-tracheal membrane,
cases of laryngitis should not be treated in company with recognised
diphtheria, and should therefore not be admitted into the diphtheria
wards of hospitals where the contagium is probably greatly reinforced
by numbers. As regards diagnosis, I admit to the full that it is often
practically impossible to differentiate severe and especially membranous
larnygitis from that of diphtheritic origin ; and I have known several
cases which during life were to all appearance purely laryngitic but were
found after death to be marked by membrane on the nose or upper part
of the pharynx. It is, further, true that although epidemic diphtheria
is nearly always prominently pharyngeal, and the usually sporadic
membranous laryngitis rarely, if ever, spreads to others, yet cases of
seemingly pure laryngitis occasionally occur side by side with recognised diphtheria of epidemic character.

As a matter of practice, therefore, in default especially of any clinically
available and definite test of specificity, I would regard, at least provisionally, all cases of membranous deposit in the fauces, nares or air-
passages as diphtheritic in nature until such time as the appearance of
symptoms significant of other diseases, or the continuous absence of those
characteristic of diphtheria, may render this diagnosis untenable or improbable. With regard to those numerous cases where the symptoms
of laryngo-tracheal obstruction are alone observed, it must be remarked
that their membranous nature is only to be inferred as a rule from our
knowledge that most severe and ingravescent cases of laryngitis are as
a matter of fact membranous ; for, apart from the rare occurrence of
expectoration, the fact of membranous deposit can only be positively
established by tracheotomy or death.

There may be some hope that the microscopical detection, by staining
methods, of the Klebs-Lœffler bacillus, corroborated by culture observations, may come into general clinical use and thus afford some means
of accurately distinguishing between diphtheritic and non-diphtheritic

membranous deposit. Baginsky has reported cases where by this means he differentiated two classes of membranous deposits, the one, marked by the presence of the *bacillus diphtheriæ*, with a mortality of 50 per cent.; the other, marked by streptococci and other organisms, being always of favourable event. The time is, however, probably far distant when the non-discovery of the presumably specific bacillus may be regarded as a satisfactory negative test of diphtheria or justify us in disregarding other possible indications of this disease.

Sources and Spread of the Contagium.—We have seen that the disease is pre-eminently one of childhood. It is very rare before the sixth or seventh month or during the period of exclusive suckling, and its frequency rapidly rises up to the fourth or fifth year and as rapidly declines till about the tenth year. Of 140 cases of all ages observed by Mr. James Dickinson at the Homerton Fever Hospital three-fourths were under two years old. It is certain that a very large number of these youthful cases, occurring, as they often do, among the well-to-do and in country districts where there is no crowding, are not to be referred to direct personal communication; and this is quite in accord with the fact that doctors and nurses in constant contact with large numbers of even the worst cases in hospital are but rarely infected, except in certain circumstances. My own experience strongly corroborates the teaching that by far the most frequent mode of infection is through the actual invasion of the mucous membrane by the exudative material from a diphtheritic throat, and in my own hospital, where for many years diphtheria was admitted into the general wards, I have scarcely ever had reason to suspect its spread in any other way. On the few occasions when other children in the ward apparently took the disease there was almost always a possibility and generally proof of their having been in close contact with diphtheritic patients, except sometimes when numerous cases of diphtheria were being simultaneously treated— a condition which I regard as probably greatly reinforcing the activity of the contagium at a distance. I believe, further, that, apart from the apparently striking predisposition of certain families to take the disease, a certain morbid condition of the naso-pharynx is a most important factor in the morbific action of the organic cause; for, leaving out of account the doubtful cases of diphtheritic sore throat with fibrinous exudation which occur in the course of scarlatina, measles, enteric fever and other affections, diphtheria with all its clinical marks is undoubtedly often seen sooner or later after these diseases, and especially after scarlatina. Examples of this connexion are very frequent in hospitals, and diphtheria epidemics are often preceded by apparently non-specific sore throat. It is indeed highly probable that an unhealthy or injured condition of the mucosa of the naso-pharynx or, possibly, of the upper air-passages is an enormously important and perhaps even a necessary factor in the production of diphtheria without, or perhaps even with, actual contact with the exudative material which bears the contagium.

Of the ætiological conditions and vitality of the contagium outside the body we know but little for certain in spite of much research and more speculation. It is said, with the support of much evidence, to remain long potentially active in clothes, bedding, soil and elsewhere, and therefore to be possibly communicable by persons themselves uninfected. Cold and damp I believe, both from authority and experience, to be certainly favourable to its energy, and there is valuable evidence to show that milk-supply from cows probably suffering from the disease in some form, and contact with animals, especially cats and fowls, infected by human or other diphtheritic matter, are more or less frequent sources of infection. It is further highly probable that the massing together in schools or other assemblies of large numbers of young children, some of whom are infective, is sometimes a powerful cause of epidemic spread. I will only add to this bare statement of ætiological belief, based on study and experience of these difficult matters, that I know of little positive evidence of the spread of diphtheria by drains or drinking water; but regard it as at least exceedingly likely that the potential contagium may exist for long in the soil and in refuse heaps of animal matter, and that its energy may be excited by certain physical conditions. Until, however, we acquire a sounder ætiological knowledge of diphtheria we shall remain mostly powerless for rational and effective prophylaxis.

The *incubation* period of the contagium in the body may be certainly as short as two days or probably shorter. Its maximum limit is difficult to fix owing to the frequently insidious onset of the symptoms of the disease. It may be at least a week. Of a longer period, probable though it be, we have no certain information.

Symptoms and Course.—The onset of diphtheria is marked in most cases by malaise, headache sometimes very severe, some fever, and a feeling of soreness in the throat; but often enough both local and general complaints are slight even when extensive deposit is seen on examination of the fauces. If the throat be examined quite at the outset, only swelling and dark redness are usually apparent, or patches of yellowish deposit indistinguishable from that seen in follicular tonsillitis or scarlet fever; but this is generally followed in a day or two by a coherent and adherent exudation which occupies the tonsils, especially at their opposed surfaces, and most often involves other parts of the pharynx. When the tonsils only are covered with a membranous exudation, and, still more, when one is affected alone or a day or two before the other, the case need not be diphtheritic in the absence of other symptoms, but may

be one of scarlatina or of follicular tonsillitis. Membranous involvement of the *uvula, soft palate or back of the pharynx*, confusion of which
with inspissated nasal mucus must always be duly avoided, is most
important local evidence of diphtheria; as also is the presence of sero-
purulent or sanious *discharge from the nostrils*. In the absence of these
two marks we are helped towards a diagnosis by remembering that cases
of sudden onset of illness with severe faucial symptoms, high fever, and
especially a history of previous similar attacks are usually not diphtheritic, however difficult of differentiation the appearances on the tonsils
may be. Much swelling or tenderness of the glands in the neck is not
frequent in diphtheria, unless either the spreading of the membrane
beyond the tonsils be well-marked, or there be nasal discharge or
severe constitutional symptoms; and when such swelling occurs suddenly we should rather think of scarlet fever. It is, with few exceptions, in the gravest cases only of diphtheria that extensive glandular
swelling, with or without areolar infiltration, is seen. In several cases I
have noted the absence of all glandular enlargement, but it is mostly
present in some degree, especially behind the angle of the jaw, as indeed
it is in other forms of tonsillitis. The nasal discharge above-mentioned
is sometimes the first observed sign of diphtheria, and must always,
when copious and continued, be regarded as of bad prognostic import.
It is usually thinly purulent or sanious and often excoriates the nostrils
and upper lip. Sometimes severe epistaxis occurs and is of grave
meaning, implying considerable diphtheritic involvement of the nares.
It is a favourable sign, on the whole, when the discharge becomes thickly
purulent. In the absence of well-marked membrane in the nose or
fauces we cannot always quite distinguish the diphtheritic nasal discharge from that of scarlatina, but apart from this the diagnostic value
of the rhinitis described can scarcely be over-rated. Occasionally
an exudation on the lips or buccal mucous membrane, resembling
confluent patches of stomatitis, is observed before other local signs of
diphtheria are manifested. In rare instances, examples of which I have
never seen, diphtheria, according to unquestionable authorities, may
begin on the conjunctiva, the external genitals or on the morbidly
or traumatically abraded skin, and in such cases the pharynx may
be unaffected.

Laryngo-tracheal symptoms are not infrequently observed early, and
may be urgent before further signs of diphtheria are manifest. Such
cases present of necessity much difficulty of diagnosis until there be
evidence of membrane in nose or fauces; and sometimes are only recognised with certainty on post-mortem examination, especially when the
membrane is confined to the upper part of the pharynx or involves only
the posterior surface of the soft palate. In my experience membranous

laryngitis is mostly an early phenomenon when it occurs in diphtheria, and the naso-pharyngeal signs much oftener follow than precede the laryngo-tracheal. In only a small minority of my cases has severe croup followed on diphtheritic disease after an interval of more than three or four days, and a large majority of those which required tracheotomy to relieve urgent symptoms were admitted as cases of "croup" from the first.

That diphtheria may exist without any unmistakable local signs or symptoms in pharynx or larynx is undoubtedly true, as is evidenced by the occurrence of paralysis of the diphtheritic pattern after very slight attacks of sore throat. In such cases the local affection is probably mainly nasal and escapes observation. The allegation, however, that diphtheritic paralysis is as a rule the sequel of indefinite or undiscovered rather than of severe attacks is incorrect, being based not on observation and the tracing out of the events of diphtheritic cases but rather on inference, owing to the absence of diphtheritic history in some cases of paralysis which at first sight appear to belong to the so-called diphtheritic category. The records of large numbers of cases of diphtheria show that paralysis is most often preceded by well-marked attacks.

The *paralyses* which are especially associated with diphtheria are for the most part of only retrospective value in diagnosis, for they most often occur after the subsidence of acute symptoms or quite late in convalescence. I shall allude further to this subject in the section on "Disorders of the Nervous System." Paralysis of the pharynx and soft palate may occur early, but the return of swallowed fluids through the nose is sometimes caused by impaired palatal movement from local swelling apart from nerve paralysis. Weakness of the ciliary muscle, shown by loss of near accommodation, is next in order of frequency, and strabismus of some kind is not very rare. Any or almost all of the muscles may be involved in turn, including those of the larynx and of respiration generally. Paralysis of the diaphragm is always a part of more wide-spread mischief and, when marked, is mostly fatal. It is evidenced by stillness or recession of the epigastrium in inspiration, increased expansion of the lower part of the thorax and feebleness of voice and cough. Dr. W. Pasteur has specially studied this condition and draws attention to its frequent association, which he believes to be causal, with pulmonary collapse and broncho-pneumonia. Affection of the heart, characterized by either slowness and irregularity or, more frequently, by great rapidity of action and often ending in almost sudden death, is an ever-present danger in diphtheria, especially when other paralyses exist. It may occur early, although it is oftener a latish sequel. I have seen two cases of sudden death from this cause in patients who were up and about ; one of whom had had, unknown to me, some slight paralysis of the velum palati ; while the other was absolutely free from all paralytic symptoms, and had been allowed to get up three days after an attack of exudative tonsillitis which had been decided on many apparently conclusive grounds to be non-diphtheritic in nature. Only one tonsil was affected, there was no albuminuria, and the boy had been long in an adult ward, suffering from chorea and unexposed, as far as we knew, to any source of diphtheritic contagion. In some of these cases the heart-muscle is found to be fattily degenerated, but certainly not in all. Collapse or death from this cause is often preceded by severe præcordial or abdominal pain.

The knee-jerks are usually absent in diphtheritic paralysis, returning, though sometimes after a long interval, when health is perfectly re-established. They are, however, frequently absent in diphtheria without paralysis, but may be well-marked throughout in cases which subsequently develop paralysis, when they may or may not disappear. When diphtheritic paralysis affects the legs the knee-jerks are, I believe, always absent. I would remark here, as possibly bearing on the vexed question of the identity of croup and diphtheria, that, in those cases of proved membranous laryngitis without other signs of diphtheria which require tracheotomy and, in children over three or four years old, not infre-quently recover, paralysis of the kinds above-mentioned is practically unknown.

Albuminuria occurs in a considerable majority of the unquestionable cases of diphtheria, and mostly in a marked degree. It generally appears on about the third or fourth day of the disease, too late to be of much help in the great difficulty of initial diagnosis. The nephritic albuminuria of scarlet fever is however usually of much later date. Diphtheritic albuminuria is almost always unaccompanied by oliguria, hæmaturia or dropsy, and is but very rarely followed by chronic nephritis. Its amount and frequency seem to vary much in different epidemics, and cannot be ranked as of very important prognostic value. I have, how-ever, observed a continuous absence of albuminuria in several undoubted cases of mild type, and it is in the severest cases that it is most copious and enduring. Nephritis is often found post-mortem but is not charac-teristic. Hæmaturia I have seen in but two cases, both fatal.

The *temperature* even in very severe cases is as a rule not high, averaging between 100° and 102.5°. The disease however is almost always dangerous when the initial or early temperature is over 103°, and a persistently high temperature is of the gravest import. It is often associated with a very frequent pulse, great lessening or absence of the first heart-sound, somnolence, loss of appetite and vomiting ; and there may be severe nasal affection without much faucial trouble.

Certain *rashes*, mostly erythematous or roseolous, occur occasionally in diphtheria. I have seen, however, only a few undoubted instances where scarlet fever could be positively excluded. In one case with protracted convalescence a wide-spread erythema occurred on the sixth day and again on the twenty-seventh day after the onset of the disease.

Vomiting of a persistent character occurs, as a late symptom, in a certain proportion of cases, mostly fatal. It is in these instances that marked *slowness of the pulse* is especially noted, and, as has recently been insisted on in a valuable paper by Dr. G. G. Morrice,[1] complete or nearly complete *suppression of urine* is a very prominent feature. This observer of large numbers of cases at Homerton Fever Hospital describes cases with an early fall of pulse-rate, some of which may slowly recover after a stationary period of some days, while in others there is a rapid acceleration, with extreme irregularity, of pulse, followed by death. A steady diminution of urinary flow coincides with this condition, bearing, however, no apparent relation to the degree of albuminuria. Vomiting begins usually about twenty-four hours before death, and during the last day only a few drachms of urine are passed. Dr. Morrice adds that there is no coma, and no relation to at least palatal paralysis; and that, very rarely, convulsions precede death. The bladder is found post-mortem to be quite empty and contracted, and the kidneys are in a state of acute nephritis. In the cases which recover increased urinary flow accompanies the pulse's return to the normal.

Mortality and Prognosis.—Fatal diphtheria is not common over ten and most frequent under five years old. Mr. Dickinson's statistics above quoted, based on cases of all ages, show, in general accordance with my own and others' records, that while nearly half of those attacked under ten die, nearly nine-tenths over that age recover. In the younger set we find an immense majority of the "croup" and "nasal" cases. I cannot fix the mortality between one and four years of age lower than 75 per cent. In the first six or seven months the disease is decidedly rare. Were we to reckon as diphtheritic all cases of apparently pure membranous laryngitis, both the incidence and the fatality of the disease in the first three years of life would be considerably greater. Involvement of the upper air-passages is always a mark of great gravity, scarcely one-fourth of the cases recovering; and a peculiarly excoriating nasal discharge is mostly seen in fatal cases. It is, however, to be insisted on that a very large number of patients dying with laryngo-tracheal involvement succumb, not to the mechanical effects of membrane in the air-passages, but to the stress of the diphtherial poison, which becomes manifest before, or more frequently after, the occurrence of laryngeal symptoms. There are indeed numerous cases where respiratory difficulty

[1] See St. Bartholomew's Hospital Reports, vol. xxviii., "On Suppression of Urine Diphtheria."

is completely relieved and immediate death prevented by tracheotomy, but which are killed by the general effects of the poison or by cardiac paralysis. Broncho-pneumonia, too, is a prominent cause of death, even when the upper air-passages are but little or not at all affected. It is possible that the great frequency of broncho-pneumonia is due to the inhalation of diphtheritic material from the naso-pharynx. However this may be, it is almost always fatal.

Bad prognostic symptoms are continued anorexia, offensive exhalations, much anæmia, great prostration, very frequent, feeble and irregular, as well as very infrequent, pulse, much albuminuria, deficient urinary secretion, great glandular enlargement giving the appearance of "bull-neck," extreme adherence of the false membranes, blackness and bleeding of the affected surfaces including the nasal mucosa, continued high temperature, and persistent vomiting. With most of these symptoms there is the greatest danger of death from the stress of the poison; and we must never forget the great risk of almost sudden death from cardiac failure, owing to paralysis of the heart or to a combination of this condition with degeneration of its walls. The constitutional symptoms are usually most prominent when the local mischief is severe, but there are many fatal cases with but slight and even indefinite pharyngeal affections and no respiratory trouble.

Of good prognosis, very generally speaking, is the limitation of the local process to the tonsils or to the tonsils and uvula, and the whiter and the less adherent the membrane the better the forecast. We have seen, however, that a whitish exudation, overlying the tonsils alone and but slightly adherent, is not necessarily diphtheritic. Moderate frequency and good quality of pulse, slight malaise and prostration, improvement of the faucial condition after four or five days, and but slight lesion of the surfaces after the membrane has disappeared, are all more or less favourable symptoms. But until complete convalescence be established our prognosis, even in the apparently mildest cases, must always be expressly guarded; for, besides the late onset of cardiac and other paralysis, all the severest symptoms of the disease may follow after a seemingly favourable course of several days.

Treatment.—In all cases strict confinement to bed in a well-ventilated room, or preferably in two rooms alternately, with a temperature not under 65° F., is to be enjoined. If there be any respiratory trouble the bed should be closed in by a tent, and a steam-kettle kept constantly in action. Food should be given frequently in small quantities. In many cases, whether there be palatal paralysis or not, pultaceous material is more easily swallowed than liquid. Milk puddings, meat-juice of various kinds, beaten-up eggs or pounded meat are each useful in certain cases. When food is refused owing to faucial swelling or pain or to the apathy and anorexia which are so frequent, or when, if swallowed, it enters the larynx, forced feeding with the nasal tube, as recommended by Scott Battams, is to be instituted. A soft catheter fixed to a glass syringe is passed through the nose into the œsophagus, and the food slowly injected at frequent intervals. Stimulants are required according to the cardiac indications which are mostly present in some degree in diphtheria, and may be prescribed as brandy, or as the tincture of cinchona, ammoniated quinine, or the aromatic spirit of ammonia. Quinine in doses of not less than one grain for the youngest child may be given three or four times a day, but its value is very questionable. In convalescence arsenic and iron are decidedly useful.

Considering that diphtheria is in all probability the result of local inoculation with the products of a specifically pathogenic bacillus, the question of local treatment is of great importance ; but, as it seems equally probable that the locally generated poison is rapidly absorbed and diffused, there is perhaps but little hope that any application to the diseased surface or destruction of the false membranes can much affect the course of the disease. Few cases are seen early enough to render such a result even *primâ facie* probable. But, not only in quite early cases but also in all others, I believe that every possible effort should be made in this direction, in spite of the questionable success that has followed on the practice. By thoroughly and frequently cleansing the fauces and nostrils with some antiseptic solution we may not only relieve discomfort, assist deglutition and breathing, and limit the hurtful effects on the respiratory tract of infective inhalation, but also may lessen or prevent the further development of the rapidly absorbable poison, however powerless we may be in antagonizing its action after absorption. With these objects the fauces should in every case be thoroughly syringed or swabbed out (swabbing being preferable when practicable, as it usually is with skilled assistance) with the solution of chlorinated soda (1 part in 20 of water) or with Condy's fluid (1 in 40) ; and, after the removal of the false membranes, as far as possible without force, all attainable surfaces should be painted over with the glycerine of carbolic acid or of borax, or brushed over with a solution of silver nitrate (15 grains to the ounce), or of zinc chloride. Subsequently the syringing or swabbing may be repeated every three or four hours, if the exudation should reappear, until the local trouble subsides ; and afterwards less frequently until the fauces take on their normal appearance. A steam-kettle with a 2 per cent. solution of carbolic acid should be kept in continuous use. The nose should be similarly syringed out with the chlorinated soda or Condy's solution whenever nasal trouble occurs, or the irrigating tube, fixed to a receptacle placed above the patient's head, may be inserted into one nostril and a pint or so of the fluid allowed to run through to the other nostril, the head being bent forwards. I have not had much experience in my cases of the action of papain as a solvent of the false membranes, but, as far as I know and can learn, its use is unsatisfactory. An alkaline spray kept constantly at work, or used for several minutes every half hour, has seemed of benefit in some cases, preparing the parts for antiseptic applications. Dr. Lewis Smith recommends the use of such a spray, consisting of 2 drachms each of eucalyptus oil and sodium bicarbonate, 1 drachm of sodium benzoate, and 2 ounces of glycerine to a pint of lime-water.

In the numerous cases undistinguished by definite membrane, where early diagnosis from follicular or other tonsillitis is not possible and especially where there are concurrent cases of faucial or laryngeal illness, or in epidemic times, one thorough application of solution of silver nitrate, followed by frequent use of a milder antiseptic lotion or spray, may be reasonably recommended. Cleansing with antiseptics is especially indicated when there is any hyperæmic or ulcerative condition of the fauces or nose, which, as we have seen, is a predisponent to diphtheritic infection. Of this both scarlatina, measles, and tonsillitis furnish good examples.

The treatment of diphtheria advocated by Seibert of New York, consisting in deep injection of freshly-prepared chlorine water into the submucous tissue of the affected parts by means of a many-pointed syringe, appears to be well worthy of trial in quite recent cases.

The management of laryngeal diphtheria is mentioned under the heading of laryngitis. For full practical directions I would refer the reader to surgical works, among which I may mention the excellent monograph by Mr. R. W. Parker. I am on the whole an advocate for operation by tracheotomy in diphtheria, whenever the marked symptoms of mechanical obstruction to breathing, such as indrawing of the soft parts of the thorax and throwing back of the head, are present, and especially when there is inspiratory intermission of the pulse; and this in spite of the few cases presumably saved thereby. Early tracheotomy in children over three years old, besides preventing the suffocation which might soon imperiously demand it, may lessen exhaustion and thus favour natural recovery; and is in my opinion guiltless of the broncho-pneumonia so often ascribed to it. Broncho-pneumonia whether due to spreading of the inflammatory process or, in true diphtheria, to infective inhalation, occurs not only in membranous laryngitis but also in purely faucial or nasal cases, and is quite as frequent in cases of laryngitis which die without tracheotomy as in those which are relieved by that operation. In view of the probably evil effects of constant inhalation of air which has passed over the diseased naso-pharynx, tracheotomy seems preferable to intubation in all cases of true diphtheria.

In the extreme prostration and collapse which usually herald death we are probably helpless. Alcohol will already have been given up to its full limit as a stimulant, and large doses are rapidly narcotic. Musk in doses of five grains placed far back on the tongue is well spoken of by some, and Henoch faintly recommends the trial of subcutaneous injections of strychnia.

In paralysis of all kinds rest is of great importance until complete recovery is attained, in view of the possibility of heart failure ; and, when improvement is delayed or there is distinct weakness of the respiratory muscles, strychnia injections, as advised by Henoch and according to some experience of my own, may be at least safely tried, beginning with a daily dose of $\frac{1}{50}$ grain for children about three years old, with gradual increase up to double the original quantity. Paralyses other than cardiac or respiratory usually recover completely, sooner or later, without any special treatment.

Diphtheritic cases should be kept continually in bed until the disappearance of all symptoms of whatever kind, and for a week after the cardiac sound and rhythm have returned to the normal, even if otherwise completely well.

Attendants on the sick should be instructed to observe strictly the ordinary rules of cleanliness in dealing with infectious diseases, to thoroughly wash their hands after touching the patients, to avoid as much as possible taking their breath, and to at once remove all particles of matter which may light upon them from the mouth or tracheotomy wound. Mr. Parker advises that all the inmates of a house where diphtheria exists should frequently rinse out their mouth with Condy's fluid and similarly brush their teeth thoroughly before going to bed ; and that the fauces of young children should be daily swabbed with glycerin of boracic acid. All excretions from the patient should be received into vessels containing a strong solution of carbolic acid or of corrosive sublimate (1–500), and all soiled linen should be soaked in corrosive sublimate of half the above strength before being washed. The room should be thoroughly disinfected after the illness. Patients may be regarded as non-infective after a week from the disappearance of all local symptoms ; but, on account of the necessity, for their own sakes, of confining them to the house for at least three more weeks, it is well to continue isolation and defer local disinfection until complete recovery be established.

CHAPTER III.

SCARLATINA.

The following remarks are intended mainly as a practical comment on some of the chief clinical points of this affection. A detailed description of scarlatina and its complications is beyond the scope of this work, and if there be room, as I think there is, for yet another essay on this subject, those only could write it usefully who have had special hospital experience of some thousands of cases.

From one point of view, and that a very practical one, scarlatina may be regarded as an infective form of sore throat, the diagnosis of which is usually difficult and sometimes impossible until the generally characteristic rash appears in from about twenty-four to forty-eight hours after the onset of the illness, or, in some cases, until the almost pathognomonic desquamation sets in either late in the fever or after its subsidence. It were well, indeed, if every suddenly occurring pharyngitis with tonsillar involvement, especially in childhood, and whether marked by swelling only, by diffused redness with swelling or by exudative tonsillitis, patchy or confluent, were suspected as scarlatinous until further observation have established a definite diagnosis. I have so often found my own and others' provisional diagnosis of "follicular" or of "herpetic" tonsillitis, or indeed of diphtheria, completely falsified by the rash, the desquamation or, sometimes, the dropsy of scarlatina that I would record a practical warning against reliance on any descriptions of the so-called "typical" scarlatina throat which, except for the purpose of satisfying some examiners, are apt to be very misleading. The deep-red colouration with swelling of the pharygeal mucosa, so common at the outset and through the course of most attacks of scarlatina even in its milder forms, is, however, a valuable diagnostic aid when we have to pronounce on the probable nature of an ill-defined eruption.

The occurrence of *latent* scarlatina, or sore throat which may be slight and apparently simple without observed rash, is too often ignored in practice, although alluded to by several and strongly insisted on by Collie and others. It is not perhaps of much importance to inquire whether or no there is always a rash in scarlatina. The rash is certainly sometimes very evanescent, appearing and almost disappearing in the course of one day or night, and in the latter case is often overlooked and always difficult of recognition. I have seen several cases of undoubted scarlatinous sore throat, as evidenced by concurrence with other and ordinary cases,

M

by desquamation, or by renal sequelæ, where no rash was revealed to repeated and careful observation, and with no distinctive appearance of the fauces. A similar absence of observed rash may occur in scarlatinous ulceration of one or both tonsils, and also in cases where there is a tonsillar exudation indistinguishable at first sight from diphtheritic affection. In connection with this we must remember that persons who have had scarlet fever, as well as some who have not, are liable to sore throat of indifferent appearance, without any other distinctive symptom, when exposed to the contagion of scarlet fever ; and that such cases may infect others with the ordinary form of the disease. I am well convinced of this fact from personal observation.

Symptoms.—Scarlatina usually begins with a sore throat accompanied for the most part by enlargement of the cervical and submaxillary glands ; and vomiting, rare in adults, occurs in an immense majority of cases in childhood, being far more frequent as an initial symptom than in any other of the acute febrile diseases. There are also as a rule headache, very frequent pulse and pains in the limbs with shivering, and sometimes there is diarrhœa. Convulsions occasionally take place at the outset. The temperature may rise to 104° or over, even in favourable cases and quite at the beginning, or may be as low as 100°, and I have seen several instances of a persistently normal temperature after the second day in cases where the rash appeared after an erroneous or deferred diagnosis. As a rule, however, the temperature and other febrile symptoms are in proportion to the extent of the rash and the severity of the sore throat. The temperature is usually highest at the acme and subsides generally with the fading of the eruption, differing herein from the fever of measles which generally falls suddenly very soon after the rash has reached its height. In very severe cases of scarlatina the temperature may rise higher than in almost any other fever, even 110° being occasionally registered. When the temperature remains up after the rash has disappeared there is probably severe faucial inflammation or one or more of the following complications :—glandular suppuration, cervical cellulitis (which is very dangerous), otitis, simple or occasionally purulent synovitis, pleurisy, pericarditis or endocarditis, or some acute lung-trouble, especially broncho-pneumonia,—most of which may be demonstrated or suspected from their proper signs or symptoms.

The *rash* may deviate from the usual type in being blotchy, papular, slightly vesicular and occasionally minutely hæmorrhagic in some parts, without necessarily serious significance ; but as a rule the severity of the attack is directly in proportion to the darkness of the eruption's hue. An extensive and fatal kind of purpura, of which I have seen one instance and others are recorded by Henoch under the name of " purpura fulminans," must, I think, be regarded as in all probability an occasional and immediate sequela or accompaniment of scarlatina. These hæmor-rhagic patches are strictly cutaneous, not involving mucous membrane or internal organs, and very tender to the touch. More frequent and far less grave are attacks of ordinary painless purpura a few weeks after scarlatina, with or without mucous hæmorrhages, and not associated with fever.

In cases of doubtful distinction from measles it is well to remember that the scarlatinal rash is not characteristically punctate on the face, but rather a bright blush on forehead and cheeks; and that the skin of the nose and its neighbourhood and round the mouth is usually pale.

The *throat* appearances, as I have said, vary much. There may be only pharyngitis of varying degrees of severity, without breach of surface, the whole pharynx with the tonsils being often much swollen and injected, and swallowing more or less impeded. More often there is marked ton-sillitis with "follicular" exudation or ulceration; or the whole of one or both tonsils may be covered with a dirty white or yellowish pellicle often mistaken as diphtheritic, but leaving a more or less excavated ulcer on removal. In some severe cases after three or four days a condition, often indistinguishable from diphtheria, is seen where both tonsils are covered with a membranous-looking deposit, on the removal of which, however, extensive subjacent ulceration is frequently found. With this there may be excessive pharyngeal swelling and purulent or hæmorrhagic discharge from the nose, and glandular and cellular inflammation and abscesses may ensue with extensive gangrene and sloughing, both internal and external. This condition, though frequently regarded as diphtheria super-vening on scarlatina, is probably to be distinguished from that disease by the marked ulceration which is its essential feature, and by being neither accompanied by laryngitis nor followed by paralysis on recovery.

True diphtheria, however, with all its characteristic marks and in both its pharyngeal, nasal and laryngeal forms, frequently follows on scarlatina, and not alone in hospitals where both diseases are admitted; this association being probably explicable by a predisposition of the injured scarlatinous throat to receive the diphtheritic germ. A still closer connexion between these two diseases is, moreover, hinted at by the occasional concurrence in the same house of ordinary scarlatina and what seems to be diphtheria, of which I have seen some examples. The whole question involved here is as obscure as it is important, and demands further experimental and clinical investigation. Dr. MacCombie of the South-Eastern Fever Hospital, with a very large experience of scarlatina, speaks of frequent ulceration of the fauces in children during the acute stage of the fever, sometimes ending in the formation of membrane on the fauces and pharynx, spreading to larynx and trachea, and mainly associated with ulceration of the posterior nares; and further tells me that the membranous affections of fauces, nares and larynx, which so often follow on scarlatina after some interval, are practically identical with diphtheria, both in their course and sequelæ.

Nasal and vaginal discharges are common in the scarlatina of childhood. Post-pharyngeal abscess occasionally occurs in severe cases, and may cause not only dysphagia but also marked dyspnœa. This complication must always be looked for, especially when there is late dysphagia or dyspnœa, and, when found, relieved by prompt operation. Cancrum oris is sometimes observed, but less frequently than in measles.

In some cases, styled malignant, the *stress of the disease* is highly dangerous or mortal without any serious local symptoms. Great prostration rapidly ensues on perhaps repeated vomiting; the pulse is very weak, frequent and irregular; the extremities are cold; the rash may be irregular or perhaps livid or may never appear; and death, preceded by coma and occasionally by convulsions, may follow in from twelve hours to a day or two after the onset of the attack.

Albuminuria is frequently found after the fever, but has no certain relation to the severity of the attack. It mostly begins in the third or fourth week, though sometimes much earlier, and generally lasts from a few days to two or three weeks. It is occasionally accompanied by hæmaturia, the presence of renal casts and dropsy, vomiting, and other clinical symptoms of acute nephritis. A slight amount of albumen may also be found, without other symptoms, during the febrile attack, and is then probably of no more significance than that which occurs in many febrile diseases. In severe cases, however, there may be much albuminuria in the first week. The true *nephritis* of scarlet fever is part of the poisonous effects of the contagium and certainly, at least in many instances, quite unconnected with chill. Frequent examination for albuminuria throughout the illness and for at least three weeks after recovery is necessary for purposes of treatment, for this affection is occasionally one of the most dangerous phenomena of the disease, involving the risk of heart failure, uræmia, and extensive pulmonary œdema. Absence of albumen, when dropsy and oliguria are present, does not exclude the diagnosis of nephritis. There is much variety, moreover, in the daily amount of albumen in many cases. It is very common to find marked albuminuria for the first time when the patient is allowed to get up, its previous absence having been established by repeated examination. This affection usually lasts some weeks, but sometimes for many months, and in a minority of cases is chronic or recurrent with slight or undiscoverable exciting causes. Uræmic convulsions are, according to Dr. MacCombie, more frequent in the nephritis of childhood than of adults.

Pains in the joints and limbs are common about the end of the first week or earlier, and usually last but a few days. Articular pain with some effusion is, moreover, apt to occur in the third week and is generally known as "scarlatinal rheumatism." Dr. MacCombie informs me that this is usually associated with albuminuria, is very intractable, lasting several weeks, is prominently characterised by fibrous thickening of the tissues round the joints, and often leads to pronounced stiffness. My own much smaller experience of scarlet fever quite bears out this description of the joint appearances; and from this, as well as the great rarity of the sweatings, heart troubles, and other incidents of acute rheumatism, and from the smaller tendency to arthritic metastasis, I am of opinion that the usual designation of scarlatinal rheumatism is misleading. Occasionally, however, without doubt the two diseases may be closely connected in time. I have more than once seen genuine acute rheumatism in known rheumatic subjects follow immediately on scarlatina.

The **prognosis** in any case of scarlatina should always, in view of the many possible complications, be very guarded, and no definite opinion should be given of its ultimate event until one month after the onset. In epidemics of a mild and non-complicated character the prognosis is *pro tanto* good, and, without severe throat symptoms, great prostration or continuously high temperature, recovery may be expected in a great majority of cases after five or six days. The disease is commonest in children between three and seven years old, and its highest fatality is in the first quinquennium and among the children of the poor. Diarrhœa is often seen in dangerous cases, and, according to MacCombie, especially in weakly children. In the absence of other severe characteristics MacCombie does not consider that diarrhœa much affects the prognosis. I have myself seen two cases of scarlatina in infants with severe initial diarrhœa and vomiting, simulating at first the acute summer diarrhœa of infancy, which were otherwise mild and made a rapid and complete recovery.

A *very frequent* pulse is the rule in scarlatina, even in some quite mild cases; but an *irregular* pulse is in my experience a very grave symptom in this disease, and necessitates a very cautious prognosis.

I have seen a few indubitable cases of true relapse of typical scarlet fever after about a week's interval. I also know from personal observation that second attacks may take place, and have no doubt of the occurrence of third attacks. It seems certain that once- or twice-repeated attacks, although infrequent, are much more common than in the rest of those diseases which usually confer future immunity on their subjects. In corroboration of these statements I may refer to several cases of indubitable re-infection within short periods, reported to me by Mr. Scott Battams and other medical friends, where definite rash and definite peeling was observed in both attacks; and I have similar information from officers of some of the fever hospitals. Dr. MacCombie tells me that a small number of cases of scarlet fever are re-infected within two months of the commencement of the first attack. He has seen such cases as early as three weeks after this date, but states that they usually take place after an interval of from four to eight or ten weeks. " These second attacks," he says, " occur in patients convalescing from typical attacks of scarlet fever. The second attack is sometimes as severe as the first, and is followed by a second desquamation and, in some cases, by albuminuria."

The **contagium** of scarlet fever has not been isolated, but is probably admitted into the body through the faucial mucosa. The poison is undoubtedly volatile and infective at some distance, but is seemingly far less widely energetic than that of measles, many unprotected children escaping although constantly exposed to the risk. I have ascertained this not only from the more certain and wide spread of measles in hospital wards, but also from many single cases of scarlatina I have known in the crowded families of the poor. There is much evidence to show that it may be conveyed in milk which either has been contaminated by persons suffering from scarlatina or is directly infective as coming from diseased cows. Although there is a lack of positive evidence of many of the accepted modes of spread, and of the duration of vitality in the poison outside the body, we cannot deny that the patient is infective from the onset of the disease, or possibly from the beginning of incubation, until desquamation has ceased. The period of the greatest infectiveness is, I believe, from the beginning to the height of the disease. Of the great infectiveness, indeed, by means of close contact in the very earliest stage, experience at a children's hospital gives ample proof. After six weeks from the onset, in the majority of cases, the patient may be regarded as harmless to others, provided thorough disinfection of the body has been observed, and no article of clothing worn for a week before, or, say, six weeks subsequent to the onset of the illness, be still in use. Every object with which the patient has been in contact during the attack should be destroyed, if possible.

If desquamation be continued after this period, as it not infrequently is, especially on the feet, there is probably no risk of infection, provided a course of hot baths and antiseptic washings have been duly followed ; but, to be on the safe side and in view of some reported cases which indicate the contrary probability, it is on the whole advisable to isolate patients until all desquamation has ceased. The incubative period of the contagium is often under two days, as I have repeatedly observed, and very rarely over four or five. Practically, however, a complete week should pass before a child who is known to have been exposed to the poison be allowed to mingle with others, and his person and clothes should be thoroughly disinfected before his release.

In ordinary cases **treatment** is simple. Confinement to bed for three weeks and to the room for three weeks more, a milk and farinaceous diet during the first four weeks, occasional purgatives if there be constipation, and daily warm baths, with all precautions against chill, during the second three weeks are perhaps all that is necessary. With high fever frequent tepid sponging should be ordered, and in extreme cases the cool bath.

When the throat affection is troublesome chlorate of potash should be given internally, and the fauces should be brushed over with glycerine of borax or syringed with Condy's fluid (1 in 40). In all cases when the fever is over, and especially when there is renal trouble, a long course of iron medicine is strongly to be recommended. When there are symptoms of laryngitis the bed should be inclosed in a tent into which steam should constantly play. The renal affection should be treated according to principles elsewhere laid down, and all other complications appropriately dealt with by medical or surgical means. Nephritis, as evidenced by albuminuria and dropsy, probably affects about 15 per cent. of all cases. It might be well, considering that the kidneys may be affected even more often than is apparent, to diet all cases of scarlet fever on as slightly albuminous food as possible for even longer than the above-mentioned period of four weeks.

Forced feeding (by means of the nasal tube, if necessary) must be employed when food is refused or swallowing is difficult, and free stimulation with carbonate of ammonia or repeated small doses of alcohol should be resorted to when indicated by general depression and feeble or frequent heart action. Especially is such stimulation required when there is irregularity of pulse. Henoch recommends camphor (grs. 1–3) as a valuable stimulant in bad cases, or a hypodermic injection of sulphuric ether (min. 15) when swallowing is difficult.

During convalescence and until the end of the sixth week the patient, besides being regularly bathed in hot water and scrubbed with soap, may be daily anointed with carbolic oil; but, when perfect isolation has been instituted, this process is unnecessary until the time comes for discharge from confinement.

It has been frequently urged by Dr. J. B. Curgenven[1] that, by the inunction of the whole body with eucalyptus oil, twice daily during the first three days of an attack of scarlatina, and nightly, after a warm bath, for the next seven days, the patient is rendered free from the poison and is not in a condition to infect others. The disinfectant, he says, should also be sprinkled over the bed and diffused in spray about

[1] For the latest exposition of his views see the *Medical Magazine* for Feb. 1893.

the room.   He advises further the internal administration of from three
to six drops of the oil three times a day.   On behalf of this plan he
claims not only far more effective prevention of the spread of the disease
than is attained by the segregation of patients in fever hospitals, but
also a high degree of immunity on the part of the sufferers themselves
from untoward complications and sequelæ.   It is only by extensive experience that the efficacy of this treatment can be established, and I can
say nothing on the matter from personal knowledge.   From the evidence
adduced, however, it seems to me that a sufficiently good case has been
made out to justify an ampler trial of this method, in view especially of
the unsatisfactory results obtained by observation of the current official
regulations.

CHAPTER IV.

MEASLES.

Symptoms and Course.—This disease usually begins with feverishness,
headache, loss of appetite, pricking of the eyes, lachrymation and photophobia, and there may be sneezing, hoarseness, cough or epistaxis.   On
the fourth or fifth day, rarely later, and still more rarely on the third
day, the characteristic rash appears, first behind the ears and then on the
forehead close to the scalp, rapidly occupying the rest of the face, and
travelling down the body.   It is at its height both in colour and extent
on the day following its appearance, and on the next day or next but one
begins to fade gradually from above downwards.   On the appearance of
the rash the early symptoms become more marked, but the temperature
tends to fall, and often suddenly, soon after the rash has fully developed.
The facial eruption is important when the previous history is obscure
and doubt may exist between measles and scarlet fever ; for the region of
the nose is almost always occupied by the eruption of the former, and
left free by that of the latter, which frequently, indeed, affects the face
but very little or only with a flush of indifferent appearance.

I have several times observed the *remission* of all symptoms and the
appearance of perfect health on either the second or third day after a well-
marked febrile invasion with rigors.   Not seldom there is a papular or
blotchy and slightly raised rash on the neck or forehead at the very outset,
followed on the fourth day by the typical and rapidly-spreading eruption.
Vomiting is not very rare at the beginning, though much less frequent
than in scarlatina, and initial *convulsions* occasionally occur in young

children of no demonstrable convulsive tendency. *Sore throat* is sometimes complained of, and, if examination be made during the first few days, we almost invariably find a patchy redness of the pharynx and some tonsillar inflammation. *Laryngo-tracheal* catarrh is frequent even at the outset. I have often seen alarmingly acute laryngitis, in some instances unaccompanied by other catarrhal symptoms, which was none other than the beginning of measles. Such initial laryngitis, albeit of great severity and suggestive of the necessity of tracheotomy, terminates as a rule favourably, and often subsides with the developing eruption. Operation should therefore be postponed as late as possible, even in severe cases. A rapid invasion of acute laryngitis should always suggest measles, especially in the absence of any presumably diphtheritic appearance in the fauces.

In some cases, on the other hand, the symptoms of the pre-eruptive stage are so indefinite and mild that no diagnosis of their true nature is possible; and in others there may be an observed absence of fever until the rash appears, when the temperature may only rise two or three degrees. Generally speaking, convalescence from all symptoms is established in about a week from the day of invasion, and the rash completely disappears and the patient is well in another week. Often, and especially when the rash has been intense, there is fine branny desquamation, mostly on the face; but this is by no means the rule, and in the majority of cases the appearance is so slight as easily to escape notice. It usually begins with the fading of the rash. The last stage of the eruption is a pale brownish-yellow staining of the skin, the result of small extravasations of altered blood-pigment.

Some amount of *bronchial catarrh* evidenced by coarse rhonchi is very often present even when the cough is but slight. Catarrh of the respiratory tract may indeed be regarded as a part of the disease even still more strictly than in the case of enteric fever; and acute general bronchitis and broncho-pneumonia, with severe symptoms, are excessively frequent in young children, especially among the poor, contributing largely to the mortality of measles. Acute broncho-pneumonia with rapid invasion in a previously healthy child may sometimes indeed, though rarely, be the first observed phenomenon, preceding the rash, which may then be slight and indistinct.

Diarrhœa in greater or less degree, probably of catarrhal origin, is very common, and is sometimes, especially in summer-time, of serious import.[1] When it is protracted, with other signs of imperfect convales-

---

[1] Dr. Hastings, formerly Resident Medical Officer at Shadwell Hospital, lays stress on the importance of always endeavouring to check this symptom in measles at once, and on the frequent difficulty of so doing, seeing that the concurrence of bronchitis often renders opium an unsafe remedy.

cence and notably with pulmonary complication, it should create some suspicion of tubercular disease and always occasion repeated examination and guarded prognosis. Even in severe non-tubercular cases with diarrhœa and melæna the result may be fatal, and all degrees of entero-colitis may be found post-mortem, attended sometimes by ulceration chiefly affecting the solitary glands.

In the later stages of measles or as more or less immediate sequelæ numerous *complications* may occur. Of such are otitis media leading occasionally to cerebral abscess, ophthalmia, inducing sometimes destructive keratitis, "noma" of the cheek or vulva, retro-pharyngeal abscess, ekzematous and pustular eruptions of the skin, stomatitis, "whooping" cough and membranous laryngitis. I question much whether all cases of these last two affections are due to the specific poisons of pertussis and diphtheria respectively. I allude again to the relationship between measles and spasmodic cough under the head of "whooping-cough;" and, as regards the membranous laryngitis which may be found either at tracheotomy or after death or may very rarely be established by expectoration during life, my own experience and belief correspond with Henoch's that it is not to be credited to the diphtheritic virus. In such cases without faucial or nasal symptoms the membrane is found post-mortem strictly confined to the respiratory tract.

Chronic bronchitis, broncho-pneumonia long in resolution, and, though less often, empyema, are familiar sequelæ, as well as *chronic tuberculosis*, especially of the lungs and bronchial glands. Measles indeed seems to prepare the ground for the tubercular process in a large proportion of children who die from tuberculosis in its various forms, whether acute or chronic; and there are very frequent instances of previously healthy children in whom wasting and chronic disorder, both in the pulmonary and alimentary tracts, and not necessarily tubercular, seem to arise directly out of severe attacks of measles.

True pharyngeal diphtheria undoubtedly occurs sometimes in close association with measles, especially in hospitals; but the connexion is far less close and frequent than that between diphtheria and scarlatina.

Paralytic affections of various kinds seem to me to bear a certain relation to measles. I refer, under the heading of nervous disorders, to some examples of a special form of muscular atrophy with this connexion; and have seen, besides a few instances of Infantile Paralysis, several cases of nondescript paresis or ataxia of the legs, apparently arising directly out of measles and with difficulty or not at all distinguishable from such as may be sequent on diphtheria. I have notes, among others, of a case of a girl of three, previously quite healthy, who three days after the onset of measles lost her speech and all power over her limbs, and had much difficulty in swallowing. After six months she had lost most of these symptoms but had no control over her legs. Three years subsequently the walk was ataxic and the speech slow and drawling. There was no wasting nor disturbance of sensibility, and the knee-jerks and electrical reactions were normal. Dr. Barlow has called attention to a myelitic paralysis following measles, and I feel convinced that the influence of this disease, if only as an exciting cause of neuro-muscular breakdown, requires further observation.

Various epidemics of measles have variously prominent characteristics; in some, acute chest affections are the rule, in others, severe diarrhœa; and I would especially mention early acute laryngitis and ulcerative stomatitis as having respectively marked by their frequency two considerable epidemics of which I had experience in the East of London. The stomatitis mainly affected the tongue and gums, and in many cases there was an apparently "diphtheritic" membrane leaving a bloody surface on removal. The fauces were unaffected except by hyperæmia.

The rash is sometimes in quite mild cases characterised by small extravasations, and more extensive *purpura* may occur with a very severe and sometimes even a fatal result. These extravasations may either be concurrent with the ordinary rash or appear when it has nearly faded. I have seen two cases of somewhat extensive purpuric eruption, appearing late, which recovered within the usual time. Of a form of measles mentioned by many, where the whole rash from the first is like that of hæmorrhagic or "black" small-pox, I have had no personal experience. Considering the present great rarity of its alleged occurrence, it seems probable that Collie's opinion that such cases are mostly, if not always, variolous is correct. Collie mentions also an occasional diagnostic difficulty between measles and small-pox, when the rash in the former is at first sight very like a certain initial and diffused eruption of the latter, especially in its confluent variety. I have occasionally met with this difficulty, and in one case, during a time of small-pox, where the rash seemed to me unusually prominent and hardish to the feel and was unaccompanied by signs distinctive of measles, I was confident in a wrong diagnosis. But Collie points out that there is almost always in such cases of small-pox a character of shottiness in some of the papules, and that the surface of the measles face is usually felt to be much smoother; and compares the small-pox and measles skin, as regards touch, to corduroy and velvet respectively. I do not think that the isolated papules which sometimes usher in the rash of measles should ever cause much difficulty of distinction from the early eruption of ordinary small-pox, for they are never of shotty character, and the different course of the early fever with the ingravescent rash in the two diseases would soon clear up any possible doubt.

I shall leave the matter of the **diagnosis** of measles from scarlatina and so-called "rubella" to be inferred from the general consideration of these affections; and would only remark here that, however observant and experienced a man may be with regard to the exanthemata, a diagnosis should never be made dogmatically on the sole basis of the rash, which is in no disease always pathognomonic. Innumerable errors, mostly avoidable, are made from neglect of this warning, and when, as may be the case in any exanthematous affection, the eruption is throughout uncharacteristic, or is never observed in virtue either of its evanescence or non-existence, it is clear that too much reliance on eruptive appearances may leave us helpless or cause us to blunder gravely.

The **contagium** of measles has not been isolated, although certain bodies are said to have been found in the blood and expired breath. It is probably contained in the breath and possibly in all other emanations from the body; is certainly most active from the outset to the height of the disease; and is, with the possible exception of that of influenza, more energetic than any other. There is no evidence that it outlives the symptoms, although there is some that its activity may anticipate them; and the patients, after a hot bath and plentiful scrubbing and disinfection of all clothes which they may have worn during the illness or a fortnight previously, may certainly be pronounced harmless to others when three weeks have passed from the day of invasion and probably as soon as the rash has quite vanished.[1] If, however, they be still desquamating, or suffering from a continuously marked chest affection with any rise of temperature, it is well to confine them longer both for their own sake and possibly that of others. We have no certain knowledge as to the conveyance of the contagium by clothes or by unaffected persons. Our precautions, therefore, must be on the side of safety. The period of *incubation* is generally believed to be about ten or twelve days before the invasion symptoms. On some occasions at Shadwell we have been able to fix on eleven days with considerable accuracy, and in several more the rash of measles has been noticed on the fifteenth day after a short exposure to infection from cases which have been discharged from the ward immediately on discovery of the eruption.

Measles may undoubtedly occur twice, even during childhood, and a second attack in adult life is not uncommon. True *relapses* of all symptoms within a week or so of convalescence are also, though rare, beyond question. For further details as to the course of measles, its symptoms and its varieties, I refer to the larger works.

Treatment.—In all cases the chief necessity is protection from chill and everything that may aggravate the respiratory catarrh which forms

---

[1] Cases have been frequently sent back to the general wards from the "Infectious Block" at Shadwell, as soon as the rash had gone, with no instance of spreading of the disease.

the chief element in the immediate gravity of the disease. This is especially necessary in children under three years of age, who, however, frequently die, sooner or later, in spite of all precautions. Chest and laryngeal affections and all important complications or sequelæ are to be treated on the lines elsewhere laid down; but slight epistaxis, slight cough and other unimportant symptoms may be let alone. I usually give a combination of carbonate of ammonia and compound tincture of camphor for a very troublesome cough in ordinary cases, and advise tepid sponging, when the fever is high, with due precaution against chill. There is no need for confining to bed a child who feels well, after the temperature has fallen to normal; but he should be kept in the room for a week longer and in the house for a fortnight more. While any cough remains, even though there be no abnormal physical signs in the chest, precautions against chill should be continuous. In all complicated cases and those with tubercular tendencies there is enhanced necessity for the greatest care.

Considering the pulmonary sequelæ and their encouragement of tuberculosis, measles is ultimately one of the most fatal of the fevers of childhood. Some regulations and, perhaps, hospital accommodation to prevent its spread and minimise its severity seem to be more urgently required, in the interests of children, than in the case of those diseases which are already dealt with in London by the government institutions.

CHAPTER V.

RUBELLA (OR "GERMAN MEASLES").

IT is now recognised by most observers that an exanthematic contagious fever of short duration, with a rash usually resembling that of measles but sometimes that of scarlatina, occurs from time to time both in sporadic and epidemic form; and that there is no mutual protectiveness between this fever and either measles or scarlatina. There is, however, a much wider discrepancy in almost all points between the many descriptions of so-called rubella, published by observers of large numbers of cases in various epidemics and in different countries, than obtains in regard to the definition of any other specific fever. This discrepancy indeed is so great as to justify the statement that rubella, apart from the doubtless important, though not crucial, test of its epidemic occurrence among those who have previously suffered from scarlatina or measles,

has no greater mark of specificity than appertains to recognised varieties
of these two diseases.

It is especially in the diagnosis of sporadic cases of this supposedly
specific disease that a generally insuperable practical difficulty arises;
and my own experience and careful study of much of the literature of
the subject forces me to teach that such a diagnosis should never be
made until such time as a subsequent series of like instances may seem
to justify it, and even then should never be pronounced at the outset
of any individual case. If rubella be a specific disease it certainly
cannot breed scarlatina or measles; but it is unquestionable that cases
which many observers, including several who have made this question
a special study, denominate rubella may, and often do, give rise to ordinary
scarlatina or ordinary measles. It is no answer to those who are frequently confronted with this difficulty and, therefore, provisionally conclude, as I do, that the question of specificity is not settled and that
there may be more than one form of fever included under the title of
"rubella," to adduce the frequent variations from the normal type in
many particulars of other infectious diseases. For it must appear to
any careful student of the widely-varying first-hand accounts of rubella
by different observers, who nevertheless alike entertain no doubt of its
specificity and attribute either want of experience or reason to those
who do, that there is nothing approaching to the average description of
a clinical type disease to be extracted from an unprejudiced collation
of the most important authorities; and, further, that what some writers
regard as almost pathognomonic symptoms are stated by others to be
very exceptional. Thus many observers regard enlargement and tenderness of the cervical glands, especially those behind the sterno-mastoid
and in the post-aural region, as almost constant, while others as positively
state that it is very rare; and there is no general consensus about the
frequency of invasion symptoms before the rash, the prevalent presence
or absence of coryzal signs, of inflammatory sore throat, of complications,
or of desquamation.

Concerning the rash, descriptions differ so much as to distribution,
duration, form and colour that it may be fairly said to be of no diagnostic value. This, indeed, is freely confessed by many who regard
rubella as one and indivisible. Often it is stated to be indistinguishable
from the rash of measles, and not seldom very like that of scarlatina,
and in certain instances it may be at first morbilliform and later scarlatiniform in appearance. Each observer, having diagnosed any given
epidemic illness as rubella, naturally regards the rash and other symptoms
that he has himself observed as characteristic; and it can scarcely be
denied, judging from the multifarious literature of this subject, that if
there be one, there are more than one clinical group of symptoms which

equally deserve an isolated position. It must be added that, while there is a general belief in the almost uniform mildness of rubella, some teach that it is often severe and sometimes fatal from laryngeal and pulmonary mischief; and, although most regard its occurrence as protective against a second attack, some urge that it is very often recurrent and may affect one individual several times. The different statements made as to the period of incubation are of less weight than other discrepancies in connexion with the question of specificity, for the limits of this period are wide enough, as far as any certain knowledge goes, in most recognised affections of this class. Such limits in rubella are said to be from a few days to three weeks, most authorities fixing about the same period as that of measles, namely about a fortnight from infection to eruption. A well-marked epidemic, however (whereof I had some knowledge and a full first-hand report), which seemed to correspond very closely to many descriptions of rubella, was certainly not scarlatina, and attacked many who had very recently had measles, had definitely in some cases, and probably in most, an incubation-period of not more than five days. From what I have seen myself in the ordinary course of practice I can say but little regarding extensive epidemics of this kind; but from certain sporadic cases that I have from time to time met with I should be inclined to regard the following group of symptoms occurring epidemi- cally as deserving of a clinical position probably separate from both measles and scarlatina :—Fever, mild and of from one to three or four days' duration, a measles-like rash from the beginning of symptoms, and slight or marked sore throat, with or without swelling of the cervical glands. The papules of the rash are perhaps smaller and tend to be more confluent in places than those of ordinary measles, but show the same order, with possibly greater rapidity, of progress from the face and neck downwards. I have also sometimes seen isolated cases which answered exactly to those described by many as "rubella scarlatinosa," there being a scarlatinous rash, with some fever, illness and inflammatory sore throat, not followed by desquamation. Some of these have had considerable coryza. Again, I have seen, more than once, several cases, in one family of children, of a rash much more like that of scarlatina than of measles, preceded by half a day's slight feeling of malaise, with no sore throat or other complaint, and very slight pyrexia; the rash beginning on the face and progressing over the body in a downward direction, the face being clear before the legs were affected, and all signs disappearing by the end of the second day. In one set of three cases of this kind there was a fortnight's interval between the appearance of the rash in each. On the whole, however, I personally regard as rare the occurrence of any cases which suggest the diagnosis of rubella, however described; and the mainly negative hospital experience of my own for

many years in this respect is corroborated by that of several of our
resident medical officers at Shadwell, a district teeming with children.
Mr. Scott Battams, who held that office for nine years and saw many
thousands of casualty cases brought for all kinds of acute diseases,
including innumerable instances of scarlatina and measles, was never
forced to the diagnosis of rubella.   One of the most positive assertors,[1]
moreover, of the specificity of rubella, while making the self-evident
remark that "should the disease preserve a *typical* course but little
difficulty will be met in the diagnosis," states that in a single case there
is no positive diagnostic guide; and similar admissions are freely made
by the majority of authorities, however widely they may differ *inter se*
as to the "type" of the disease.

Without therefore denying the probability of the existence of some
such specific disease as the "rubella" of many authorities, which may
conceivably indeed be ætiologically one in spite of an indescribable and
almost limitless variety of clinical expressions, I feel sure that as a matter
of practice it should be diagnosed in individual cases with the greatest
hesitation, and that we are rarely justified in positively excluding both
measles and scarlatina in any isolated case of supposed rubella.   By this
precaution we may be saved from many a blunder at the small expense
of a frank confession of ignorance.   It is beyond doubt to any one of
experience that many cases of afterwards unquestionable measles or
scarlatina vary so much from the "type" or prevalent form that they
can only be rightly judged of from their close association with cases
of the universally recognised type-disease; but in the matter of rubella
the typical image, if indeed it exist, still awaits the medical artist to
liberate it from its conglomerate matrix of clinical material.   Its alleged
varieties are all that meet the eye.

The difficulties of diagnosis will probably not be set at rest without
the discovery of a specific organism, of which criterion of individuality
no contagious disease stands more in need than so-called rubella.   Of
the theory that rubella is a hybrid between scarlatina and measles I can
but say that, if this means that the subject is infected simultaneously
by the two poisons, the epidemic occurrence of the disease is not thereby
rationally explained; whereas, if the term "hybrid" be inaccurately
predicated of the hypothetical germ of rubella supposed to be in an
imperfectly differentiated or transitional condition, the notion, if superficially plausible, is certainly fanciful and unsupported by analogy, either
clinical or biological.

Of treatment there is nothing for me to say.   In my own experience
only ordinary care with no special medication has been indicated in any
case which I would regard as neither measles nor scarlatina, or in which

---

[1] Dr. Edwards in Keating's *Cyclopædia of Diseases of Children*, s. v. "Rubella."

I practically excluded those diseases on the ground of an epidemic prevalence of an exanthem among those who had already been their subjects. But there is good evidence that cases of a serious nature occur from time to time, which have as much claim as the milder ones to be regarded as " sui generis," although it does not yet appear whether they belong to the same or a different category.

CHAPTER VI.

CHICKEN-POX, MUMPS AND INFLUENZA.

Chicken-Pox (*Varicella*).

On this exceedingly common disease, which is mainly incident on children from one to ten years old, I have little comment to make, and shall not attempt more than a brief description, with cautions as to diagnosis.

The affection is very contagious, its poison being in all probability readily air-borne and carried also in clothing; its fever lasts but a few days and is usually very slight; and recovery is for the most part complete without sequelæ. Epidemic prevalence is very frequent.

Although as yet the specificity of the poison of chicken-pox has not been micro-biologically established, we are justified in believing in its existence, and the clinical evidence of its distinctness from that of small-pox is overwhelming. Nevertheless there are, doubtless, a few cases, both in children and sometimes in adults, which at first occasion insuperable diagnostic difficulty even to expert observers. I have myself known more than one case definitely diagnosed as small-pox by men who have had great hospital experience of that disease, but whose opinion has been subsequently proved erroneous, not only by the course of the individual case, but also by the concomitance in the patient's family of several cases of typical varicella.

It may be freely conceded to the few authorities who may still regard these two diseases as due to one and the same poison that, at least clinically, it is absolutely impossible in some cases to differentiate between varicella and that modified form of small-pox, so familiar in children during epidemic times, which is known by the title of varioloid.

The *incubatory* stage of the disease may be practically stated as from two to three weeks, although some observers give the minimum period as eight days. In my experience fourteen or fifteen days has been the most frequent period. Whatever slight symptoms of illness there may

N

sometimes be before the rash appears are quite indistinctive. In very rare cases there is premonitory fever of considerable severity. I have myself seen two cases where the characteristic eruption of vesicles was preceded for nearly two days by a bright red rash almost, if not quite, indistinguishable from that of scarlatina. In both these cases fever was high with marked illness, and the varicellar eruption was very profuse. No desquamation ever took place nor was there any sore throat. As is well-known, the eruption begins as small red spots, slightly elevated, which in the course of a few hours as a rule become vesicular. The vesicles sometimes retain a surrounding red areola, but at other times this is absent, causing the appearance popularly known in some districts as " glass-pox." In a small minority of cases, otherwise normal and benign, vesiculation may be retarded, certainly for many hours if not for a whole day, and these are the instances where there is the greatest difficulty of diagnosis between varicella and varioloid. Successive crops appear during two or three days, after which the fever falls or disappears and the skin affection begins to decline with drying up or crusting of the vesicles. Sometimes the vesicles become enlarged and show a slight depression (or "umbilication") in the centre. The larger vesicles and others which have been injured by scratching or otherwise often leave behind them a persistent round white cicatrix. The duration of the disease until the falling off of the scabs is from about eight to eleven days. The eruption may occupy the whole body as well as the oral and faucial mucous membranes, and usually proceeds from above downwards.

In some cases, otherwise slight and ordinary, some of the vesicles have purulent contents.

I have seen a few cases where some sore-throat has been complained of, with signs of moderate pharyngitis.

There is a form of varicella, not very rarely observed since attention was first called to it by Mr. Hutchinson, where some of the vesicles enlarge and ulcerate, or may become black and gangrenous with deep ulcers underlying the scabs. This process, according to Dr. Crocker, does not always occupy the seat of the varicella vesicles, but may attack other parts. It would appear that this form or rather complication of varicella occurs especially, if not entirely, in weakly children or those who are tubercular, and should scarcely be regarded as other than an accidental epi-phenomenon. In its graver form it is frequently fatal, especially, if not always, in tubercular cases; but in its lesser degrees, which alone can be considered as not very rare, recovery may take place with comparative frequency. Continued pustular eruption not seldom, and pemphigus and urticaria sometimes, are met with as sequels of varicella. Very occasionally acute nephritis follows in a few days after the subsidence of the disease.

It is unnecessary to dwell on the differences between varicella and ordinary variola, but we must always be on our guard, whether in epidemic times or not, against the dangerous, albeit sometimes unavoidable, diagnostic confusion of varicella and varioloid. When a definite period of three days' prodromal fever, especially with back-ache and chill, has existed, we must not diagnose varicella, however characteristic of this affection the eruption may be, but isolate the patient at once on the suspicion of small-pox ; if, however, as is by no means very rare in varioloid, there be no clear period of symptomatic invasion, but slight illness and fever only and no pustulation, we have nothing to depend on for distinction from varicella but the possibly shotty feel of the papules. If well-marked this is a valuable positive sign as a rule ; but I have observed it at least once in true varicella, with much retarded vesiculation, where it deceived not only myself and several others, but also a careful connoisseur of small-pox of many years' hospital experience. As regards the distribution of the eruption in these affections it has been observed by Dr. MacCombie and others that in small-pox the extremities, in chicken-pox the trunk, is most affected.

In spite of the almost constant benignity of chicken-pox all patients should be isolated if for no other reason than the possibility of the rare and grave complication above noticed. No special treatment is required in the vast majority of cases. There is no reason to believe that the patient is infective to others after the disappearance of all scabs.

Mumps (*Contagious Parotiditis*).

Of this disease with its well-known characteristics my experience has furnished me with no grounds for comment on the classical description given in the text-books. The diagnosis is rarely difficult, except in cases where the swelling is but indistinctly localised in the parotid region or more especially involves the submaxillary glands, and in those where the parotid swelling remains unilateral for more than two or three days. These difficulties are of course more prominent if the case be a sporadic one, or the first encountered in a time of epidemic.

I have seen cases of lymphatic glandular abscesses which have been at first sight taken for mumps, as well as some of swelling in the parotid and submaxillary regions which were secondary to periostitis of the jaw.

It is frequent for the swelling of one side to precede by a day or two that of the other, and not rare for the submaxillary glands and even the lymphatic glands of the neck to be so much involved in addition to the parotids as to modify considerably the typical appearance of the subjects of this disease.

We occasionally meet with inflammation of the parotid glands in the course or as the sequel of other diseases, such as enteric fever (though very rarely in children), scarlatina, measles and small-pox. I have also seen it several times, but in adults only, in patients with gastric ulcer who were being fed exclusively by the rectum. In such cases the swelling is generally unilateral throughout, and may proceed to suppuration. Any diagnostic difficulty hereby presented will usually be dispelled during the progress of the case. The swelling in true mumps is almost always at last bilateral, although the involvement of the second side may occasionally be long delayed; suppuration is extremely rare; and there is scarcely ever any cutaneous redness or much tenderness on pressure over the enlarged glands. The chief complaints made in mumps are more or less dull aching pain in some cases, and, in nearly all, difficulty in opening the mouth and considerable pain in attempting to masticate. In most cases that I have seen, pain on movement of the jaw was apparently the first symptom of the disease. Deafness is sometimes complained of, and there is occasionally either dryness of the mouth or, for a while, increased salivation. The symptoms of prodromal feverishness may precede the local complaint for a few days, but are very often absent or unnoticed, and the temperature at the height of the attack is rarely more than 102°. The swelling usually reaches its height about four days after its first appearance, and then gradually recedes for a similar or somewhat longer period. Having never had occasion to study any number of cases during an epidemic I can say nothing personally either of the so-called "metastasis" to the testes, which all authorities agree to be very rare, at least before puberty, or of that to the ovaries or mammæ, which is perhaps problematical. I have once seen pericarditis, otherwise unaccountable, following immediately on an attack of mumps, but no stress can be laid on this coincidence.

The period of *incubation* seems to be from one to three weeks. One case of mine pointed almost conclusively to an infection of the latter date. The event of mumps is almost always favourable, with no sequelæ. Although the disease is probably highly infectious the questions of isolation and quarantine may, in my opinion, be left to the discretion of those concerned.

Nothing is usually required in the way of *treatment* besides confinement to the house or room, or to the bed when there is considerable fever, and hot local applications when there is much discomfort. Ordinary remedies for the febrile condition may be given should relief of symptoms seem to require them.

This disease very seldom attacks infants, and is rare in early childhood.

Influenza.

Although influenza is neither mainly nor most severely incident on childhood, my experience of the recent epidemics of 1891 and 1892 leads me to write very shortly of some of its manifestations in young children, which are on the whole less distinctive and thus much more often overlooked than is the case with adults. It is not within my scope to discuss or detail the characters of this disease. I shall, therefore, state that, from the experience I have had, I regard it as a specific contagious fever, mainly of epidemic character, with acute onset and varying duration ; liable to relapse ; and not seldom recurring in the same individual in one or more subsequent epidemics. The actual fever lasts usually for not more than three or four days, often for a much shorter, and occasionally, although without discoverable complication, for a considerably longer period. The attack begins almost always with either a definite rigor or, more often, a feeling of chilliness, as expressed by almost all patients old enough to describe their sensations. It is attended by general discomfort and usually by distinct pain, mostly in the back and legs, with as a rule more or less severe headache, great prostration, and drowsiness. Some cough, generally frequent, hard and painful, though sometimes very slight, marks most cases at all ages, and is often so paroxysmal in its nature as to be quite indistinguishable from a typical attack of whooping-cough. Of this I have seen several examples in adults as well as children. In my own experience of the recent epidemics coryzal symptoms were decidedly rare both in children and in adults. Diarrhœa was frequent ; nausea or actual vomiting by no means uncommon even long after the subsidence of the fever ; and the attack was often followed by giddiness, neuralgia in various parts, and many other symptoms of nerve disturbance enduring for a long and indefinite time, with much prostration and often with obstinate anorexia.

In a certain number of cases, proportionately very great in children, there were inflammatory attacks of the respiratory tract over and above the laryngo-tracheal, tracheal, or large-bronchial catarrh which, in some degree, may be regarded as part of the clinical picture of the type-disease. Such attacks were acute general bronchitis, often of the so-called capillary form, or demonstrable broncho-pneumonia. Occasionally also, though in my experience much less often than was apparently the case in that of others, what was seemingly ordinary pneumonia occurred, strictly limited to one side, beginning acutely, and ending, after a typical course of signs and symptoms, with a critical fall of temperature attended by sweating. Judging, however, from the numerous cases I have myself seen at all ages, including some which were fatal and were examined post-mortem, I must emphatically state that the lung-inflammation proper to influenza is, in an overwhelming majority of instances, unquestionably broncho-pneumonic. In children, indeed, a severe broncho-pneumonia, beginning acutely but running an indefinite course with signs of much bronchitis, may be the chief and perhaps only definite indication of influenza. Such cases occurring in children over three or four years old, with no history whatever of antecedent illness either acute or chronic, were exceedingly frequent during the influenza epidemics of 1891 and 1892, and, considering that acute broncho-pneumonic attacks of so-called simple or catarrhal nature are almost unknown at this and later ages, there seemed to be no doubt of their truly influenzal origin. An additional reason for regarding these cases as specific, even in the frequent absence or slight prominence of other symptoms of the disease, was the concurrence in the same family or house of typical instances of adult influenza. Without this latter factor the diagnosis of influenza in infants is often very difficult, and I am inclined to think that this was the chief reason for the prevalent belief, in which I strongly shared, during the epidemic of 1889-90—the first in the experience of most of us—that children as a rule escaped the disease. In the following years, however, I observed an unusual prevalence of acute bronchitis and broncho-pneumonia in previously healthy infants with good surroundings, which disappeared coincidently with the subsidence of each epidemic.

It must also be remarked that in several cases in infants and quite young children, which for the reasons above stated I regarded as influenzal, drowsiness, starting during sleep, sudden screaming and occasional squinting, with fever, were observed ; and that the diagnosis of meningitis was frequently suggested and even sometimes adhered to by medical observers. There was, however, no paralysis nor spasm, other than occasional slight convulsions in one or two cases, nor any other evidence of meningitis ; and the patients always recovered.

I have not made careful observation of enough cases to enable me to say whether or not there are other distinctive points worthy of notice in the influenza of young children. Ear-ache was certainly frequent even in otherwise uncomplicated instances. Sometimes tuberculosis, either of the lungs or brain, made its appearance as a sequel of what was apparently an influenzal attack. I have, however, several times seen phthisis in adults starting with the lung-inflammation of influenza.

Apart from the occurrence of severe broncho-pneumonia, I believe influenza to be rarely fatal in children. Sometimes, though far less frequently than in adults, there is weakness and protracted convalescence.

Concerning medicinal *treatment* there is nothing special to say. I have usually given quinine, with strychnia or nux vomica ; and often antipyrin or salicylate of soda, with apparently good effect, when there was much pain. There is, according to my experience, no reason to believe that either the salicylate or salicin itself has any effect on the morbid process; and the latter drug seems to be here, as it certainly is in acute rheumatism, a much less certain anodyne than the former. For the rest, confinement to bed at once and for some days after the fever is over; abundant food as soon as the appetite returns, and plenty of concentrated liquid nutrients when it does not; and alcoholic stimulants, in small doses frequently repeated, when there is much prostration, are the chief points to be observed. If convalescence flags, change of scene with plenty of fresh air and other hygienic remedies, assisted by a course of arsenic and iron with or without cod-liver oil, is strongly to be recommended.

CHAPTER VII.

ENTERIC FEVER.

THIS disease is mainly incident on childhood and youth and is especially frequent between five and fifteen years of age. Under two years of age it is rare. In the second quinquennium its fatality is less than either before or afterwards, and after the eleventh or twelfth year differs but little from its average at all ages. From my own experience and a study of the records of the fever hospitals I am convinced that the danger of enteric fever in childhood is usually much under-rated; and it would seem that the lesser death-rate under twelve years of age, which is about four to five per cent. below the average at all ages, is mainly due to the comparative rarity of deep ulceration with its frequently consequent perforation.

Although our central conception of enteric fever must be that of an infectious disease with special symptoms, due, in all probability, to the action of a specific bacillus, and marked post-mortem by inflammation or ulceration of Peyer's patches and the solitary glands of the ileum, we nevertheless sometimes meet with cases, clinically indistinguishable from the type-form, which either show no very characteristic lesion after death or may be marked by ulceration of the large intestine alone. On the other hand, the well-known ulcerative lesion and even perforation may be found in cases which have run a short course with no specially diagnostic symptoms. It is, according to my experience, only in adult life hat ulceration exclusively occupying the large intestine accompanies a fever which has all the clinical marks of enteric; but it is common in childhood to find only swelling and softening of Peyer's patches and of the solitary glands, without ulceration, and this not only in cases where early death may have anticipated ulceration but also in those of prolonged duration. Swelling and even softening of Peyer's patches is certainly not confined to enteric fever in young children ; and, when we reflect on all these facts and on some clinically anomalous cases of fever which we occasionally meet with at all ages, unmarked by the characteristic post-mortem lesion, we must either entertain doubts that enteric fever, as generally diagnosed during life, is always one and the same affection, or question the claim of disease of Peyer's patches to be a strictly integral part of its definition.

The **contagiousness** of enteric fever is still a matter of debate among clinicians. Without entering here into a detailed support of my own views, which are largely based on my experience at a children's hospital as well as among adults, I would state my agreement with the opinion of Collie and other authorities that the disease is frequently conveyed directly from the sick to those in attendance upon them, and that young nurses are especially apt to suffer. I regard it as proved that the active contagium is not confined to the decomposing fæces, but, at the same time, as still somewhat doubtful whether it is carried by other emanations than the intestinal discharges. It is of course almost, if not quite, impossible to prove, in any given case of apparent contagion, that some fæcal matter has not been conveyed directly to the mouth after handling the patient ; but I have seen a sufficiently large number of examples of infection of nurses and others in attendance on enteric cases when there has been no diarrhœa, and when all precautions with respect to cleanliness were with the greatest degree of probability observed, to cause me to regard, for practical purposes at least, the possibility of contagion apart from fæcal convection as by no means disproved. I am further of opinion, owing to some important experience I have had, that, however seldom enteric fever may spread to other patients in a hospital ward, the massing together of several cases may reinforce the contagium and be in all probability a source of direct infection through the air. Hence I would discourage the simultaneous admission of more than a very few cases even to a large general ward, and, in private cases, forbid all children and young people access to the patient's room. I cannot dwell here on the other recognised sources of infection from drains, drinking water, milk and the like ; but would insist on unremitting attention to personal cleanliness on the part of the attendants on the sick, and on the disinfection of all discharges and soiled linen according to well-known methods.

Assuming the reader's knowledge of the typical phenomena and course of the disease, I shall confine myself here to the notice of such points in

symptomatology as seem to be more or less peculiar to its examples in childhood.

The *duration* of the actual fever is certainly much more often under three weeks than in adults, and is sometimes under two. If there be absence of diarrhœa, rose-spots and splenic enlargement, as well as of the typical temperature-curve, some diagnostic doubt may continue even after recovery; but this difficulty is not often met with in practice. There are but few cases of fever which run a course of even ten days or a fortnight without some characteristic signs, and it is but seldom that we have to diagnose enteric fever on purely negative grounds. I have hitherto failed to recognise, as some authorities do, any considerable class of cases of apparent febricula or simple continued fever which, from the lack of distinctive symptoms alone, deserves to be called enteric.

In spite of the frequent difficulty of ascertaining the date of onset of the disease, I am fully satisfied, from otherwise typical cases where this date could be accurately fixed, that at least in early childhood the whole course of the fever is sometimes considerably under a fortnight.

When a case lasts longer than four weeks there is, generally, progressing ulceration of intestine with more or less continuous diarrhœa. Such cases are grave, but I have seen several recover perfectly even after two or three months' course with no apyretic interval.

The *temperature* shows greater variation from the typical curve than in adults, and this not only in cases of short duration. In scarcely more than half of my cases are the charts so significant as in a vast majority of adult patients. This phenomenon is in keeping with others dependent on the readily modifiable action of the nervous system in early life. Not only does the temperature-curve, even in severe cases, often touch the normal line in the earlier part of the fever day, and show a much greater comparative rise in the later part than in adults, but also the general range of temperature is frequently higher. It is pre-eminently true in childhood that the severity of the disease is by no means always proportionate to the average height of the temperature, and the remembrance of this fact will often be a check on meddlesome or harmful treatment. In convalescence, too, the temperature, far more often than in adults, rises unexpectedly for a short time. Such a rise in some cases is certainly associated with constipation and disappears with its relief. Constant subnormality during early convalescence to the extent of one, two, or even three degrees, and also irregularity of temperature are observed in a majority of cases, especially in those where there is rapid recovery after great wasting. This phenomenon may occur in adults, but is far more frequent in children.

Sudden onset of the fever is very often seen in childhood. Out of a series of 62 cases under 14 years old, the exact day of onset could be fixed in 24 by definite symptoms, the child having been previously to all appearance perfectly well; and in many more the beginning of malaise and anorexia was very nearly dated. In adults insidious onset is far more frequent, although, according to my experience, by no means so prevalent as is usually taught. Among the symptoms referred to, headache almost always occurs, shivering or rigors usually, and vomiting in most cases. Pain in the belly, sometimes severe, is often complained of, especially in cases where diarrhœa is an early symptom; and there is frequently much aching of the back or legs. Occasionally tonsillitis is an early symptom.

Diarrhœa, in the sense of three or four loose motions in the twenty-four hours, occurs at some period in most cases. It is, however, still more rarely in children than in adults that very large, frequent and watery motions are noted in cases which recover; and excessive diarrhœa at any age may be regarded as both an exceptional and grave symptom. *Constipation* after the first week is observed in perhaps about one-fourth of all cases in children, and in a much smaller number is more or less persistent throughout. This condition is somewhat more frequent than in adults, owing probably to the less severe inflammation and rarer ulceration of the intestinal glands; extent of intestinal involvement being roughly proportionate to amount of diarrhœa. Cases with persistent constipation are, however, by no means always of the mildest, for both in adults and children they may be marked by high and protracted fever, by severe pulmonary and other complications, or even, as I have more than once seen, by deep though very limited intestinal ulceration with fatal peritonitis from perforation. I would state here that among fatal cases in childhood, with or without much diarrhœa, perforation does not appear to be proportionately less frequent than in later life, although its symptoms may be more obscure. Certain epidemics of enteric fever appear to be marked by frequent instances of constipation, while in others diarrhœa occurs in a large majority of cases.

Splenic enlargement can usually be established by careful palpation after the first week; but it is rarely detectable before the eruption appears, and is therefore of no great diagnostic value except in cases where the eruption is very late or altogether absent. It may persist for a week or more after convalescence sets in; or may soon disappear, as far as palpation can prove, even in cases which subsequently relapse. Determination of the spleen's size by percussion is, I think, seldom accurate enough to be of much practical use. In children the sign of splenic enlargement is even of less value than in adults, for their spleens are apt to become palpably enlarged in other febrile conditions. Tenderness in the splenic region is very common, and more characteristic when it can be satisfactorily isolated from the general abdominal tenderness so often observed.

Rose-spots are but seldom absent. They were noted in 48 out of 62 cases after admission to hospital at varying periods of the disease. Careful observation after the fifth or sixth day will establish their existence in an immense majority of cases. Often, however, they are infrequent and may occur in single or only two successive crops; they may last but one or two days, and are sometimes seen only on the back. Very occasionally there is a scarlatiniform rash on the body in the early time of the fever, especially in cases which begin suddenly. This rash may endure for several days, spreading from the face downwards. At first it may be difficult in such cases to exclude the diagnosis of scarlatina, but the throat is not characteristically affected nor is there any subsequent desquamation.

Bronchitis, with or without much cough or dyspnœa, and often so slight as to be detectable only on careful examination after deep inspiration, is the rule in enteric fever with, I think, fewer exceptions than in adults, and may be regarded more as one of the expressions than as a complication of the disease. Even when there is no cough slight bronchitic signs are exceedingly often observable. Sometimes the gravity of the illness is mainly due to this bronchial involvement, which may mask other symptoms and cause difficulty in diagnosis almost from the beginning. Occasionally in protracted cases it seems to be the cause of death. Evidence of bronchitis can very often be established by examination of the front of the chest alone. Broncho-pneumonia is not very rare, but is less often detected by signs during life than demonstrated post-mortem.

The *heart* almost always suffers in enteric fever to a degree discoverable by examination. In numerous cases with perfect recovery I have observed complete absence of the first sound for many days, and in still more the first and second sounds were for long indistinguishable from one another. We often hear a rapid succession of short sounds, the normal pauses being abolished. The impulse is proportionately diminished, or impalpable. In protracted and severe cases bilateral ventricular dilatation may be made out; and I have sometimes heard a soft but well-marked systolic murmur at the apex, which has afterwards disappeared with the return of the heart's dulness to its normal dimensions. This cardiac feebleness and dilatation are of great clinical importance, and are evidenced, symptomatically, by the extreme rapidity of pulse and occasional syncope which follow on movement or on suddenly sitting up, and, anatomically, by the softness, flabbiness and thinning of the heart's walls which have been frequently found post-mortem. Actual myocarditis may occur, of which I have seen one well-marked instance.

Not the radial pulse alone, therefore, but also the heart should be constantly and carefully examined before allowing patients to rise from bed or sit up. The morbid condition of the heart may long outlast the fever, and is a strong indication for the greatest caution in convalescence.

In many of the complications of enteric fever there is nothing special to childhood. Laryngeal ulceration ; tonsillitis with exudation, not easily distinguishable from diphtheria and causing much dysphagia which should always suggest examination of the fauces ; epistaxis ; otitis ; glandular and cellular abscesses ; bullous eruptions ; parotiditis, and venous thrombosis are certainly rare, and probably, with the exception of tonsillitis, rarer in children than in adults. In one case, aged $7\frac{1}{2}$ years, which ultimately made a good recovery, peritonitis set in during the fifth week, and after a while there was swelling and purulent discharge from the umbilicus. The swelling was incised and drained. There were also bed-sores, multiple abscesses, pericarditis and pleural effusion. Occasionally we meet with cases of destructive arthritis, peri-ostitis, and necrosis of jaw.

Special *nervous symptoms* are not common, although deafness, without otitis, and some amount of delirium, are very frequent. I have known one case of definite meningitis following on purulent otitis, one case of typical hemiplegia, and a few instances of indefinite paraplegic weakness with ataxia. In some cases which perfectly recover there is retraction of the head, with other apparently cerebral symptoms such as the "tache," photophobia, and great drowsiness. Marked mental changes, both in the direction of mania and imbecility, are rare in childhood, though met with by most observers from time to time in adults. I have, however, seen several cases where children, after a long and severe attack, became forgetful and silly for a considerable time and were unable or unwilling to speak.

Relapses in the sense of fresh febrile attacks with distinctive symptoms, occurring several days after the establishment of apparent convalescence, are from all accounts not so frequent in children as in adults. I have, however, seen many cases. More common is the recrudescence, often repeated, of symptoms, with or without apyretic intervals of a day or two. Owing to these relapses and recrudescences the whole course of the disease may be prolonged for months, and yet perfect though tedious recovery ensue. That true relapses are mostly milder than the primary attack is perhaps a statistical truth, but, in my experience, of no practical moment in prognosis. I have seen several instances of far greater severity during relapse. True relapses are but seldom traceable to dietetic errors, such possible exciting causes being positively excluded in most cases ; but return of diarrhœa, with heightened fever and other symptoms which constitute recrudescence, is sometimes in all probability referable to careless feeding.

Diagnosis.—Among the diseases which may be confused with enteric fever either early or late in the course of the affection I would mention pneumonia, meningitis, bronchitis, phthisis, and acute tuberculosis.

Pneumonia is sometimes evidenced by only very late physical signs or by none at all; and may involve but slight respiratory trouble. Beginning with similar symptoms to those of some cases of enteric fever, this disease may then be very difficult of diagnosis until a definite pneumonic crisis may remove doubt. The diagnosis of enteric fever is at the best, during the first week, a process of exclusion. *Meningitis*, especially tubercular, may closely simulate enteric fever, from which it is by no means always to be separated by the infrequent pulse which is insisted on by many. The cleaner tongue, however, the normal, doughy or retracted abdomen and the intense dislike of disturbance will greatly aid us in excluding the probability of enteric fever. *Bronchitis* has often been diagnosed as the primary affection in the second week or later in this disease, when distinctive symptoms have disappeared or are masked. Remembrance of the rarity of simple and extensive bronchitis after infancy has more than once led me to suspect and, subsequently, correctly diagnose enteric fever from the mere presence of severe bronchitis with fever of some duration. *Phthisis* is sometimes diagnosed in cases of enteric fever coming late under observation. Much wasting, with bronchitis and perhaps doubtful pulmonary signs, and other phthisical symptoms are often salient phenomena in late enteric fever; and in some cases the correct diagnosis can only be arrived at by a careful retrospect on the recovery of the patient. Cases of extreme wasting due to enteric fever which has never been recognised have several times come under my notice and caused great diagnostic difficulty. *Acute general tuberculosis* may be mistaken for enteric fever after even careful observation; but the rarity of this form of tuberculosis, especially after infancy, is much greater than the frequent mention of this possible diagnostic difficulty would lead us to expect. With the exception of some temporary hesitation over a few cases of tubercular peritonitis with some diarrhœa, I have but rarely met with this source of confusion in practice after a careful study of any case in question.

There are several other affections, among which is influenza, that may be mistaken for enteric fever in the first week or even later. I shall, however, say no more on this head than that enteric fever may begin in almost any way, and that, without the most careful observation and thought, it should never be positively excluded from the diagnosis of any case of illness where it may be suspected with the faintest show of reason. We must always provisionally treat a fever as enteric until the lapse of time or the appearance of other symptoms remove our suspicion.

Treatment.—In many instances of enteric fever in children perfect rest in bed from the earliest possible time, and assiduous feeding with milk and meat-juices, in proportions varying according to individual requirements and indications, are all that is necessary or advisable for the successful conduct of the case. Anorexia and refusal of food is exceedingly common. While adults will usually, except in the severer cases, take nourishment fairly well throughout, it is very rare, except in the mildest cases, to meet with no difficulty in this respect with children. It is often indeed necessary, even in the absence of digestive disturbance or pharyngeal difficulty, to feed the patient by the nasal tube. When there is constipation, which is often associated with tumidity and tenderness of the abdomen, the amount of milk must be reduced, and more beef-tea or other meat-juice substituted; or a raw egg beaten up with a little brandy may be given once or twice a day. With persistent diarrhœa the diet should be mostly or exclusively milk and barley-water. The amount of food must be regulated by the age and, as far as possible, by the individual digestive capacity of the patient. But we are often in the dark on this latter point, and I am sure, from some cases I have seen, that the gravity of the illness is often increased by an excess of food, especially milk, which is apparently well digested. In such cases there is usually constipation which, even when attended by abdominal discomfort, is often neglected on principle. Such a condition, however, sometimes calls for the greatest clinical acumen in deciding upon measures of relief. In my opinion constipation should always be treated by small and simple enemas when it is accompanied by distension or discomfort, by the escape of flatus, or by the presence of palpable scybala; while mere absence of bowel action without these accompaniments may be left alone even for many days. In deciding to act on the bowels, however, while the fever lasts, the greatest caution should be observed; for deep ulceration and even fatal perforation may occur with persistent constipation. Marked nervous symptoms with tremor should give us pause before interfering with constipation. After convalescence is established the constipation, which is so common, frequently causes discomfort and slight rises of temperature, and should always be relieved, though cautiously, by enemas or, occasionally, by mild aperients. In this context I cite a case of illness where constipation, caused by an excessive diet of milk and by abundance of astringent medicines given on the theory of the disease being enteric fever, seemed to induce, or at least encourage, by pressure the occurrence of very scanty urine highly charged with blood and albumen in excess, and a highly dangerous degree of the so-called "typhoid" condition. The patient whom

I first saw on the eleventh day of this illness, which ultimately was proved to be scarlatina, had been fed throughout on six pints of milk and two pints of beef-tea daily. The abdomen was enormously distended with solid matter. Very large evacuations produced by repeated enemata were followed by a complete subsidence of all symptoms and disappearance of all blood and albumen, with a copious flow of urine, within thirty-six hours; rapid recovery ensuing on a scanty diet with scarcely any milk.

Abdominal pain and tenderness, with or without constipation, and particularly when accompanied by tremors, restlessness and delirium, should always be treated by opium in sufficient doses to produce drowsiness; and, if there be more than three or, at the most, four copious motions in the twenty-four hours, I always give opium, beginning with small doses. With prolonged diarrhœa, which so often means ulceration of considerable extent, opium should be given persistently and freely for its healing action. I have frequently seen many grave symptoms quickly recede, and both appetite and vigour increase, after the bold administration of this drug. Bismuth and other astringent remedies are also useful in diarrhœa, but in severe cases time should never be wasted in the trial of these drugs alone. The contra-indications to opium are as a rule, though not without exception, extreme nerve-prostration and excessive repugnance to food, and, always, marked respiratory trouble.

Frequent, feeble and, especially, dicrotic pulse, or marked weakening and, still more, absence of the first heart-sound, necessitate alcohol. So also does tremor with abdominal discomfort. It is always well to omit the alcohol for a while after a few days, for if we persist with it after the desired effect is produced we run the risk of paralysing the nerve centres and preventing the tendency to natural recovery. I have sometimes seen, on omission of this drug, a sudden amelioration of symptoms which were doubtless due to the obscuration by alcoholic narcosis of the natural improvement that had all the time been taking place. In convalescence alcohol is generally advisable, and sometimes necessary when the heart has suffered much.

Antipyretics are in my opinion scarcely ever needed in childhood, and but rarely at any age. After many trials of both quinine and cool baths I have almost abandoned both, and from some experience of antipyrin am of opinion that it is generally useless and sometimes harmful. I have seen some cases rapidly improve on its discontinuance, although the temperature immediately rose after its artificial fall. It is, however, perhaps permissible and, possibly, useful in some degree in cases where, without any other untoward symptoms, the temperature remains persistently very high. Full doses should never be given, on account of the frequently depressing effect on the heart, and, when the pulse is dicrotic or the first cardiac sound is markedly feeble,

this drug should be entirely withheld.   Frequent tepid or even cold
sponging is, however, often very useful in treatment, soothing the patient,
inducing sleep, and lessening delirium.   In my opinion it is scarcely
ever imperative at any age to treat the temperature *per se*.   In some
severe cases, nevertheless, with persistently high temperature, or when
hyperpyrexia occurs, I give quinine in large doses in preference to cold
baths or to antipyrin.   The temperature can thus be reduced in many
cases ; but the utility of this procedure is open to much doubt.

Solid food should not, as a rule, be allowed at all until ten or twelve
days after the subsidence of the fever.   Unformed or even liquid stools
sometimes follow on the resumption of ordinary diet, owing to the unac-
customed stimulation of the bowels.   We should always satisfy ourselves
of the nature of such stools by personal inspection, before venturing to
continue or deciding to alter the diet.   For solid food I usually begin
with bread and butter or fish, and occasionally with a very little pounded
meat.   Only small quantities of solid diet should be allowed for the first
week ; subsequently, if all goes well, but little restriction need be placed
on the satisfaction of the appetite.

Patients should never be permitted to exert themselves until examination shows considerable improvement of the power of the heart.   The
longer the rest, the more rapid, as a rule, is the convalescence.   Arsenic,
iron, and the best hygienic conditions may all aid in establishing perfect
recovery.

The treatment of enteric fever can be written about in but very general
and inadequate terms.   Each individual case demands constant attention
and often varying treatment, and no bad case can be treated in the best
way by any rules whatever.   The doctor's acumen and the nurse's skill
are of primary importance and are frequently taxed to the utmost.

Concerning the theoretically rational attempt to treat enteric fever
according to its causal indications by endeavouring to produce intestinal
asepsis and thereby check the production of the pathogenic material, I
can speak but little, as yet, from my own experience, and have thus left it
to the last.   Judging, however, from the possible or even probable good
effects of the internal administration of " naphthalene " or " naphthol "
in the septic diarrhœas of infants, and from some reports of cases of
enteric fever thus treated, I am inclined to think that an extensive use
of the drug from the commencement of symptoms in this disease is well
worthy of trial.   From one to ten grains of naphthalene, in divided doses
according to age, may probably be given daily, without any drawback
other than the unpleasant taste of the drug, which, however, is not of
much moment, considering the dulled sensibilities of most enteric patients.
The very slight solubility of this medicine seems to favour antiseptic
operation at all parts of the intestinal canal.

CHAPTER VIII.

RHEUMATISM AND ARTHRITIS DEFORMANS.

By the term rheumatism, which I use synonymously with rheumatic fever and as inclusive of both acute and subacute rheumatism, ignoring whatever distinction may be implied in these latter words as practically useless, especially in the case of children, I denote the well-known general affection with more or less pyrexia, which is marked in almost all instances, at least after very early childhood, by some arthritis and sweating; by inflammation of the heart and pericardium, at least before puberty, in a considerable majority; and, very frequently, by various other affections, especially of fibrous and serous structures.   To this disease only, including certain recurrent or chronic symptoms in some of its subjects, I apply the term "rheumatism," which, however slight and obscure some expressions of the disease it denotes may be, is nevertheless to be employed with no vague meaning.

Without discussing the still open question of the ultimate **pathology** of rheumatism or reviewing the numerous hypotheses of its origin, whether exploded or extant, I must here merely state that my own experience and study of this disease at all ages inclines me to regard it as in all probability essentially due to a faulty condition of the nervous system, for the most part inherited, which expresses itself in various inflammatory and other modes in ready reaction to diverse impressions upon the nervous periphery, among which "chill" is probably to be regarded as the chief.   At present, indeed, neither chemistry, bacteriology, nor morbid anatomy has supplied us with any approach to an ultimate explanation of the source and phenomena of what most agree in recognising as rheumatism ; and it must be admitted that its clinical facts and conditions seem highly unfavourable, if not contradictory, to any hypothesis of germ-origin.   A certain small number of cases in all appearance, if not actually, rheumatic, and often marked by endo- or peri-carditis and some sweating, are well known to follow, sometimes immediately, on scarlatina, apart from those less shifting joint-swellings without cardiac involvement which are alluded to under the heading of the last-named disease.   It may scarcely be denied that scarlatina is the exciting cause of these attacks, which are probably truly rheumatic.

Symptomatology.—Although the heart-affection of rheumatism is much more common in childhood than when the disease is first developed at a later age, occurring indeed with a frequency almost, though not

O

quite, inversely proportionate to the years of its subjects, so that those who may be first attacked in their maturity most often escape heart-disease altogether, yet without doubt the younger the patient the less severe and enduring, though not, in my opinion, much the less frequent or at all the less recurrent, are the arthritis and the pyrexia. Tonsillitis again, and both dry and liquid pleurisy, which are frequent in rheumatic subjects of all ages, are perhaps seen oftener in early life; as also may be certain skin affections, such as erythema nodosum, urticaria, and other rashes, sometimes simulating scarlatina, and occupying chiefly the arms and legs. These phenomena and, still more prominently, endocar-ditis and pericarditis may, indeed, occur from time to time as the only manifestations of rheumatism afterwards evidenced by an attack of arthritis. Many writers include chorea as often or indeed almost always an indication by itself of the rheumatic tendency, but of this relationship I treat elsewhere. The fibrous nodules, of subcutaneous site, hereafter to be shortly described, are almost wholly confined to the rheumatism of childhood, occurring especially or exclusively in association with heart-disease.

Typical acute rheumatism, with profuse sour-smelling sweat and severe arthritis, with much swelling and redness, shifting from joint to joint, is undoubtedly rare in little children, and but seldom persists in childhood, although untreated by salicin or the salicylates, for that period of two or three weeks or more so common in adult cases even when masked or modified by medicines. I have, however, seen one instance of this kind, strongly characteristic in all particulars, in a child of $2\frac{1}{2}$ years with definite rheumatic heredity, as well as several other less severe but still very typical cases under five years old. Out of a series, moreover, of 70 cases in the wards at Shadwell Hospital between the ages of four and fourteen, all with more or less arthritis, I find 44 noted as having much sweating, many of them with the familiar sour smell; 18 with slight sweating; and 8 only with no sweating at all while under observation. In the parents' history of some of these latter cases there is an account of much sweating at the onset, with remarks on a peculiarity of smell. I mention these facts because there seems to be too great a tendency on the part of some writers, who rightly urge the oft-times slight and elusive characters of articular rheumatism in childhood, to underrate largely the frequency of these well-marked cases, and to multiply and emphasise unduly the distinction between the symptoms as incident on childhood and on later age. As a result, too, of these views rheu-matism in childhood becomes so wide in extent as to be almost indefinite in content; and single phenomena, of probably or certainly multiform causation, such as chorea, tonsillitis, erythema and others, are liable to be erroneously attributed to a necessarily rheumatic origin.

In commenting on some of the qualities of rheumatism in childhood I shall quote largely from the above-mentioned 70 cases, registered as "acute rheumatism," which I have taken consecutively from my ward-books; but would repeat that I am in full accord with those who urge the frequency of less well-marked instances, especially in young children. This is amply attested by many further cases of my own registered as "heart-disease," with a history of slight and evanescent joint-pains without observed redness or swelling; and by some others which, by reason of concomitant symptoms or of a rheumatic family history or both, were almost certainly instances of rheumatic heart-disease although lacking evidence of any arthritis. There are, indeed, comparatively few cases of pericarditis, and still fewer of endocarditis, which are not either certainly or with the greatest probability of rheumatic origin. Over and above the said 70 cases of rheumatism I have noted the positive or extremely probable occurrence of this disease in 77 out of 98 cases of endocarditis in my wards, as well as in 17 out of 26 registered as "pericarditis." I would add here, as bearing on the statement that rheumatic heart-affections alone are frequent in children without any articular symptoms at all, that in several cases of apparently idiopathic pericarditis and endocarditis, seen outside hospital and closely observed by the mothers, there was a clear report, on inquiry, of antecedent joint-pains which would have certainly escaped the parents' notice in the case of most hospital patients.

As regards incidence on *sex and age* the 70 cases may be briefly tabulated as follows, the ages being given in all cases at the date of the first attacks. It will be seen that the greater liability of girls does not begin until after the age of 10 years.

	Boys.	Girls.
Under 5 years old	1 (æt. 4)	...
Between 5 and 10	17	17
„ 10 and 14	13	22
	31	39

All but 9 out of these cases, ranging between 6 and 14 years of age, were the subjects, while under observation, of either old *heart-disease*, of active peri- or endo-carditis or of apical systolic murmurs. In 9 of these, including 4 with both mitral-valve murmurs (single or double) and signs of pericarditis, and 5 with single or double murmurs only, all signs and symptoms of cardiac affection disappeared before discharge.

There were signs of pericarditis in 19, all with valve murmurs as well; and the rest had murmurs mostly mitral, many having signs of heart-enlargement, but 12 being marked by no enlargement or other sign or symptom of cardiac disorder than a soft systolic murmur at the apex. It would thus seem likely that actual rheumatic heart-affection may not seldom disappear in childhood, leaving no morbid result ; and it is certain that many well-marked systolic murmurs at the apex pass away, whether due to valvulitis or to temporary regurgitation through the mitral orifice by reason of ventricular dilatation.

Such disappearance of blowing murmurs at the apex is familiar to us in the case of adults as well ; but it is especially in childhood that we must hesitate, and sometimes for long, before inferring heart-disease from the presence of an apical bruit alone. So-called hæmic murmurs at the base and over the ventricles are met with fairly often in rheumatic children, especially when the anæmia is clinically well-marked ; but I think they are more common in adult cases. Further consideration of the symptoms and treatment of heart-disease finds place under its proper heading.

The painful *onset* of rheumatic attacks is sometimes accompanied by rigors, occasionally by vomiting, and not very rarely, especially when there is cardiac involvement, by marked præcordial or epigastric pain. Tonsillitis, too, is often an early symptom. Many cases, however, begin with very slight joint-pain which afterwards increases, and in others there may be at first only slight fever and sweating.

The *arthritis*, which we have already seen to be often slight and some-times overlooked or possibly non-existent, is certainly of much shorter duration in children under 12 than later on, and the swelling and red-ness are usually much less in proportion to the pain. I have often observed acute joint-pain with tenderness, equal to that of severe adult cases, when there has been neither redness nor tangible swelling. In very many cases the arthritis is confined to the lower extremities ; in several to the knees alone.

The *temperature* is usually of an actually lower range than in adult rheumatism, and hyperpyrexia is very rare. I have met with but one probable and seemingly well-marked case, which I now cite, of the so-called cerebral rheumatism in childhood. A necropsy, however, which might have definitely excluded other disease was not obtainable.

A boy of 12 fell ill with pain in the back, severe headache, and rigors. On the fourth day of his illness I saw him lying bathed in sour-smelling sweat, profoundly apathetic, with frequent twitching of face and fingers ; the respirations were 60, the pulse 180, and temperature 106°. He was roused only by movement of his limbs, when he screamed with evident pain ; but there was little or no visible swelling, nor any physical sign of thoracic mischief. A few years previously the boy had had a severe attack of definite articular rheumatism with occasional pains, sub-sequently, in limbs and back. In spite of cold bathing sedulously carried out the temperature fell for only a very short time, and next day death followed on increasing coma with continued sweating.

The occurrence, in various parts, of the subcutaneous *fibrous nodules*, first described by Meynet of Lyons in 1875 and, though noticed by Hillier, first brought prominently before the English profession by Drs. Barlow and Warner at the International Medical Congress in 1881, is probably a definite symptom of rheumatism, is very generally associated with arthritis, and almost always connotes heart-affection, which is said to be often of a severe and progressive character. These nodules are frequently very small, can sometimes be felt when almost invisible, and may be of any intermediate size up to that of an almond. They are but rarely tender on pressure, are mostly movable under the skin, and usually occur in the neighbourhood of joints, especially about the elbow, the knuckles, the malleoli and the edge of the patella, and, less frequently, over the vertebral spines, the iliac crest, the occiput, and various other parts. They may be few or many, and often make their appearance in successive crops. Apparently they develop very rapidly and often increase in size after their discovery, but subside much more slowly. I have, however, seen a few cases of the complete disappearance of many in about a week. Their limit of duration is usually stated to be from three days to five months or more. Pyrexia is certainly no regular accompaniment of their evolution, and, according to Dr. Barlow, its presence is probably due to some concomitant inflammation.

The difference between observers as to the frequency of these nodules is probably the outcome of varying degrees of care in the quest of them, as I have learnt from several demonstrations, by my colleague Dr. Coutts, of minute nodules which I had quite overlooked. Probably they occur to some extent in about 20 per cent. of all cases of acute rheumatism with heart-disease, but large nodules are much rarer. Besides their possibly prognostic bearing on the progress of heart-disease, these nodules have a diagnostic value in sometimes establishing the true origin of otherwise doubtful rheumatic affection of the heart and other symptoms, or the rheumatic association of a given case of chorea. But, as I have already said, it is known that these fibrous nodules are much more often seen in cases which have suffered or are suffering from arthritis.

In two out of the only four fatal cases, from my list of 70, where death seemed due either to severe pleurisy or heart-disease during an acute attack of rheumatism, there were numerous and well-marked nodules.

It must, however, be noted that nodules have, though rarely, been seen where other symptoms of rheumatism were absent. Dr. Hadden has reported [1] a case in point where there had been neither arthritis, chorea, nor heart-disease.

Pleural effusion, single or double, or *pleuro-pneumonia* are noted in at

[1] Clin. Soc. Transactions, vol. xxiii.

least eleven of these cases, and in others there were signs of old pleurisy. Extensive pleurisy is a bad prognostic symptom in whatever form it occurs. Localised dry pleurisy is very common; but, apart from its frequent tendency to recur, and thus to indicate a deep rheumatic taint, it has no grave significance.

Acute *tonsillitis* occurred in 7 of the 70 cases while in hospital, and there was a ·history of previous and severe sore-throat in many more.

Erythema was observed in only two of the cases, but no inquiry was made as to this point while taking the histories; and *chorea*, either previous or concurrent, was noted in three. The well-known relationship, however, between rheumatism and chorea is treated of under the heading of the latter disease. It must be remembered that pleurisy, pericarditis, endocarditis, chorea and other probably rheumatic symptoms may be seen, from time to time, in patients who sooner or later afterwards have their first attacks of articular rheumatism.

With respect to the *family history* of acute rheumatism, although careful inquiry was made in each instance, no facts, either negative or positive, could be gained in very many—a common failure in hospital cases; but there was a perfectly definite account of one or more subjects of this disease in the immediate families of twenty of the cases, a probable history in many others, and a satisfactorily negative one in four. Considering these statistics, however, and, further, the great frequency of a history of rheumatism in the families of rheumatic patients in private practice, the prevalent hereditary character of the rheumatism of childhood is well attested.

In the **prognosis** of rheumatism we must consider chiefly the state of the heart and pericardium and other so-called complications, such as pneumonia and pleurisy. When there are marked signs of failing heart, with or without definite evidence of valve mischief, extensive pericarditis or pleurisy, the case is always grave, although, even then, but rarely fatal in first attacks. In the very rare instances of purulent pericarditis, or the somewhat less rare ones of empyema, the immediate prognosis is bad ; and an acute pleural effusion on both sides may be rapidly fatal. For the rest, there is nothing special to children, in whom acute rheumatism by itself is of even better immediate prognosis than in adults. But, seeing that the heart suffers so often and so severely in the rheumatism of childhood, the ultimate forecast of most early attacks is at the best but very doubtful. A very small proportion of children become the subjects of chronic rheumatism proper, by which I mean more or less enduring pains, stiffness, swelling, or deformity of joints ; but many suffer from marked and frequent recurrence of articular pain and swelling, and more from repeated flying pains, either neural or "muscular," in

various regions. Torticollis, too, must be classed as an occasional sequela
of true rheumatism. In the few definite instances which I have seen
in children of advanced and permanent deformity of joints, such as
is usually described as "arthritis deformans" or "rheumatic arthritis,"
there has been no history of primary acute articular rheumatism or any
of its usual accompaniments.

Emotional and other *nerve disturbances*, such as night terrors and,
according to Dr. Goodhart, enuresis, are very common in children subject to rheumatism; and frequent headache also illustrates the neurotic
relationships of the disease, which are often so marked in adults as
well.

The **treatment** of rheumatic fever in childhood has no claim to detailed
special consideration. For the acute manifestations, as long as the
heart is working well, even though there may be abnormal auscultatory
signs, salicin or the salicylate of soda should be given. These drugs,
and especially the salicylate, are successful, as in adult cases, in proportion to the severity of the pain and the height of the fever; and, according to the best clinical experience, seem, valuable though they are, to be
merely symptomatic remedies. But, considering that in children both
these symptoms are often slight and evanescent, they are by no means so
often required as in the case of adults; and further, in view of the more
common cardiac troubles of childhood, they are often, in my opinion as
in that of some others, inadmissible. I am sure that in several cases in
adults I have known the heart functions impaired and improved by the
alternate giving and withholding of salicylate of soda, or, in some
instances, of salicin, which was less often tried. In such a difficulty
pain and restlessness, and other cardiac symptoms as well, may be much
relieved by small and repeated doses of opium; or, when the joint-affection
is at all severe, small blisters may be used, frequently with good effect.
In these cases, as well as in others which are slight and relapsing, I still
think that there may be at least symptomatic relief from giving, as I
often do, bicarbonate of potash or other alkaline medicine, according
to older custom; and the more chronic or recurrent any rheumatic symptoms without fever are, the more should we insist on tonic treatment,
with passive movement of the joints and systematic rubbing. The diet
during the acute attack should be milk with some farina. During convalescence it should be light, but, especially when there is much anæmia,
not necessarily without meat. If lithates appear to any extent in the
urine, fever diet should be returned to for a while. The child should
of course be kept in bed until the attack is over, and carefully preserved
from chill at all times, whether well or ill. It is, in my opinion, neither
necessary nor advisable to force either a child or an adult to lie between
blankets; for I have often seen much discomfort, increased sweating and

cutaneous irritation produced thereby. Arsenic, iron and cod-liver oil are all very useful medicines in convalescence.

The treatment of the cardiac troubles of Rheumatism is considered under the heading of Heart-disease.

Arthritis Deformans.

It is necessary to treat shortly of this disease, for, although not frequent in childhood, it is scarcely as rare as might be believed from its usual omission from the text-books. Some cases under two years old have been reported. Owing to its clinical likenesses and still greater contrasts to rheumatism it seems practically better to deal with it here than under the possibly more appropriate heading of diseases of the nervous system.

In all essential characteristics "arthritis deformans" or, as many style it, "rheumatoid arthritis" is the same in children as in adults, and appears to arise in similar conditions. Its origin is as a rule insidious, without fever; and pain with stiffness in the joints is at first a much more prominent symptom than swelling, which indeed for a while may be unobserved. Some cases, however, begin more or less acutely with an attack of joint-pains of definite date. The hands are usually first affected in the small joints, and much time may pass without involvement of the larger joints of the extremities. The larger joints however, judging from the cases I have seen and from those reported by others, are liable to be affected sooner or later; and not only the wrists and ankles but also the knees, elbows, shoulders and hips may be quite disabled. The swelling of the joints is followed by displacement of the bones; and, after a period when much grating and creaking on movement may be felt or heard, permanent ankylosis may at last take place. I had a case of a girl, now under observation at Westminster Hospital in the ward for Incurables, in whom the symptoms began at the age of 11 years with pain, swelling, and redness of the right knee and right ankle soon after she had been wet through two or three times. The knee soon became semiflexed, and the swelling and pain increased. Gradually most of her joints were affected and distorted, the knees becoming flexed and the feet permanently everted almost at right angles to the leg. The patient's paternal aunt and grandmother were similarly affected, and her father had had rheumatic fever when he was nineteen. There was also a probable history of acute rheumatism in other relatives not of the immediate family.

A marked characteristic of this disease, well shown in the case above referred to, is the atrophy of the muscles in connexion with the affected joints: a condition quite inexplicable by the theory of disuse, and unparalleled in other joint-affections, always excepting those which occur in association with other nervous phenomena such as the symptoms of tabes dorsalis.

The anatomical condition of the joints, as far as clinical examination shows, is apparently the same as in the adult disease. This is marked by chronic inflammation affecting all the joint structures, leading to erosion of the cartilages, bony outgrowths from their edges, and at last to considerable absorption of the ends of the bones and fibrous ankylosis.

In the four or five cases I have seen of the disease before puberty there was mostly a history of acute rheumatism in the family, no febrile movement with the attack, and no evidence of any cardiac affection. Some cases, however, as recorded by Sir A. Garrod and others, and as illustrated by the instance above quoted, appear to begin acutely, and sometimes with fever. Mr. Hutchinson has shown that the joint-disease may anticipate all other symptoms for some time. That "arthritis deformans" may have some more or less remote alliance with true rheumatism cannot be denied. Some few cases are preceded by definite attacks of the latter disease; but in so common an affection as acute rheumatism this sequel is extremely rare.

The favourite subjects of the adult disease are seemingly the ill-nourished and anæmic, the physically exhausted, and the mentally depressed. The most common excitants are, with equal or greater probability, damp and cold. As far as my small experience of the disease in children has taught me anything, I would say that a neurotic constitution and exposure to cold and damp are prominent antecedents. In early cases the good effects of generous diet, tonic treatment, warmth, and dryness of climate are very marked at all ages.

The important observations of many, tending to show that in the adult disease the joints most used are the most likely to suffer early and severely, cannot, as far as I know, be regarded as markedly applicable to the affection as seen in childhood. The joints of the toes however, with the exception of the ball of the great toe, seem as a rule to escape, at least for a very long time, in all forms of the affection.

The disease in adults is liable, as is well known, to periods of symptomatic latency, and to exacerbations of pain and swelling in varying extent and degree. Sometimes a few joints only are affected, and, after more or less damage done, no recurrence takes place. I have not been able to find any published data on which to ground any practical rules of prognosis as regards the disease in childhood.

The earlier the case is recognised, and **treatment** undertaken, the better the chance of arrest or approximate cure. Arguing from adult cases, we are justified in looking for improvement from good nutrition, warmth and dry air. In severe cases I have seen marked benefit result from the simple addition of arsenic to the other therapeutic means used.

I am of opinion that this drug should be persistently given from the outset. Cod-liver oil is also valuable, and probably iron as well. If there be no great subsequent pain, passive movement of the joint and daily prolonged rubbing of the limbs should be part of the routine. The patient should always be encouraged to use the joints as much as possible. When pain is considerable guaiacum is sometimes useful; and, although I would only advise its temporary use, a combination of colchicum, potassium iodide and bicarbonate of potash has not seldom, in my experience, seemed to give much relief to the patient's feelings. Counter-irritation over the joints, by means of the liniment of iodine, may be of some use in early cases. If possible, the sufferer should be sent in the winter to such a climate as Egypt. In England the treatment at Bath at any time of the year seems worthy of trial; considerable benefit having appeared to me to result therefrom in two cases below the age of puberty, as well as in several adults.

CHAPTER IX.

WHOOPING-COUGH.

By this name is denoted a peculiar spasmodic cough, occurring in paroxysms of sometimes excessive frequency, often associated with catarrh both bronchial and nasal, and largely tending to spread among children.

The whooping inspiratory sound so often heard is a result of laryngeal spasm coincident with the inrush of air to the lungs, which have been emptied by the violence of the expiratory paroxysm. This is, as a rule, absent in the earliest period, may never appear at all, continues sometimes long after the affection has ceased to be communicable, and recurs frequently for months, or occasionally for years, with every fresh attack of catarrh. The essential clinical mark of the affection is the sudden onset of severe paroxysmal coughing, with varying degrees of suffocative symptoms and frequent vomiting, rather than the laryngeal noise which so often follows of necessity on expiratory violence. During the recent epidemics of influenza it was not uncommon, especially in the earliest stage of the disease, to note a paroxysmal and suffocative cough accompanied by well-marked whooping in no way distinguishable from this familiar symptom of whooping-cough in childhood; and it is certain that such a sound is not confined to these diseases, but not infrequently occurs with violent coughing otherwise and suddenly excited.

Objections to the generally accepted view of the germ-origin and

specificity of all cases of cough with whooping have occurred to many thoughtful observers, owing to the differences presented, in many instances, from the usual characters of most affections of the infectious class. It is not to be doubted that clinically typical and uncomplicated whooping-cough is often quite free from fever from first to last, and is accompanied by no other sign whatever of illness. At the outset indeed, before any characteristic cough is heard and when the existence of catarrhal symptoms only makes an initial diagnosis impossible, thermometric observations of temperature are rarely made ; and there can be no justification for the positive statement, re-iterated by most writers, that an early febrile stage is the rule. At the height of the disease, moreover, there is no fever unless there be pulmonary or other complications ; and I have amply satisfied myself, by searching inquiry into the history of very many cases, that there is probably no initial fever or even malaise in a large number. Those who are feverish and ill at the beginning and, often, for some weeks afterwards are mostly infants or quite young children ; and these are nearly always the subjects of more or less extensive bronchial catarrh or broncho-pneumonia. The purest examples of whooping-cough are non-febrile cases in children beyond infancy ; and it is in these that the disease is most justly studied. In such cases the interparoxysmal state is usually unaccompanied by any abnormal physical signs in the chest.

The **course** of whooping-cough is of very variable duration, as any one may learn from experience without referring to the widely discrepant statements made on this head by writers. The usual division of the disease into three stages is artificial, and useless for practical purposes ; for the so-called first stage can be but rarely studied and is very variously described, the second, with the characteristic cough, may set in at very different periods and last indefinitely, and the third is nothing other than gradual convalescence. Again, it is not by any means rare to find *single* instances of what is clinically whooping-cough, and quite indistinguishable from contagious cases, occurring in families where the other children have not previously suffered. I have frequently established this fact by observation and inquiry extending over some months, so as to more than include the longest alleged period of incubation ; and at Shadwell Hospital, where whooping-cough for many years was formerly admitted into the general wards, spreading of the disease but very rarely occurred. The striking difference in this respect between the **spread** of whooping-cough and measles, which has been noted by Henoch and others and observed though not recorded by many more, is not to be accounted for by more children being protected by previous whooping-cough. Our extensive records at Shadwell Hospital show that less than **23** per cent. of children admitted under six years old—the favourite period for whooping-cough—had previously suffered from that disease, while 34 per cent. had had measles. Some have endeavoured to explain the infrequent spread of whooping-cough in wards by the contention that actual contact or very close proximity is necessary for the contagium to work—a condition which is more likely to obtain when the children are up and about than when separated by inter-cubicular spaces; but it must be remembered that many children in a ward are not kept in bed; that when whooping-cough does spread among children in bed it spreads sporadically, not from neighbour to neighbour; and, further, that a very large amount of the evidence of the contagiousness of the disease falls to the ground without the assumption of the greatest diffusibility and subtlest activity of the poisonous principle. But none of these facts seem to me to weigh materially against the view of the germ-origin of the cases of whooping-cough which are confessedly contagious, especially in the light of our extended conception of micro-organic pathology; for neither febrility, nor definite course, nor striking contagiousness apart from inoculation are necessary elements in microphytic disease. I am strongly inclined to believe that, in what we must clinically and for practical purposes call whooping-cough, we have really to do with many cases which are neither specific nor catching, albeit at first virtually indistinguishable from those which are; and I further think that, although true contagion by means of the sputa or other emanations is scarcely to be questioned in many instances, the spread of whooping-cough as a whole from the sick to the unprotected is far less ready and much more sporadic or fitful than that of measles and other exanthems. In other words, I believe that a paroxysmal cough with whooping may be engrafted as a purely nervous incident on ordinary catarrh, apart from the influence of any specific organism; and, further, that in the clearly contagious and epidemic catarrh, which we recognise as "typical" whooping-cough, there may be very varying degrees of infective power. The great difficulty of regarding "whooping-cough" as always one and the same disease is also somewhat enhanced by its frequent and immediate sequence on, or development out of, the catarrh of measles, even when a fresh source of infection can be with all probability excluded. Among many other striking instances of this concurrence with catarrh I have seen whooping-cough follow on broncho-pneumonia in two children suffering from scarlatina, who were strictly isolated, and for several weeks inhabited alternately two rooms daily ventilated by widely opened windows and frequently sprayed with carbolised vapour.

As to the contagious form of whooping-cough, there is no direct evidence as yet generally recognised of a specific germ, however probable its existence may be, although certain bacilli have been found and believed by some to be pathogenic; nor is there any agreement among

authorities as to the period of incubation. About fourteen days has
been established with much probability in some cases; but, if we were
to exclude instances where the incubative period must have been much
shorter or much longer, the evidence of any contagion at all would be
considerably weakened. We must believe, however, at present, after due
consideration of hard facts and conflicting commentaries, that there are
contagious catarrhs, with perhaps diverse microphytic sources, which
specially affect the upper air-passages and cause an exceptionally
paroxysmal and violent cough usually entailing the characteristic whoop;
and that the contagium is probably contained in the sputa and breath
of the sufferers. For practical purposes we should therefore isolate
patients with "whooping-cough" as long as there are distinct signs of
nasal or bronchial catarrh; and, in all cases, empirically, (for we have
absolutely no knowledge of the limits of the infective period) for six
weeks. If there be, as I believe, very numerous cases of non-contagious
whooping-cough, we have certainly no means of diagnosing them at the
beginning; we must, therefore, practically regard all cases at first as
coming under the category of infectious disease. But in instances, such
as I have frequently met with, where a child whoops with every fresh
catarrh for months or even sometimes for years, we may usually be
certain of their harmlessness.

The paroxysmal and suddenly explosive nature of the cough will, in
many cases, tell the experienced observer of the nature of the affection to
come, even before any suffocative or severe symptoms set in or any
whoop be heard; and the diagnosis is much strengthened, in spite of the
absence of evidence of laryngeal or bronchial catarrh, when the cough is
very frequent, especially at night. The prevalence of an epidemic, as in
other cases, may give valuable aid; but this occurrence is far less frequent
and definite than in most other contagious diseases. In the established
disease, a full symptomatic description of which I purposely omit as
familiar to my readers, great diagnostic stress must be laid on the
redness of the face, which is marked even at the beginning of the
paroxysms of repeated coughing; and on the puffy and venously congested face of many of the sufferers even during their inter-paroxysmal
states.

Whooping-cough, as a whole, is most frequent in children *over six
months and under six or seven years old*. The second and third years
are perhaps its favourites. From the exhaustion due to the frequent
coughing and vomiting, and the frequent co-existence of extensive bronchitis, broncho-pneumonia and pulmonary collapse, the *mortality* of the
disease in children under two years old is enormous. Hæmorrhages
into the conjunctivæ, epistaxis, hæmoptysis from faucial or laryngeal
sources, rupture of the ear-drums, ulceration near the frænum linguæ

from friction with the teeth, and convulsive seizures may be more or less frequent incidents of the disease; severe headache is very often complained of; and hoarseness is not uncommon. The cough is apt to be especially severe and frequent at night. In some few cases meningeal hæmorrhage has been observed as the probable result of a violent paroxysm; and Henoch reports a case of hemiplegia referable to a similarly caused cerebral hæmorrhage.

Diarrhœa with much mucus in the stools, indicative of intestinal catarrh, is doubtless frequently seen with whooping-cough, and is probably a part of general catarrh; but, having for many years made special inquiries and observations on this point, I am convinced that in the majority of cases intestinal disorder plays no part.

Convulsions, if frequently repeated, are usually of serious import. A previously convulsive child, however, will have its fits more frequently in whooping-cough; and the paroxysms of cough are, in their turn, most readily excited by the slightest disturbance in convulsive or very nervous children. The nervous element in whooping-cough, indeed, is not only always prominent but also may be considered, apart from the question of the nature and mode of contagion, as the very essence of the affection. Space forbids me to say more on this subject, which has been so thoroughly treated by Sturges. Some convulsions seem directly induced by the suffocative paroxysm, and others by the subsequent exhaustion. Those which may be due to embolism or thrombosis of the cerebral veins or sinuses will be of considerable prognostic gravity, and their nature may be suspected from concomitant paralysis or a strictly unilateral distribution.

The co-existence of other diseases, especially of rickets with its yielding thorax, or of tuberculosis, largely aggravates the dangers of whooping-cough; and it must never be forgotten that a vast number of cases of chronic pulmonary disease, tubercular and otherwise, with enlarged and cheesy bronchial glands, seem to take their origin from this affection. Instances of chronic bronchitis, broncho-pneumonia, emphysema, asthma, phthisis and other disorders, where the history is given of continuously bad health after whooping-cough, are simply legion.

We have seen that the duration of the disease is very indefinite because of its probably varying origin. Perhaps in the truly contagious cases an average may be struck of about three months, from the onset to the disappearance of all incidental symptoms; but the contagious period is probably much shorter. It has already been remarked that whooping-cough and measles frequently occur in very close connexion; and I would further state, after extensive inquiry on this point, that in an overwhelming majority of cases of such connexion the whooping-cough is sequent upon measles. In many instances of this kind the spasmodic cough is doubtless a nervous epi-phenomenon in no way connected with specific infection; and this consideration, together with the fact of whooping-cough being far more common in girls than boys after the first few years of life, points strongly to the conclusion that in so-called whooping-cough we have not always to do with one and the same disease.

The **prognosis** in whooping-cough depends on the facts of the individual case, and little can be gained by generalisation. Uncomplicated cases are seldom fatal except in children under two years old, many of whom die in or from convulsions or from exhaustion. For the rest, the main dangers lie in pulmonary complications, coincident rickets, and the untoward sequelæ above mentioned.

Treatment for this disease is as multiform as it is important. Although we have certainly no means at present of cutting short the attack or of antagonizing the action of the germs which we believe to be the source of many cases, we can often successfully lessen its most troublesome symptoms and palliate some of its complications.

Immediate isolation of children suspected of whooping-cough should be ordered; and especial care should be taken to remove infants and younger children from possible sources of infection, for in these the most untoward results are to be dreaded. The patient should be kept indoors in an equable temperature of about 63° F., for chill is especially liable to induce a paroxysm of coughing; but it is equally important that the room should not be too hot and that the air should be frequently renewed. For this purpose a frequent change between two rooms is to be advised. In uncomplicated cases, indeed, our chief effort should be to prevent all known excitants of cough, among which vitiated air is prominent. The patient should also be kept quiet, and free from excitement which, unquestionably, often determines a paroxysm. When there is fever, and especially when there is marked bronchial or pulmonary complication, the patient should be confined to bed. For the purpose of antagonizing the cause of the disease or checking cough many drugs have been used both theoretically and empirically. My experience is that we can often control the cough with marked success by many different medicines, but that there is no one drug of which we can predict any considerable degree of efficacy in any given case. Inhalation of diluted carbolic acid spray or of other tar-products, and of antiseptic remedies generally, widely practised and much lauded by many, I have tried in many cases; sometimes, as in other modes of treatment, with apparent success, and sometimes with signal failure. In several early cases, recalcitrant to other remedies, I have ordered constant impregnation of the air of the room with carbolic acid or creasote by means of the steam-draft apparatus, on the plan of Dr. Robert Lee; but have rarely seen any notable result.

In the later stages equal improvement may apparently follow on very diverse remedies. Among the innumerable drugs recommended, either as specifics or palliative remedies, I think I have seen benefit arise in different cases from the bromides, chloral, cannabis indica, and possibly sometimes from the extract or tincture of belladonna; but as regards the latter drug, as well as the solution of atropia, I am convinced, by very numerous trials to which out of deference to authority I have again and again recurred after repeated disappointments, that it ranks very low even as a palliative; and I have frequently pushed it to the production of its physiological effects. For many years I have found no remedy in uncomplicated whooping-cough so often useful as opium, in the form either of the compound tincture of camphor or of laudanum; and I now scarcely ever treat a case of ordinary whooping-cough without it. This drug can be given with safety, in very small doses, even to infants; and the only contra-indications are dyspnœa from pulmonary obstruction and evidence of renal failure. By opium, far more than any other remedy, both the force and frequency of the paroxysms of whooping-cough are often markedly lessened. I do not think that quinine in any dose has any effect on the course or symptoms of the disease. For the frequently accompanying bronchitis and broncho-pneumonia the ordinary treatment proper to these affections is applicable. The bromides are often useful in neurotic, and especially in convulsive, children.

After improvement has set in, as evidenced by a notable reduction of the number of paroxysms, and when there is but little remaining bronchial catarrh, as shown by physical examination, the patient may be allowed to go out, being kept away from other children; and a change of scene or, it may be, a change of air will often be followed by an almost sudden cessation of all symptoms, as I have personally observed in several instances. I have, further, sometimes seen apparently ingravescent and early cases of whooping-cough almost suddenly recover, not only with change of place, but also even while remaining at home or in the hospital ward. Many instances of this have been characterized by some pulmonary trouble for an indefinite and sometimes long time before the paroxysmal cough, with reddening of the face and whoop, has been observed; and are perhaps to be regarded as belonging to the non-specific group alluded to above. The nervous element in whooping-cough, evidenced by the paroxysmal cough and its frequent excitement and stoppage by psychical means, is, as we have seen, often very prominent; and in protracted and recurrent cases is probably the only fact to be dealt with. There is abundant evidence that a child may have occasionally typical paroxysms of cough, and be perfectly well and free from contagious influence on others; but our ignorance of the time-limits of contagion gives rise to a vast practical difficulty in deciding on the necessary period of isolation for the protection of others, and we are thus driven back on the barest empiricism. My practice, believing though I do that the contagious period is probably not more than two or three weeks, is to isolate all patients for six weeks ; and, after that term, to let them go free whether they whoop or not, provided always that they feel well and have neither symptoms of nasal catarrh nor any expectoration. We may hope, if not trust, that the time may come when micro-biological investigation will supply us with means towards accurate diagnosis, scientific prophylaxis, and possibly efficacious treatment of this familiar though still indefinite disease.

APPENDIX TO SECTION III.

ABSTRACT OF THE CONCLUSIONS GIVEN IN THE "REPORT OF A COMMITTEE APPOINTED BY THE CLINICAL SOCIETY OF LONDON TO INVESTIGATE THE PERIODS OF INCUBATION AND CONTAGIOUSNESS OF CERTAIN INFECTIOUS DISEASES." Supplement to Vol. XXV. of the Clinical Society's *Transactions*. London: Longmans, Green & Co.

Small-pox.

Incubation Period.—The interval from exposure to the appearance of the initial symptoms is commonly 12 days; but not infrequently it is a day more or a day less. Occasionally it is as short as 9 or 10 days, and sometimes as long as 14 or 15 days.

Infectious Period.—The patient remains infectious from the onset of the initial symptoms until all scabs have cleared off. The infection is, however, much more intense during the height of the active stage than during the initial illness.

Isolation may, for the reason just given, be practised as late as the time of the appearance of the rash with some expectation that the spread of the disease may thus be checked. The infection is easily carried in clothes, and in the hair of a person in attendance on a small-pox patient.

Chicken-pox.

Incubation Period.—The incubation period is usually 14 days, but may be a day less or 4 or 5 days more.

Infectious Period.—A patient is infectious at least as soon as the rash appears, and remains so during convalescence. The infection can be conveyed by fomites.

Measles.

Incubation Period.—The most usual period is 9 or 10 days. Occasionally the period is as short as 5 or even 4 days, and sometimes as long as 14. A susceptible person who has been exposed to infection must be found free from fever and catarrh at the end of a fortnight before it can be said that the disease has not been contracted.

Infectious Period.—A patient is very infectious during the prodromal stage, and, probably, not less so during the acute attack; thereafter infectivity declines rapidly and has ceased altogether three weeks after the rash. As it is probable that the infection can be retained, for a short time at least, in fomites, it is necessary to practise disinfection before terminating the period of *isolation*, which should be for three weeks after the appearance of the rash.

Rubeola (*Rubella or "German Measles"*).

Incubation Period.—The most frequent period is, probably 16 to 18 days; but it may be as long as 21 days or as short as 8, possibly even less. In 62 out of 69 cases it was some period between 12 and 18 days.

Infectious Period.—A patient begins to be infectious 2 or 3 days before the rash appears, and continues so during the height of the disorder. Infection rapidly declines thereafter, and ceases in a week in mild cases ; but it probably persists in cases where desquamation occurs until that process is over.

Quarantine.—A person who has been exposed to the infection must be kept under observation for 23 days ; and, as the disease in very infectious in its earliest stage, it is desirable to isolate on the least suspicion of catarrh or malaise.

Scarlet Fever.

Incubation Period.—The period is, as a rule, more than 24 hours and less than 72 hours. It has not been shown ever to exceed 7 days.

Infectious Period.—A patient is infectious from the onset of the earliest symptoms, and remains so until long after convalescence is established. A second desquamation may be infectious.

Isolation.—Isolation, to be effectual, must be commenced at the onset of the disease, and continued for seven or eight weeks, or until all desquamation has ceased. As the infection is easily preserved in fomites, disinfection should be practised with great care, and by the most efficient methods.

Quarantine.—A person who has been exposed to a source of infection should be kept in quarantine for 7 clear days. If at the end of that time there is no elevation of temperature and no indication of sore-throat he may be pronounced to have escaped infection.

Unrecognised Cases.—The symptoms of scarlet fever may be very anomalous, or very little marked, and, especially in the adult, may consist only of sore-throat. The infection is often spread by such cases.

Surgical Scarlet Fever.—The production of a traumatism, surgical or other, may determine the onset of scarlet fever in a person who has been exposed to infection, but who, previous to the traumatism, appears to have resisted the infection. It is, therefore, unadvisable to perform any operation which can be deferred upon a patient who has recently been in an infected house or ward.

Influenza.

Incubation Period.—The usual period is 3 or 4 days ; but it may be as long as 5 days, or as short as 1 day or a few hours less.

Infectious Period.—A patient is infectious from the onset of symptoms until convalescence has been sufficiently established to enable him to return to his ordinary avocations.

Diphtheria.

Incubation Period.—The period is usually 2 days, and seldom exceeds 4 days. Seven days is the longest period for which there is trustworthy evidence.

Infectious Period.—A patient begins to be infectious in the incubative stage, and no term can be certainly assigned to the subsequent duration of infection as long as any unhealthy condition of throat endures. The danger certainly persists for one or two months. The infection can be retained and conveyed by fomites.

Quarantine.—Seven clear days is probably sufficient, if at the end of that time the person who has been exposed is subjected to careful medical examination.

Unrecognised Cases.—The infection of diphtheria may be conveyed by cases so mild that they never come under medical treatment, or so anomalous that their true nature is not recognised. In schools, and other places where large numbers of susceptible persons are gathered together, all cases of sore-throat ought to be isolated as though they were diphtheria, at least during periods of epidemic prevalence.

Enteric Fever.

Incubation Period.—The period varies very much, and is probably in large measure determined by the " dose " of the virus received. In children who have drunk freely of infected milk it may be as short as 7 or 8 days or even less. The usual period is 12 or 14 days. In rare cases it may be as long as 23 days.

Infectious Period.—Infection lasts from the onset of the illness until convalescence has been established for at least a fortnight. Infection can be retained in fomites for two months at least.

Mumps.

Incubation Period.—The prodromal stage of mumps is of very uncertain duration, and frequently passes unperceived. The interval between exposure and the onset of parotitis is most often three weeks, a day more or a day or two less. It may be as long as 25 days, or as short as 14.

Infectious Period.—The prodromal period of mumps may certainly last as long as 4 days ; and during the whole of this period the patient is infectious, though he may make no complaint of illness. Infection diminishes progressively from the onset of the parotitis, and ceases in a fortnight, or at most three weeks.

Quarantine.—It appears that the infection of mumps is not easily conveyed for even a short distance. Separation of a patient in a single room in a house containing many susceptible persons may be effectual. Quarantine should last 25 days. A susceptible person first seen 10 days after exposure to infection may be placed in quarantine with every prospect of preventing infection ; and it is worth while to resort to quarantine if exposure has taken place even three weeks earlier.

[*P.S.*—Regarding whooping-cough, the facts submitted to the Committee were not sufficiently definite or numerous to afford material for any useful conclusions. —H. D. D.]

SECTION IV.

DISORDERS OF THE NERVOUS SYSTEM.

SECTION IV.—DISORDERS OF THE NERVOUS SYSTEM.

DISORDERED nerve-function forms a highly important part of the maladies of childhood, and most of the affections of this kind which may be regarded as special to our subject are largely referable to the double fact of the higher cerebral centres in early life being imperfectly developed and at the same time rapidly developing.

This is evidenced not only by inference from functions but also by examination of the new-born child's cerebral cortex which, both macroscopically and microscopically, is of far less structural complexity than in the adult. Co-ordination exists but in its lower grades, instability of the nervous mechanism is conspicuous, and it is verily out of disorder that nervous order gradually comes into being. Hence the pre-eminent helplessness of the infant and the abounding tyranny of its environment. The distinctively human qualities of the individual are thus late in development, even as was man himself in the biological series; and the inchoate organism of the child must run the gauntlet of innumerable untoward surroundings, dependent upon others for the higher nerve-control in all its aspects. We may thus expect to find, among the nervous disorders of infancy and childhood, instances of arrest and of vices of development; multiform aberrations from the normal, both temporary and permanent; and many evidences of falls by the way during the progress from instability to co-ordinated perfection of nervous function. I cannot here enlarge as I would on this point; but content myself with remarking that, when we consider the double function of the nervous system in regulating the whole individual organism and in bringing it into relation with the external world, we cannot but recognise that this formula of imperfect nerve-control may be applicable to many other disorders of childhood than those which are technically styled "nervous." It may at least be said that, wherever the hypothesis of a nervous origin for a complex of morbid symptoms may be rationally applied, it has mostly a double force when the matter in question is disease in early life.

The field, however, of what for practical purposes are known as diseases of the nervous system, is less wide in infancy and childhood than in later life ; for with many affections whose predominant symptoms give them a place in this category we have little or nothing to do. Such for instance are numerous diseases of the adult, due to degeneration and marked changes in tissue either within or without the nervous structures, and causing symptoms of functional failure of the brain or spinal cord.

It is especially, as we have seen, in connection with the brain in childhood that nerve-affection manifests itself, and under this heading both mental and physical disorders should be considered. The pathological part played by the spinal cord and the peripheral nerves is much less in childhood than in later life. I shall, however, treat of the several nervous disorders of childhood neither in anatomical nor ætiological order, for no classification of this kind can be practically useful, involving, as it does, numerous cross divisions ; but shall rather use the method of simple enumeration under the best known clinical titles, whether such titles be taken from structural change or from predominant symptoms.

I have been forced, by reason of want of space, to forego the discussion of mental disorders proper.

CHAPTER I.

SPASMODIC DISORDERS.

THE affections of children which are especially marked by spasm, with the exception of those occurring in association with prominent paralysis and hereafter to be noticed, may be clinically grouped under the following headings of Infantile Convulsions and Tetany, Epilepsy, and Localised Spasms.

Infantile Convulsions.

Convulsions may be described as paroxysmal attacks of involuntary muscular contractions, generally accompanied by insensibility. In a slight degree the convulsive tendency is inherent in infancy, by reason of the cortical incompleteness already mentioned ; and is evidenced by the spasmodic twitching of limbs and sudden breathlessness often observed in cases where neither typical convulsions occur nor further neuroses follow. This condition I have seen arise in healthy young babies who from some cause or other, as, for instance, artificial feeding, have temporarily failed in nutrition, but completely disappear when flesh is regained by appropriate treatment. Many intermediate grades may exist between this quasi-physiological spasm and the ordinary convulsion which must be recognised as pathological. There is, moreover, no purely symptomatic distinction between infantile convulsions and epilepsy. The typical and fully-developed attacks of both are identical in appearance, and epilepsy appears in various guises. The differentiation of the epileptic convulsion in all its aspects depends, as we shall presently see, on other considerations than the character of the actual fit.

Infantile convulsions may be general or partial or strictly unilateral in distribution. The trunk is usually involved in the spasm as well as the limbs. It may be said that severe general convulsions in early infancy differ from those in later life in that they most often lack the orderly march of the spasm from the smaller and more specialised muscles downwards to those of the trunk. Yet we may not seldom observe the fingers and thumb of one hand twitching almost simultaneously with the twisting of the mouth and turning of the eyes, which are so frequently the first observed indications of the coming fit. In the height of the attack the breathing is shallow and often arrested, the face blue, the saliva may be frothed by the maxillary movements, and there is frequently passage of urine and fæces. The separate attacks last but a few minutes ; but they may be repeated in extremely rapid succession for even days at a time, the intervals being often occupied by varying grades of apathy or by deep coma. It is in cases of this kind that there is imminent risk of death through engorgement of the brain and lungs. In but few instances does a single marked convulsion occur; but in many there are long intervals between the fits. General convulsions, whether slight or severe, are the rule in infancy. Unilateral spasm, however, is often noticed, but certainly has not the same significance of organic mischief in the brain as in adult life, unless it be frequently repeated or accompanied by lasting hemiplegia. Even recurrent unilateral convulsions I have seen, both in cases which recovered without any drawback and in others which died and showed no cerebral lesion ; but here, when making our diagnosis and forecast, we must always think of the probability of a lesion such as tubercle or other growth, or of vascular mischief on the side of the brain opposite to the spasms.

Laryngismus is a very frequent phenomenon in connexion with general convulsions, especially in rickety children; it may indeed, when marked, be regarded as almost exclusively an indication of rickets. Of this, however, I speak more in detail under another heading.

Very often the attack is, like epilepsy, represented by nothing which can be called a convulsion, there being but a temporary loss of consciousness, a catching of the breath, a sudden waking with a start, or a falling forwards of the head and stertorous breathing for a few seconds.

Regarding the **ætiology** of infantile convulsions I shall say nothing but what is special to the subject and confine myself to what appear to be the clinical conditions, both predisposing and exciting, out of which they arise; referring the reader for the general pathogeny of spasm, including of course epilepsy, to larger works, and especially to the teaching of Dr. Hughlings Jackson. An admirable article by Dr. James Anderson, clearly stating Dr. Jackson's views, is to be found in Tuke's *Dictionary of Psychological Medicine.*

Nervous heredity of many kinds and *rickets* are the two conditions which, one or both, underlie an enormous majority of all cases of convulsions in childhood, whatever the immediately exciting or reflex causes of the attacks may appear to be. This rule applies, I think, to most fits which usher in acute disease, as well as to those which strictly concur with difficult dentition, with markedly deranged digestion, or with definite psychical disturbances such as fright. My colleague Dr. Coutts has shown that the occurrence of fits at the onset of acute disease is much rarer than is usually believed; and I entirely agree with his contention that most of the cases even thus occurring are connected with either rickets or pronounced neurotic heredity. But nevertheless I would urge the practical caution that, when fits take place for the first time, especially after the earliest infancy, and there is any rise of temperature, we should never forget the possible sequence of pneumonia or other acute febrile disease, however good the personal and family history of the case may be.

Among organic diseases of the brain, and other affections which may occasion the nervous discharge resulting in convulsion, we must bear in mind all kinds of *meningitis*, and especially the tubercular form, in which convulsion may occur at any period and is not seldom the first observed symptom; cerebral thrombosis and embolism; injuries to the head; otitis media; tumours; and sometimes hæmorrhages, both meningeal and cerebral. Convulsions also frequently take place at the outset of *chronic hydrocephalus;* in wasting diseases involving, perhaps, arterial anæmia of the brain; with venous hyperæmia of the brain as instanced in pertussis; and sometimes, seemingly, in direct association with high temperature. I enumerate these conditions as concomitant, but not necessarily causal. In exhausting diseases, such as diarrhœa and vomiting and many others, convulsion is the immediate herald of death, attended, not infrequently, by a considerable rise of temperature.

It must not be forgotten that general convulsions not seldom occur in connexion with coarse disease limited entirely to one side of the brain; that they are antecedents of many cases of infantile hemiplegia and sometimes of aphasia alone in varying degree according to the stage of the child's development; and that they may be associated with renal disease. We must therefore search with all care for concomitant symptoms of brain or other affection, examine the urine for albumen, and exercise due caution before pronouncing upon the possible nature and import of any given case of convulsions.

When one-sided convulsions occur without local disease of brain they are not constant in seat, but often shift from one side to the other. Local paralysis, and marked inequality of pupils, following on convulsions, point strongly to the probability of brain-disease. Protracted strabismus is also suspicious, although this symptom may last for long after convulsions, with no other indication of organic mischief. Above all, localised convulsions of recurrent character and fixed seat, and apparently unaccompanied by loss of consciousness, indicate with the greatest probability the existence of organic disease of the cerebral cortex.

Besides those **sequelæ** of repeated convulsive attacks which have already been glanced at, more or less permanent mental deficiency may be observed, and other marked neuroses may ultimately follow. According to Dr. Coutts, who has studied the question of the sequelæ of infantile convulsions, there is much evidence that a considerable proportion of the subjects of this affection in its ordinary form suffer in later years from definite nervous trouble. He found[1] that 40 out of 85 subjects of infantile convulsions were in later life victims of epilepsy, somnambulism, insanity, chorea or migraine ; and that of the remaining 45 most were either eccentric and irritable or below their brothers and sisters in intelligence. Although it seems necessary to discount these figures to some extent, owing to the method of inquiry which in many of the older cases must have proceeded backwards from the fact of present neurosis to the history of past convulsion, a perusal of Dr. Coutts' arguments and further illustrations will afford us substantial reasons for regarding infantile convulsions as a very probable danger-signal to the future nervous health of their subjects.

The later the period at which the convulsions of early childhood set in, the more probable is their epileptic nature ; and in many cases there seems to be no breach of continuity between the fits of infancy and the epilepsy of a lifetime. It is known that the origin of many cases of epilepsy is traceable to the first year, and that a large majority have a markedly bad neurotic heredity.

The most favourable elements of **prognosis** in the convulsions of infancy are the absence of organic brain-disease and of discoverable neurotic heredity ; infrequency of attacks; and the presence of any strongly presumable excitant of reflex action, such as marked alimentary disturbance or shock in close association with the fits in question, and thus perhaps evidencing external stress sufficient to impress even a fairly stable specimen of the infantile nervous system. I may remark here, however, that I believe the cases to be rare where fits can be referred

[1] See art. "Convulsions in early childhood," *Medical Magazine*. Aug. 1892. London.

to parasitic worms as an immediate excitant. I have certainly met with a case or two where fits occurred just before the vomiting of a "lumbricus," but I have no knowledge of convulsions being occasioned by oxyuris or tænia. It is but seldom that I have been forced to recognise alimentary disorder as a determining cause, except in cases of markedly rickety or neurotic children. No diagnostic importance can be granted to the contention of some that in young infants who are well nourished the fits are reflex, while in those who are wasted they are due to some intra-cranial lesion. Experience alone, not to mention the probable pathology of convulsion, falsifies this misleading statement at every turn. Repeated convulsions, however, in wasted and apathetic infants are always of grave prognostic import, being often, as we have seen, a mode of dying.

Post-mortem examination in cases which have suffered from convulsions usually reveals nothing. Pronounced venous congestion of the brain or small extravasations are, however, found in many cases of repeated convulsions immediately preceding or coincident with death. There is but little evidence of localised lesion of brain resulting from convulsions ; nor, in consequence, can we refer the permanent epileptic condition, which doubtless often follows on fits in children, to such an hypothetical cause.

The **treatment** of convulsions is a simpler matter than the acknowledged difficulty of ætiological diagnosis would suggest. Practically we have, in severe and repeated cases, to do what is possible towards checking the convulsions, seeing that they are not only alarming but may also be sometimes fatally injurious by causing engorgement of brain and lungs. For this purpose inhalation of chloroform, cautiously administered, is, I think, one of the best remedies, on which, in my experience, good results have sometimes followed. It should not, however, be used when there are already signs of lung- or heart-failure or serious collapse. The cyanosis produced by the fit is in my opinion, as in Henoch's, no contra-indication to the use of chloroform. Another decidedly useful method is the rectal injection of a few grains of chloral hydrate, according to age, with or without bromide of potassium. Of this treatment Mr. J. Scott Battams speaks highly. Administration of opium or, when swallowing is impossible, subcutaneous injection of morphia is often very useful in preventing the recurrence of fits. One thirtieth to one twenty-fourth of a grain may be given to a child of a year old. In prolonged cases the bromides are very serviceable, and I have found, both in children and adults, that the ammonium salt is as efficient as that of potassium. I also believe that for continuous use it is far preferable. Chloral in small doses and belladonna may be tried in chronic cases. Nitrite of amyl inhalations may be advantageously used during the attack, according to some authorities ; but in my own small experience of this remedy it has been ineffectual. I always order the hot bath, partly from convention, and partly because it sometimes seems to do good. In all fits, whether severe enough to indicate such checks as above-mentioned or of a milder character and less frequent occurrence, we should search for any possible excitant and endeavour to remove it, however doubtful we may be of its causal rôle. For this end emetics and purgatives may be given without hesitation and sometimes with success, and in cases where there is obvious swelling and redness about the gums there can be no objection to free lancing.

Tetany.

I refer to this affection under a separate heading mainly in deference to usage, and partly because the symptoms to which this name has been given are said to be quite as common in later childhood and early youth as in infancy. In my own experience of patients of all ages tetany has been mostly seen in infants and quite young children. It is chiefly marked by tonic contractions of the hands and feet (carpo-pedal); the thumbs being stiffly bent across the palms, the metacarpal joints flexed, and the fingers extended. The soles are arched, and there is generally evidence of pain with the occurrence of the cramps which may last for long with intermissions of varying duration. There may be slight twitchings of the facial and other muscles, either spontaneous or produced by irritation of the cutaneous nerves ; but the clinical picture generally is one of tonic, not clonic, contraction. In infancy, certainly, tetany arises in the same ætiological conditions as the convulsions of which I have spoken above, being notably connected with rickets and exhausting diseases, especially intestinal ; and it not seldom exists in the intervals of clonic attacks. In some cases the wrists and elbows are tonically flexed, and the ankle- and knee-joints may be rigid in flexion or extension. This state may last for days or weeks, and occasionally there is œdema on the dorsal surface of the hands and feet. I have seen one very typical case of extensive tetany, involving the large as well as the small joints, with redness and œdema of feet, where the child, aged 18 months, remaining tetanised and perfectly conscious for nearly a fortnight, was thereafter seized with violent clonic convulsions which were nearly fatal. It soon made a good recovery. Previous to this attack the child had been perfectly well, but there was some evidence of rickets. In patients beyond infancy what may be called the purest examples of this so-called tetany are found, quite unaccompanied at any period by clonic convulsions, and unmarked by loss of consciousness or other cerebral symptoms. There is sometimes in these older cases more wide-spread tonic spasm including the trunk muscles : and, as it is said and as I have seen in one case, contraction of the muscles of the jaw. Confusion is possible here between tetany, which is in itself harmless, always of good prognosis and, as we have seen, closely allied to other conditions of functional spasm, and true tetanus. Many reported cases of so-called chronic idiopathic tetanus have been doubtless instances of tetany. Tetany has been observed by some as a kind of nervous epidemic among girls, and, as such, is doubtless closely allied to the various spasms of hysteria and hystero-epilepsy. The affection is said to be often connected with taking cold. For clinical purposes I think it is better to separate, as Henoch does, these tonic contractions, which attack infants in close connexion with convulsions, from the more unmixed forms which occur in later age ; but there is no hard and fast line of division either in appearance or ætiology, and the term tetany is now generally used in the most comprehensive sense.

In infancy the *treatment*, both symptomatic and general, is similar to that of convulsions. Calabar bean has been recommended in doses varying from $\frac{1}{30}$ to $\frac{1}{3}$ grain, but the duration of the attacks so treated, including one case of my own, in no way points to any curative action. In older cases general treatment for neurotic disorder, as alluded to under the heading of hysteria, is the best line to follow.

Epilepsy.

Under this term I include the recurrent attacks of impairment or loss of consciousness, often attended by convulsions of varying degree and distribution, which are usually known as "idiopathic" epilepsy; excluding the unilateral or strictly localised spasms, known as Jacksonian epilepsy, which are mostly unattended, at least at the outset, by loss of consciousness and are connected with localised disease in the cerebral cortex. A not uncommon form of convulsions, however, sequent on hemiplegia in infants and young children, where the spasms, though most often confined to the paralysed side, are frequently general and in all respects indistinguishable from idiopathic epilepsy, must be dealt with in this context.

Epilepsy as above indicated shows itself in childhood, as in adults, in various forms, and in all grades from short affections of consciousness or "petit mal" to the well-known convulsive attack or "grand mal." It is pre-eminently a disease of early life ; three-fourths of all cases beginning before twenty, and more than one-fourth before ten years of age. Its onset is most frequent in the first year, and at or about the period of puberty.

Owing to the incomplete development of the infantile brain and the imperfect powers of expression in early childhood we often meet with less clearly defined order of spasm than in later years; and the "auræ," though occurring in many varieties, both physical and psychical, are less easily discovered and described. In children beyond early infancy the attacks rapidly approximate to those of later life, neither the spasms nor the auræ needing special description. I have seen cases illustrating many of the auræ referred to in monographs, the multiform epigastric sensations being the most frequent.

Two considerable difficulties meet us respecting the **diagnosis** of epilepsy in infancy and later childhood respectively. On the one hand infantile convulsions are often indistinguishable from those which by the sequel are proved to be chronic epilepsy; and on the other both the mental and physical phenomena of the condition known as hysteria may closely resemble epilepsy, and, indeed, both alternate with and sometimes obscure undoubted epileptic seizures. The questions of the diagnosis of epilepsy and the prognosis of convulsions in infancy are practically the same, and, with the exception of most cases connected with organic brain-disease, are very often insoluble at the time when their importance is greatest. Frequency of attacks, absence of discoverable excitants, the exclusion of rickets, marked neurotic and epileptic heredity and, perhaps, good general health may be, more or less, comparatively indicative of epilepsy; but none of these points can be absolutely relied on in the diagnosis of any given case. If frequent losses of consciousness without convulsions can be discovered by observations of altered facial expression, of sudden quietness, or of spells of noisy breathing, the suspicion of epilepsy is thereby strengthened. I cannot agree with the view that the actual presence of rickets enables us to make an absolute distinction between ordinary convulsions and epilepsy; for, although infantile convulsions frequently cease with receding rickets, they are often persistent; and an hereditary tendency to epilepsy and other neuroses is especially marked in many cases of repeated convulsions with rickets. It is, moreover, frequently found that epilepsy, apparently beginning in late childhood or adult life, has been preceded by convulsions in infancy. This takes place, according to Gowers, Hughes Bennett, and other authorities, in from 7 to 15 per cent. of all cases of epilepsy, and is probably still more frequent; for infantile convulsions are very often overlooked or forgotten. In most cases, then, of recurrent convulsions in children under two years old, lapse of time alone, as we have seen, will be a useful guide to prognosis. After this age the diagnosis of epilepsy rests on a firmer basis.

Our second difficulty meets us in cases of later date, which are often marked by strange alterations of conduct and a multiform hysterical display. The epileptic seizures, which are really present but perhaps detectable only by careful observation, especially when of the kind known as "petit mal," are often obscured by or alternate with convulsive attacks, which, however violent, are marked by movements akin to voluntary, the eyes being closed and high grades of consciousness shown by various signs.   There may also be anæsthesia and other phenomena of deep hysteria, which are demonstrably beyond the control of the will. I shall further treat of such symptoms under the head of hysteria, merely alluding to them here to emphasise the frequently close alliance between epilepsy and hysteria, and the prominently psychical elements of epilepsy known as epileptic "vertigo" and "mania," marked instances of which occur even in young children.   The lesson to be learned is that, masked by so-called hystero-epilepsy and hysteria and sometimes by acts of apparent imposture and by general depravity, true epilepsy may not seldom be met with in children as well as in adults.

The clinical **ætiology** of epilepsy in childhood, apart from such cases as are connected with rickets, involves no special consideration.   *Neurotic heredity* is manifest in many cases, as shown by the history of epilepsy itself, chorea, insanity, migraine and other disorders in near relatives, and some of these affections may concur in the individual patient.   It is unnecessary to discuss the question of the exact connexion between infantile convulsions and epilepsy ; but it may be said that, although the chronic condition may be the direct result of this repetition of impressions left by the early attacks, the very repetition in question is a strong indication of special nervous instability.

Epilepsy often marks *congenital idiocy* in its various grades, but may make its appearance at different periods, one of my cases, a microcephalic, having had his first fit at five years old.   In several others there was marked and rapid mental impairment after the epilepsy had declared itself ; and it is a general rule that the more frequent the fits, especially when of both forms, the more decided and constant is the mental disturbance.   I had once a mentally deficient patient of seven years old with marked epilepsy continuous with infantile convulsions, all of whose six brothers and sisters had suffered from convulsions which, in most, had ceased in infancy.

Exciting causes of the first fit in children over four or five years old, whether or not they have had convulsions in early infancy, can sometimes be made out with great probability.   A boy who had had a few convulsions in infancy had his first subsequent fit, at the age of eight, immediately after breaking his arm.   I saw him six weeks afterwards, when his attacks were typical and frequent, with intervals of tetany, mania, and filthy habits.   He gradually improved with large doses of bromide, but remained very stupid.   Another boy, aged 9, previously healthy, had suffered for two years from epileptic convulsions beginning immediately after he had been nearly drowned ; and I could mention many similar instances of apparent excitation of first attacks. Disturbance of this kind, I believe, may determine epileptic seizures in childhood more readily than in adults; and careful hospital treatment even without drugs, or a similar régime, is pre-eminently likely at this period to cause a prolonged remission of even very frequent attacks.

I have seen several cases which began during *fever* or acute pulmonary attacks or in early convalescence therefrom, and often between one and two years old. In a boy of six years, with innumerable fits and marked epileptic heredity, the first attack was at three years of age on recovery from scarlet fever. Many cases start just after a *fall or blow on the head*, and I think that at all ages a subordinately causal nexus must here be recognised.

There is a frequent connexion between *hemiplegia* and epilepsy in infancy and early childhood, apart from those definitely one-sided fits due to localised cortical disease, as, for instance, tubercular or other tumours, which I have excluded from the present consideration of epilepsy. The diagnosis of the organic cause of such fits rests mainly on their frequent recurrence in unvaryingly unilateral or mono-spastic form, unaccompanied by the psychical elements of epilepsy, and mostly attended by such symptoms of cerebral disease as headache, vomiting or localised paralyses. Almost all cases of infantile hemiplegia begin with convulsions, and are frequently followed by permanently recurrent attacks, which in many instances are general as regards the spasm and in no way distinguishable from ordinary epilepsy. There is good reason, as we shall presently see when considering hemiplegia, for attributing the hemiplegia in many of these cases to venous thrombosis in the cortex or lower cerebral centres, and to regard the convulsions as the result of brain disturbance consequent on the lesion. Most of these cases occur in the first few years of life; several recover partially, and some completely. When the spasms are exclusively unilateral, involving the paralysed side only, the lesion has probably been considerable; but in the chronic cases with general convulsions, which especially concern us here, the original lesion has to a great extent disappeared, leaving its effects in permanent instability. The oldest case of the kind, out of many that I have seen in children, took place at ten years of age. In one instance of well-marked general epilepsy, in a girl of thirteen years old, the fits had been constant since the first year when a severe convulsion had occurred, lasting twelve hours, and followed by permanent paralysis of the right arm. Many of my cases were hemiplegia of typical distribution, involving arm, leg, tongue and lower facial region, and thus indicating the neighbourhood of the internal capsule as the seat of the lesion. In some of the right-sided cases there was also marked aphasia.

A temporary hemiplegia or comparative paresis of one side is often seen in children, as in adults, after violent or repeated epileptic fits. This is to be regarded as due to exhaustion of the nerve-centres. It is recognised on the side which has been chiefly or alone convulsed, and usually affects the limbs without the face.

Before leaving the subject of the symptoms and diagnosis of epilepsy in childhood I must emphasise the frequent mistake of taking attacks of "petit mal" for syncope. This diagnostic error is often made in the case of adults as well, but the very frequent occurrence of epileptic attacks without convulsion is especially overlooked in children. I have repeatedly met with such cases, marked by temporary loss of consciousness, with or without staggering or falling, and many of them subsequently developing the convulsive form, which have been called fainting fits by both medical and lay observers. Simple syncope, be it always remembered, is extremely rare in childhood, and is almost confined to cases of great prostration or exhausting disease, and to conditions where, as in enteric fever or in pronounced anæmia, there may be temporary or chronic cardiac dilatation. It is moreover, unlike the far more frequent "petit mal" which is most often quite unheralded by "aura," preceded by nausea, or sweating, or a distressed feeling of "going to faint;" and is generally caused by exertion or excitement. An immense majority of so-called fainting fits at all ages, especially when recurrent without recognisable cause and attacking otherwise healthy people, are without doubt epileptic. I must also mention night terrors, especially when frequent and accompanied by hallucinations, as often co-existent with and indicative of epilepsy; and in many cases habitual somniloquence and somnambulism have a similar association.

The **prognosis** in the epilepsy of childhood must be especially guarded, considerably fewer cases recovering, according to Gowers who deals with large numbers of cases at all ages, than of those which begin in later years; and the acknowledged diagnostic difficulty as regards convulsions in infancy renders a really useful forecast at this period for the most part impracticable. Doubtless, however, a certain number of cases, apart from all treatment, tend to improve or recover as years go on, however inveterate they may have appeared. The longer the disease has lasted, and, generally speaking, the more frequent the attacks, the worse is the ultimate prognosis; but I know of no good criterion to apply to any individual case. Gowers shows that hereditary cases recover or are arrested far more often than those without such a history. He also gives reason to believe that the prognosis is worse when the attacks occur both in sleeping and waking than when confined to either of these states alone.

The best result we can look for, as a rule, is a diminution or arrest of the fits under treatment, and this takes place in varying degrees with the use of the bromides.

As regards **treatment**, the epileptic child must be guarded from all demonstrable or suspected exciting causes of the attacks, and sedulously nurtured and taught, with avoidance of all mental strain. Plenty of easily digestible food, abundant fresh air and sunlight, and the medicinal tonics, are all indicated. When the fits are as frequent as one or more in a fortnight I tentatively give the bromide of ammonium or potassium, persisting with the administration of this drug in proportion to the frequency of the attacks. But when the intervals are longer than a fortnight I am strongly opposed to the routine prescription of the bromides, and especially of the potassium salt which is so frequently given on the slightest suspicion of an epileptic tendency. It may be possible that cases beginning in quite early life with frequent convulsions may be checked from further development by the use of the bromides, such an hypothesis being incapable of either proof or disproof; and it is un-questionable that fits can be lessened in frequency or kept in abeyance during continuance of the drug. But beyond this there is nothing certain; and I have no doubt, judging from carefully observed cases both in adults and children, that, notwithstanding much that has been said on the other side, the constant use of the bromides is eminently depressing and injurious to the nervous system and the mind. Such evil effects, if not of very long standing, are doubtless mostly recovered from on the discontinuance of the drug; but there is at least a danger of setting up a permanent listless habit, and anorexia or dyspepsia difficult to overcome.

In severe and inveterate cases other drugs may be tried, but few will be found of any value. I have frequently tried both belladonna and strychnia with at the best but very doubtful effect, and can say but little in favour of borax.

In cases where the bromides are to be given I begin with the ammonium salt, as less depressing, and seldom now have to change it for that of potassium. I give from five to seven grains for a dose to children of a year old, and often up to twenty or thirty after the age of 10 or 12. I cannot call to mind a case where the potassium salt has succeeded in checking fits where the other has failed, though there have been many where both proved nearly useless. In all cases where the drug seems to be doing good, it should be stopped for a while, and instantly resumed if the fits recur. By this method we shall soon find individual indications for continuance, remission, or complete omission of the medicine. In cases where the fits always return on omission of the bromides the drug should be given for at least six months continuously, even when the fits have altogether ceased. After this period the doses may be diminished in amount and frequency during another six months, when omission may

Q

be once more attempted. I have often tried the combination of arsenic with bromides in those frequent cases where the nodular skin-eruption appears, but have scarcely ever even suspected any good effect. The inhalation of amyl nitrite is apparently successful sometimes in preventing the accustomed convulsive attack from following on the aura; and there is some evidence to show that nitro-glycerine may be useful even in cases of "petit mal." I have, however, had no experience myself of the use of this drug in children. In the rare but scarcely questionable cases which are described by some as reflex epilepsy, arising from such occasions as eye-strain, &c., we should search for any probable exciting causes; correcting, for instance, marked visual errors, if possible, or endeavouring to relieve genital irritability by careful watching or by removal of a tight prepuce.

Localised Spasms.

Under this title I shall consider briefly certain spastic symptoms of more or less common occurrence in children, which are as a rule local in expression and often in origin.

Nystagmus, or oscillations of the eyeballs in either a lateral or, more rarely, a vertical or rotatory direction, is in childhood mainly seen in connexion with blindness or much impaired vision, as in congenital cataract, corneal opacities or optic-nerve defect; with intracranial tumours, especially those involving the cerebellar and pontine regions; with chronic hydrocephalus and sometimes acute meningitis; with many cases of the "nodding-spasm" presently to be noted; with some apparently epileptic convulsions; and with the rare affections known as "disseminated sclerosis" and "Friedreich's disease." When constant and increased by fixation of the gaze on a distant object, it is generally a sign of organic nerve-disorder, with the probable exception of its concurrence with nodding-spasm or with some cases of convulsions. More or less rapid lateral nystagmus, however, is sometimes seen in normal-eyed persons, including children, when they are interested or excited, and may then be almost constant; but, as far as I know, this form is always to be checked by voluntary fixation of the gaze. I am informed by Dr. Hughes Bennett, to whom I mentioned this observation, that he has noticed this phenomenon repeatedly in students under oral examination. I have notes of one marked case of nystagmus in a child of two who had suffered from very frequent convulsions, without discoverable exciting cause, since the age of three months. The child was slightly rickety, but had no further symptom of nervous disorder. She improved much after a month in hospital, the convulsions ceasing and the nystagmus becoming much less.

Nodding-spasm, or **head-jerking**, is in my experience a somewhat rare affection. Dr. Hadden however saw twelve cases, during two years' practice at the Hospital for Sick Children, according exactly with such as were described by Henoch in 1851 and subsequently, and by other observers. The affection is marked by constant or intermittent nodding, lateral, or rotating movements of the head, rhythmical or jerky, and usually accompanied by nystagmus of one or both eyes, either vertical, lateral or rotatory. The nystagmus is of far more rapid rhythm than the movements of the head, and is generally increased when the head is forcibly held. Henoch attributes this affection to nerve-irritation from teething and probably other reflex causes. Strabismus is but rarely observed. Nearly all the subjects are under two years old, and most are first attacked between the ages of six and twelve months; but the symptom has been sometimes noted long before the earliest teething-time, and Henoch quotes a case of twelve years old. From four out of five of Dr. Hadden's detailed cases, and a few which I have seen myself, in which there were either convulsions or losses of consciousness with, sometimes, lateral deviation of the head and eyes, it would seem that this affection has alliances with the epileptic condition, or at least points to great irritability of the nerve-centres usually well-organised in early life. I am also inclined, with Dr. Hadden, to include in the same clinical category a few instances I have seen of unexplained nystagmus without head-jerking. For a more detailed account of these cases I refer to Henoch's Lectures on Diseases of Children and to Dr. Hadden's paper in the *Lancet* for June 1, 1890. The prognosis is usually good as regards this symptom, and I know nothing better to suggest for medicinal treatment than the use of the bromides.

There is some distinction to be made, though not perhaps so great as taught by the above-quoted observers, between this affection and that which has long been known as "eclampsia nutans" or the "salaam convulsions," where the movements take place in distinct paroxysmal attacks, consciousness is lost, and the upper part of the body as well as the head is bent forwards with rapidly successive jerks. Henoch says that these cases always end fatally.

I may mention lastly that cases of frequent slow swaying forwards of the body are often seen in quite young children not necessarily otherwise affected. Some of them have been referred to uneasiness of the genital organs; but in several which I have observed I have found no local cause for any irritation.

Torticollis or "wry-neck" is almost always of the tonic form in childhood, the permanent clonic variety, or ordinary spasmodic wry-neck, being practically a disorder of later life. The usually temporary jerking of the head, from contraction of the sterno-mastoid or other muscles as a part of nervous habit, may conceivably, if continued, account for a few chronic cases of torticollis otherwise unexplained. Of this I have seen one possible instance. The most usual cases, other than surgical, of this disorder in children, are either of the acute form due to cold, and then to be regarded as an extreme form of "stiff neck;" or of the chronic variety, which may result from the acute, or may perhaps be due to such reflex causes as carious teeth, but often appears to be idiopathic or a simple neurosis. I question whether the affection is ever directly caused by rheumatism properly so-called; for, although I construe the symptomatology of rheumatism in childhood as widely as possible, I cannot from my experience regard torticollis as either an incident in acute rheumatism or as an indication, from any of its associations, of the rheumatic diathesis. Congenital wry-neck is always tonic and chronic, due, probably, to abnormal position of the head in fœtal life or to some faulty development of the vertebral or neuro-muscular structures. It is said that the so-called "sterno-mastoid tumour," sometimes seen in infants obstetrically injured, and due to extravasation of blood, may lead to torticollis. Such cases, however, usually recover spontaneously. Caries of the vertebræ, glandular abscesses in the neck, and other injuries, such as cicatrisation after burns, must be remembered among extraneous causes of torticollis.

Treatment must be directed towards removal or diminution of any visible or suspected cause of the affection. The congenital cases may often be cured by manipulation or tenotomy. Acute cases as a rule recover quickly with rest and warmth. Chronic cases from whatever cause arising are usually of bad prognosis, though on the whole more likely to improve than in the adult, especially when apparently idiopathic. There is nothing either in the medical or surgical treatment of torticollis which is special to childhood.

Other spasms in great variety are seen in children. They are generally to be regarded as signs of nervous instability; are often, though by no means always, associated either individually or hereditarily with hysteria, chorea, epilepsy, migraine or other neuroses; and sometimes last through life. Of this class are twitchings of the eyelids, scalp and nose, of muscles of other parts, and multiform jerkings of the head and shoulders often described under the inaccurate name of "habit-chorea." Sudden and oft-repeated expiratory acts, often with a short gruff cough, sometimes occur, of which I have seen several examples mostly connected with a neurotic history. One was in an epileptic boy of ten, otherwise quite healthy. These phenomena are frequently occasioned by ill-health, general nervous disturbance or local irritation, and may pass away on the removal of the excitant. Sometimes in similar clinical association are seen sudden contortions of the extremities and even of half the body, or still more extended convulsive movements, unaccompanied by any disturbance of consciousness. These cases have been named by Henoch "chorea electrica" or "lightning spasm." The knee-jerks and other reflexes in such children are usually excessive, and are sometimes attended by more wide-spread spasm. Closely allied to these spasms in children are occasional cases in adults which I have seen, marked by multiform contractions and highly-excitable reflexes amounting often to clonus, unassociated with any evidence of organic disease or with the mental phenomena of hysteria, and often ending in complete recovery. The "para-myoclonus multiplex" of Friedreich, which is mainly an affection of the large muscles in connexion with the trunk, seems to be of this genus; as also does a case published by Hughes Bennett in the July number of *Brain* for 1886.

The *prognosis* in most of these various cases of local spasms in childhood is good in proportion to the slightness or absence of other and graver nervous disorders, and the possibility of suitable treatment. Discoverable causes of local irritation of the part affected often exist; such as, for instance, discomfort from clothing, or, as in the case of the winking-spasm, a tendency to conjunctivitis, which is not seldom the effect of late hours and noxious gases. From experience of many cases of all grades, I am sure that, as a rule, the child's attention should not be frequently directed to the spasms; and I quite agree with a writer on this subject who teaches that, when the movements are more or less controllable by the will, reward for improvement is much more successful than punishment for continuance of the habit. These children are mostly neurotic in some form and degree, and must be *treated* on the best principles of mental and physical hygiene. Regulated bodily exercise, as much out of doors as possible, and gymnastics suited to the child's age are very helpful; and daily school lessons should be given, which should be chiefly oral and not long continued. Prolonged rest at night and an hour or two of sleep in the day should be encouraged, but not by drugs; close and gas-lit rooms must be tabooed; and all excitement carefully avoided. In weak and anæmic children the mineral tonics, especially iron and arsenic, are of much use, as also is cod-liver oil in those numerous cases where fatty foods are hardly taken. It is but in the severest and most inveterate cases, and those complicated with some general neurosis, that sedative medicines are necessary or advisable. The bromides then are sometimes of signal service in breaking the nervous habit. I have in many different cases seen great improvement, and cure with no subsequent relapse, as the result of a short course of these medicines. Two boys, aged about ten, suffering from very frequent expiratory spasm of several months' duration, the one slightly epileptic, the other of double asthmatic heredity, but both otherwise in good health, permanently lost their trouble after daily taking for a few weeks thirty grains of ammonium bromide.

In a few cases of long-continued jerky spasms of the neck and face or other muscles I have had complete and permanent success by the method of removing the patient from home and applying some irritating treatment, such as a blister or seton, to the part affected. Such treatment, I believe, operates by means of a mental impression.

Retraction of the head is most often, when of any persistence, the signal of serious cerebral disease; and is sometimes combined with tonic spasm of the trunk-muscles, causing excessive opisthotonus. It occurs very frequently in basic cerebral meningitis from any cause, with or without extension to the spinal cord, the symptoms in the purely basic cases being probably due to the prevalent ventricular effusion; in many advanced instances of intra-cranial tumour, especially in the cerebellar region; and in some of chronic hydrocephalus. I have seen it in a marked degree in connexion with some ill-explained cases of general rigidity with wasting, which may recover, although usually with impairment of mental functions. In one case, however, of a child of two years old the occiput was retracted on the spine for many weeks, and there was extreme wasting with great rigidity of all extremities, followed by a very gradual recession of all symptoms and ultimately perfect recovery with intelligence apparently intact.

As a rule lasting head-retraction means organic disease, and my experience of autopsies in many cases where this symptom existed tends to show that the pathological constants are either basic and cervical meningitis, tubercular or otherwise, or excess of fluid in the cerebral ventricles. In many cases no other morbid condition is found than marked ventricular effusion. Chronic ventricular effusion, however, at least when of insidious origin, may certainly exist without retraction of the head, as common experience amply shows.

It must be remembered that the head is often retracted in cervical caries, and sometimes in cases of glandular swellings in the neck or of post-pharyngeal abscess pressing on the larynx. I have also seen marked retraction several times after falls, without any other nervous symptoms than those due to general shock. It is here, probably, the result of muscular strain. Endeavours to replace the head are in all such cases accompanied by pain; and as a general rule, when a local cause is in question, the diagnosis is not difficult.

Simple but rigid retraction of the head in babies, without spasm, and apparently quite unaccompanied by pain on attempts being made to overcome it, is often observable for a short time in cases of temporary disorder, whether pulmonary or alimentary, as well as in many wasting infants and those subject to convulsions. I have sometimes made a provisional diagnosis of meningitis in cases such as these (the temperature being raised in many instances), where perfect recovery took place in the course of a few days. Therefore, without any other cerebral symptom or evidence of further disease, head-retraction is by no means to be regarded as such a grave or even fatal sign as it appears to some, even when all local causes have been eliminated as far as possible.

CHAPTER II.

THE PARALYSES OF CHILDHOOD.

IMPAIRMENT or loss of motor power in infants and young children may arise, as in adults, from lesion in any part of the motor tract, whether cerebral, spinal, neural or neuro-muscular. The paralyses, however, which are due to chronic degeneration in brain or cord are mostly without their counterparts in early life ; while others, common to all periods, such as many of those caused by tumours and abscesses or by traumatic and other extraneous lesions of the nerves and nerve-centres, have no clinical or pathological characters special to childhood.

Infantile Hemiplegia.

Paralysis on one side of the body, either complete or partial in extent or degree, and arising from lesion or defect of the cerebral motor tract, is often met with in early childhood, and mainly differs from the classical hemiplegia of adults in being less often of that typical form, marked by involvement of the arm, leg, face and tongue, which results from lesions in or near the internal capsule, and in being much more frequently associated with both initial and subsequent convulsions. The reason for these differences lies in the rarity in childhood of hæmorrhage or thrombotic softening in the basal ganglia, and in the more frequent situation of the lesion over the motor surface of the brain.

One of the chief known **causes** of hemiplegia in infants is *meningeal hæmorrhage*, which, however, from its irregular distribution over the surface of the convolutions, frequently produces bilateral symptoms. It is mostly traumatic, and occurs at or soon after birth owing to protracted labour or the use of instruments. In this context I may perhaps mention a case of hemiplegia with rigidity of the right arm and leg in a child of seven years old who was suffering from well-marked purpura hæmorrhagica. The paralysis improved, and the diagnosis of superficial hæmorrhage seemed very probable. *Tumours* in the motor tract, whether in the cortex or lower down, *localised meningitis*, vascular changes from *thrombosis, embolism* or *cerebral hæmorrhage*, and, rarely, *abscesses* of the brain are also established causes; and *imperfect development of the motor cortex*, either congenital or sequent as atrophy on meningeal hæmorrhage or other trauma, has been several times observed.

Owing to the rarity of recent necropsies in hemiplegia in children, the causes in numerous cases are matter of conjecture. Tumours, especially tubercular and cortical, are doubtless the origin of some cases otherwise at first unexplained, as was evidenced by the case of a boy of two years old under my care in hospital, with left hemiplegia beginning with convulsions, who was readmitted after a year with recurrent convulsions and the signs of meningitis. At the post-mortem we found small caseous masses of tubercle in the upper Rolandic region on the right side, and recent tubercular meningitis. Arterial embolism is a well-recognised though not frequent cause of hemiplegia in children, and may sometimes be confidently diagnosed in recent endocarditis, or conjectured in some cases of marked stasis in the pulmonary circulation. I have notes of a case of rheumatic fever, with endocarditis and chorea, where sudden hemiplegia, occurring first on the right side with aphasia and soon after on the left, allowed no doubt of the diagnosis of double embolism. Hæmorrhage other than meningeal is but rarely proved in childhood, but it may occur from syphilitic disease of vessels, and in a growing tumour; and either hæmorrhage or thrombosis is the possible explanation of those cases of hemiplegia which take place during or soon after measles, scarlatina and other acute infectious diseases. Among other examples of this I may quote a typical hemiplegia of capsular origin, occurring three weeks after the onset of severe measles in a child of fifteen months. There was some rigidity, but all symptoms disappeared in four months. Thrombosis of the right middle cerebral artery, without heart-disease or embolism, was found by Dr. Abercrombie in a case of sudden left hemiplegia in a boy of six years old with diphtheria; and thrombosis of the vessels from syphilitic disease has frequently been reported. The sudden hemiplegia sometimes occurring in whooping-cough may be due to hæmorrhage from increased vascular pressure, with or without anatomical change in the walls of the vessels. Many cases of infantile hemiplegia with convulsions may probably be explained, according to the ingenious and reasonable theory of Dr. Gowers (who argues from the frequent sinus-thrombosis of infancy), by the occurrence of thrombosis in cortical veins; and the hypothesis of Strümpell, who reasons from the sclerosed and atrophied patches not seldom seen in the young brain and quotes that class of hemiplegias which begin, as infantile spinal paralysis often does, with febrile symptoms, that primary localised inflammation of the grey matter may cause cerebral palsy, is not to be disregarded, although lacking anatomical proof.

It is possible, therefore, to diagnose the cause of hemiplegia in a few instances, and to guess at it in many others from its mode of onset; but, seeing that in adult cases with much more definite pathology we frequently mistake hæmorrhage, thrombosis and embolism, I do not regard this question as of much practical importance.

Clinically infantile hemiplegia begins in most cases with convulsions, generally of the unilateral type, and with loss of consciousness. Rigidity of the paralysed limbs sets in usually very soon, the joints of the upper extremity being most often flexed, and the foot in the position of equino-varus. Convulsions, both unilateral and general, tend to recur; they are in many cases strictly epileptic, and may remain after the paralytic symptoms have largely or quite disappeared. Mental deficiency of various degrees is very frequent, especially in the earliest and congenital cases; so also is marked abnormality in the shape and size of the skull, many of these infants being microcephalic. After a more or less lengthened period wasting of the rigid and paralysed limbs frequently sets in, and there may be marked surface coldness and venous congestion. The growth of the limbs is also often arrested. In cases other than congenital sudden onset is accompanied sometimes by marked pyrexia without further specific signs; and, as we have seen, the attack may arise in the course or sequel of many acute diseases. I have several times seen true aphasia in quite young children; in one case at the age of two years. Left hemiplegia with aphasia, stated by some and supposed by others to be more frequent in children than in adults, I have not as yet observed. The rigidity seen at first usually disappears in sleep, and is notably increased by forcibly moving the limbs. Later on in bad cases there is often permanent contraction. Fine tremor, or ampler spasms on movement of the affected limbs, and the constant movement generally called "athetosis" are not infrequently seen. These latter movements, slow, irregular, and unlike either choreic or convulsive spasm, affect the upper extremity, and mostly the fingers, which are constantly working in a vermiform manner. They have also been inaccurately named "post-hemiplegic chorea," and, far more appropriately, as preventing confusion with other disorders, "mobile spasm." In adults this symptom has generally been found in connexion with softening in or near the optic thalamus; but it appears certain from general pathological knowledge that the various forms of mobile spasm with hemiplegia may be due, like the chronic rigidities, to lesions in any position which interfere with the fibres of the pyramidal or motor tract.

Cerebral paralysis is, broadly speaking, marked off from the spinal paralysis of infancy by increased reflexes; slight and slow, instead of rapid, wasting; rigidities; normal electrical reactions; hemiplegic distribution; and frequently convulsive origin.

The **prognosis** in hemiplegia depends largely on the diagnosis of the cause, which, we have seen, is often a matter of much difficulty. Congenital cases occurring with convulsions mostly involve idiocy; and frequent convulsions in any case may pass into epilepsy and be accompanied by some mental impairment. Slight congenital cases, however, may improve or perhaps recover; and I have seen at least two non-congenital ones recover where convulsions had been frequent and rigidity marked. Frequent convulsions with monoplegiæ point to cortical lesion, which, being often tubercular, may be ultimately followed by meningitis. The typical form of capsular hemiplegia is, both as to its causes and prognosis, similar to that which affects adults, and the forecast is here more favourable than in the larger class of cases due to affection of the cortex; for the lesion may be but small and the mental faculties little if at all impaired. It is common to find the paralysis of face and leg disappearing rapidly, affection of the arm alone remaining. In acute cases prolonged apathy and coma are of the worst augury. I may mention here that the temporary paralysis of the limbs seen after many and severe epileptic attacks does not occasion prolonged difficulty either in diagnosis or prognosis.

It follows from the prevailing permanency of these affections, from whatever cause arising, that the field of **treatment** is very small. It is rare that medical aid is at hand before the initial convulsions are over; but, when possible, these should be checked, according to the methods already mentioned, without waiting to make a diagnosis. The child should be kept perfectly quiet, and bromide of ammonium or potassium should be given. There is no objection to ordering iodide of potassium for some weeks, even in the absence of positive evidence of syphilis; for morbid processes due to syphilitic vascular disease may possibly thus be checked. Contracture may be treated by frequent shampooing of the limbs, but rarely, in my experience, with more success than by the probably useless faradic current. In two cases of mine, of doubtful origin, which made marked improvement, no medicinal or manipulative treatment was employed.

Spastic Paralysis.

There is doubtless a small class of cases, in infants and young children, reminding the observer at first sight of that now well-known disease of adults, which, owing to the affection being chiefly or wholly confined to the legs, at least in the early stages, is described under the name of "spastic paraplegia." The chief clinical characteristics of this malady in adults are loss of power accompanied by startings and rigidity of the limbs, increased knee-jerks, and ankle-clonus; while there is usually no wasting, affection of the sphincters, or impaired sensibility. These phenomena are frequent enough as a sequel of myelitis with descending degeneration of the lateral columns of the cord; but, when they occur in an apparently pure and idiopathic form, their pathology is at present uncertain. Some cases, progressive or stationary, are referred by many to a primary lateral sclerosis; while others, retrocedent or recovering, are sometimes classed with functional disorders. It is in my opinion to be regretted that the cases now to be considered should ever have been described under the name given to the adult affection; for, as we shall see, in but few of them are the symptoms of paraplegic distribution only, and it is not in the cord but rather in the brain that we must look for the underlying lesion or cause of the symptoms. Spastic paralysis in childhood, not only in its narrower but also in its wider sense, must be regarded as due to lesion or disturbance of function in the cerebral motor tracts; and is thus fittingly considered in connexion with infantile hemiplegia.

Hemiplegia, as we have already seen, is most often accompanied by chronic spasm in infancy; and one at least of its causes, namely meningeal hæmorrhage, is often productive of bilateral symptoms. It is known, both from clinical and post-mortem evidence, that many cases of congenital or very early spastic paralysis on one or both sides of the body are due to cortical injuries and wasting of the motor convolutions sequent on the lesion. Further, it is established by recorded cases[1] that the same symptoms may result from imperfect development of the motor convolutions on one or both sides. These considerations have an important bearing on many of the spastic paralyses of infancy, including those which have been named "spastic paraplegia." Post-mortem examinations of the least complicated cases are few. I have myself seen but one, in a markedly microcephalic idiot of nearly three years old, where the motor convolutions were strikingly small and abnormal in appearance.

Spastic paralysis, other than hemiplegic, in infancy is generally of *congenital or very early origin;* the arms mostly, and the trunk-muscles sometimes, are affected as well as the legs; there is frequently marked wasting of the body and anæsthesia of varying degree; and, owing doubtless to cerebral causation, the sphincters are often spontaneously relaxed. Microcephaly and congenital idiocy are frequent in all grades, as also are convulsions, which in non-congenital cases are often the first

[1] See Sharkey's "Lectures on Spasm in Chronic Nerve Disease," Churchill, 1886, and Ross's case in *Brain,* vol. i. p. 477.

noticed symptom and may be succeeded by mental disorder. The large majority of the whole, and nearly all the congenital cases, show signs of some mental deficiency. Strabismus, nystagmus, marked tremors on movement, and "mobile spasms" of the arms as well as the hands have all been observed in one or other of my own cases. The knee-jerk is usually much increased, but the rigidity is often so great as to prevent the phenomenon of ankle-clonus. In the slighter cases, where standing or walking is possible, the rigidity is still conspicuous, the gait is hopping, the thighs adducted, the feet inverted, and the pointed toes frequently catch the ground. Most cases, however, are from the first unable to stand, and the rigidity is excessive, the legs often crossing one another. Sometimes the limbs are all in rigid flexion; in others the legs are extended while the arms are flexed.

The difference in the extent of the paralysis and its accompaniments in various cases must depend on the nature of the original lesion or defect, as also the improvement or recovery seen in some instances. Some cases waste extremely, others but little or not at all; and it is conceivable that in some of these anomalous cases we have to do with spinal infantile paralysis of a chronic form. The electrical test, which would be of great value here, is often quite impracticable, and always difficult of application and interpretation in young children. It is possible, in cases which show improvement, that there is slight inflammatory or hæmorrhagic effusion in the meninges, syphilitic change in the arteries leading to softening, retrocedent tubercular growth, or even late compensation for previously arrested cerebral development. One very marked case I saw in a boy aged fourteen months, who for six months had had rigid spasm of all extremities, strabismus, retracted head and frequent convulsions. After many weeks in bed with excessive wasting he gradually improved, and after six months seemed well in all respects. I saw him again after ten months more, when careful examination failed to detect any abnormality of mind or body. In another case of an undoubtedly syphilitic infant the symptoms apparently began at the age of six months. After about five months in hospital the limbs recovered, and the child seemed well; but there was marked mental deficiency.

Sometimes the paralysis is accompanied by more or less rhythmical movements. I saw one marked example of this in a girl of six years old, with a good family history, where the affection had been first noticed a short time after birth. She had "athetosis," without rigidity of arms, and rigid extension of legs. Another case with athetosis of arms and complete inability from birth to maintain equilibrium, though with no rigidity or other symptoms, may be referred, I think, with many of the spastic cases proper, to some widespread defect of cerebral development.

Occasionally I have seen cases with all the symptoms of the disease known as "multiple" or "insular" sclerosis, which may undoubtedly occur in early childhood; but post-mortem examinations of young children, showing the disseminated lesions in brain and cord connected with these symptoms in adults, are not numerous. Many cases, however, have been published under this title, some showing the "typical" symptoms, others with difficulty distinguishable from cerebellar or other tumours. At any rate the subject of definite and demonstrated "multiple sclerosis" has no special claim, in my opinion, to be discussed at length among diseases of childhood.

It is probable that in children these chronic spastic affections, owing to their multiform associations and differences of extent and degree, have a multiform causation; but it seems certain that in all cases they are due to deficiency, destruction or pressure in some part of the motor tract in the brain, as is evidenced also by numerous cases of brain-tumour, especially of the pons or of the cerebellum pressing on the pons, with spastic paralysis of limbs. In such cases, however, as are unaccompanied by any evidence of localised brain-disease, by microcephaly or by mental deficiency, ætiological diagnosis is very obscure.

In spite of several anomalous cases of tremors on movement, without rigid spasm, which I have often thought are probably slight examples of the somewhat heterogeneous class of affections we have been considering, as well as of a few others with marked and wide-spread rigidity, and of the improvement of some of the numerous congenital cases usually referred to under the title of "spastic paraplegia," the *prognosis* on the whole must be very grave, especially when the affection is of long standing. It is very rarely that this set of symptoms, dating from infancy, is seen in adults; the patients tending to die young from various causes.

It need scarcely be said that *treatment*, in consequence, must be almost vain. We may try antisyphilitic remedies in all doubtful cases, but with scarcely any hope of success. In those which approximate in appearance, however much they differ pathologically, to the spastic paralysis of adults, but are however not progressive and may last indefinitely, exercise should be enjoined; and repeated passive flexion of the limbs may be of some slight use.

Infantile Spinal Paralysis or Poliomyelitis Anterior.

By this title we denote a form of paralysis which, though occasionally seen at all times of life, is especially incident on early childhood, most cases beginning during the period of the first dentition and but few after the age of three or four years. The limbs are especially affected, the lower more often and, as a rule, more gravely than the upper; the muscles are flaccid, and soon begin to waste; and the paralysis is mostly of sudden onset, attaining its highest degree almost at once. Tested by faradism, the excitability of the paralysed parts is found to be much diminished or quite lost; and the galvanic current soon shows the reaction of degeneration to be present in greater or less degree, the muscular contractions being slow, and either equally responsive to both poles or reacting more readily to anodal than cathodal closure. When the legs are affected, the knee-jerks are lost; and there is nearly always surface coldness, and often an appearance of venous congestion over the affected parts, with a tendency to chilblains and ulcers and ready reaction to injuries with slow healing. The functions of the bladder and rectum are usually not affected; but in one extensive case, among others, that I have seen, in which motor power was lost in all extremities and in the neck muscles as well, there was complete incontinence of urine and fæces for a week, perfect control over the sphincters not being regained until another week had passed, by which time the neck muscles had recovered and the paralysis had almost disappeared from the arms. The legs remained paralysed and wasted in different degrees. Anæsthesia is rare, occurring only as an occasional complication, and pain and tenderness, though not seldom present, are not often prominent and' never prolonged. In many cases where little or no improvement is shown the whole limb is stunted in growth; and there are also various deformities owing to the stretching of the weakened muscles by the weight of the limb, and to unantagonized contraction of those which are unaffected.

Sometimes but one limb or certain groups of muscles, at others several limbs or even the trunk or neck or the muscles of respiration may be involved; and probably in most cases the paralysis is greater in extent at first than afterwards. Some improvement usually sets in very soon, and one or more limbs or groups of muscles may rapidly recover while others remain partially or wholly paralysed. Recovery more often begins in the arm than the leg when both are paralysed. In fourteen cases in my wards, all more or less advanced or incurable, of which I have full notes on this point, both legs were affected, though unequally, four times, both legs and one arm four times, one leg four times, one arm once, and an arm and leg on the same side once. The detailed records of others, dealing with extensive statistics, and my own unrecorded experience of many cases formerly seen as out-patients, sufficiently attest the predominance of leg paralysis.

Occasionally, as in two of my noted cases and in a few others that I have seen, the paralysis begins insidiously and gradually increases; it may also spread after a while to fresh parts after the manner of what has been described by some as the chronic or subacute form of this affection occurring in adults. Dr. Hughes Bennett and Professor Erb have published instances of this,[1] and it is probable that such cases may be from time to time overlooked. One of Dr. Bennett's cases recovered, but the prognosis is said to be generally bad in this chronic form. It is, moreover, very doubtful whether cases of this nature are pathologically identical with the acute form of the disease under consideration. I have seen several instances, exclusive of hospital in-patients, which answered to the so-called "temporary paralysis of childhood," where perfect recovery took place in the course of three or four weeks; and a few, seemingly typical, which got well in a few days. It is, however, quite open to doubt the diagnosis of such cases; for, without wasting, coldness of surface or the electrical test, important elements of distinction are wanting. It must be remembered that, even with cases in other respects typical, the electrical test, otherwise so valuable, and especially the galvanic current, is extremely difficult both to apply and interpret in young children.

Frequent as the cases are which correspond more or less closely to the received description of infantile paralysis, it must be admitted that we are much in want of more accurate observation, both as to the clinical conditions and modes of onset, before we can form a perfectly definite conception of the disease as due to one and the same pathological process. In private as well as in hospital practice most patients escape trained observation at the outset, our chief reliance for information as to the beginning of the attack being perforce placed on more or less vague and imperfect reports; and, as we shall presently see, the ætiological knowledge we have from morbid anatomy, highly important though it be, is mainly based on the examination of long-standing cases.

The following are the chief **conditions** and **accompaniments** of the onset of cases answering to the usual description of infantile paralysis.

Pyrexia of duration varying from a few hours to several days, and unassociated with any distinctive symptoms, very frequently precedes the discovery of the paralysis; and some very exceptional cases of a severe and even fatal character, accompanied by convulsions and sweating, have been reported as occurring in an epidemic form.[2] The typical paralysis also occurs in the course or as the sequel of measles, scarlatina and other specific diseases. In three out of sixteen recorded cases in my wards there was a definite history of an initial feverish attack of a few days' duration; in a fourth, of severe vomiting the night before the paralysis was noticed; and in a fifth the paralysis appeared on the

---

[1] See *Brain*, vol. vi., 1883.
[2] Cordier, *Lyon Médical*, Jan. and Feb. 1888.

third day of scarlet fever. I have also in my memory several cases following directly on measles; and one, of an apparently complete although mainly temporary character (one leg only being affected and almost recovering within three weeks), which occurred in a child of four years old who, with three others, was just convalescent from influenza. It is possible and, according to some, highly probable, that unnoticed fever may frequently occur.

Convulsions with or without fever are often reported as ushering in the paralysis, and sometimes there is a semi-comatose condition at the onset. I have seen two cases, in one of which the paralysis followed on protracted convulsions recurring during a fortnight, and in the other immediately on a single fit. *Pain and tenderness* on pressure in the affected limbs and, indeed, sometimes in the body generally are not very rarely reported, but seldom last long. Doubtless considerable general pain exists at the outset in some cases; and this symptom, especially marked in the back, is very prominent in many instances of the disease in adults.[1] My experience, however, teaches me that marked pain and especially tenderness, evinced mainly or entirely on movement of the affected limb, are valuable diagnostic signs, pointing away from the affection we are considering. *Falls*, or *blows* on the back, are often quoted as exciting causes; but the frequency of these accidents in early childhood renders a causal nexus even more doubtful here than in the cases of antecedent exanthems. In four of my sixteen cases above quoted a fall, and in one, a blow on the back, was the alleged cause; in one, however, a feverish attack of two days' duration intervened between the fall and the paralysis, and in another there were convulsions; while in one only, where after a severe and stunning fall an eventually typical paralysis was observed on the child's regaining consciousness, could the connexion be deemed with much probability other than coincidental. "*Catching cold*" and *rheumatism*, properly so-called, are among the alleged or suggested conditions out of which this paralysis arises. There appears to be some evidence sometimes of a rheumatic connexion, either in the personal or family history of the patient; and doubtless many recorded cases have followed immediately on a definite chill. A child of two years old, whom I saw several times, had a typical paralysis of paraplegic distribution a few days after suffering from severe leg pains immediately following a prolonged sitting on wet grass. In this context we may bear in mind the vulnerable condition of the nervous system in early childhood, and the vascularity of the cervical and lumbar enlargements of the cord, which are the chief seats of the lesion found post-mortem. *Over-exercise* has seemed occasionally to be the precursor of

[1] See reports of two cases of my own in vol. ii. of **Westminster Hospital Reports**, 1886.

the paralysis, and we may remember here that the disease is probably rare in the first year of life and frequently attacks robust and active children. Caution, however, is required when attempting to draw any conclusion as to the nature of a case from consideration of age, for the younger the subject the more difficult is accurate diagnosis. Lastly, *heat* has been suggested as one of the conditions of the affection, the majority of cases, according to most authorities who have collected large numbers, occurring in the summer months.

No alleged or probable excitant is found in a large number of cases. I have seen several where it was positively stated that the paralysis came on suddenly when the child was in perfect health, and we frequently hear of children being put to bed well and found paralysed in the morning. In this context we must remember also the chronic cases above mentioned, where weakness, at first scarcely noticed, gradually develops into typical paralysis with wasting and the reaction of degeneration. Such cases as these, and especially the sudden ones, are very striking, and must clearly be reckoned with when we endeavour to formulate a rational ætiology.

Such being the apparently multiform conditions and modes of origin of this affection, the lesson taught us by **necropsies** of many old-standing and a few recent cases is a valuable contribution to the question of ætiology. It is almost certain that in the typical cases with wasting the necessary lesion is more or less destruction of the large ganglion-cells in the anterior cornua of the spinal cord, and that the morbid process, though it be found in some degree in the whole length of the cord or may even involve other columns as well, is especially concentrated in the cervical and lumbar enlargements. Corroborative evidence of this is afforded by cases in adults with similar symptoms and post-mortem lesions. In most necropsies there has been sclerotic overgrowth of the neuroglia in the anterior cornua, more or less invading the antero-lateral columns or other parts as well; but there is but scanty evidence from necropsies in recent cases as to the nature and cause of the primary lesion. Occasionally, and notably in Dr. Charlewood Turner's case at the London Hospital examined six weeks after the onset of the paralysis, hæmorrhagic foci have been found, involving the special regions above mentioned. In this case, however, there was anæsthesia as well as motor paralysis of the lower parts of the body, and the lesion was found to have invaded the posterior columns in the lumbar enlargement. In a case of my own, aged 2½ years, at Westminster Hospital, sudden, permanent and typical paralysis of the left leg, followed in a fortnight by paralysis of the right leg nearly recovering after seven weeks, preceded death from tubercular meningitis by about eighteen weeks. Post-mortem examination showed, besides the cerebral meningitis, marked wasting of the left

R

anterior cornual region of the cord, especially in the lumbar region, where
the cells were replaced by fibrous tissue; and the left sciatic nerve was
smaller than the right and contained a large quantity of withered tubules.
There was also marked affection, observable by the naked eye as well as
by the microscope, of the anterior cornual region all down the left side ;
the cells in the dorsal region having disappeared, though there had been
no marked symptom during life of affection of parts above the lumbar
region. The multipolar cells in the right side of the cord were also
much diminished in number.   The specimens were shown by Dr. Hebb
at the Pathological Society of London in April 1889.

As regards the **ætiology**, then, of a large number of cases of "infantile
paralysis," we are on firm ground when we assume that the primary
cause of the symptoms is disease of the large ganglion-cells in the anterior
cornua of the cord, which are known to preside over both motor and
nutritive functions.   In endeavouring, however, to connect the various
modes of onset with the cord lesions the question of the nature of these
lesions, whether primarily hæmorrhagic or inflammatory, at once meets
us.   The hypothesis of a primary wide-spread and acute inflammation
involving a large extent of the cord would cover all those cases which
begin acutely with fever and extensive paralysis, including those which
are marked by sensory symptoms, the fever being regarded as symptomatic; while the partial improvement and remaining weakness of limbs
would be accounted for by the predominant and permanent affection
of the cervical or lumbar enlargement of the cord.   On the other hand,
the cases which begin suddenly, unmarked by any other symptom than
paralysis of a limb or limbs, including those which may be possibly
due to traumatic causes and over-exertion, would be better explained by
such a lesion as hæmorrhage into the relatively vascular region of the
anterior cornua.   It may be suggested, too, that the changes in the
blood and vessels induced by the febrile process which, whether
specific or not, so often precedes the paralysis, are predisposing causes to
hæmorrhage.   We are not, however, in a position, owing to our very
imperfect knowledge of both the clinical and anatomical phenomena
at the outset, to formulate a comprehensive theory of the causation of
this disease, if indeed it be always one and the same.   We must remember that the assumption of a primary poliomyelitis leaves much of
the ætiological question unsolved ; while that of a primary hæmorrhage,
though harmonising with a far greater proportion of all the known
clinical phenomena and therefore possessing great claims on our consideration, is but little supported by the teachings of morbid anatomy,
however much we may be inclined to insist upon the probably greater
liability to such an accident of the actively growing cord of childhood.

While we are justified, therefore, in regarding infantile paralysis as generally due to disease of the spinal marrow, we must remember that it is almost impossible to exclude peripheral neuritis as a cause, especially in cases accompanied by pain aggravated on movement and by much tenderness in the affected limbs, which may increase gradually from the beginning. The same atrophy and altered electrical reactions result from both cord and nerve lesions. From what we know of peripheral neuritis, however, we should mainly suspect it as a cause in childhood in cases where a single limb or group of muscles is affected. As possibly bearing on the question of the spinal pathogeny of this affection I would call attention to two cases of infantile paralysis, reported by Dr. Coutts, where there was swelling of the ankle-joint in the affected limb nearly coincident in time with the onset of the paralysis. Dr. Coutts is inclined to connect these cases with the recognised spinal arthropathies described by Charcot and others.

The **diagnosis** in most cases seen some days or weeks after the onset is fairly clear ; but, before atrophy or coldness of the limbs is pronounced, or when the electrical test is either inapplicable or gives doubtful or nearly normal results, the difficulty is often very great. The fever which frequently precedes attacks has no distinctive marks, so that we depend for diagnosis on the paralytic phenomena alone. Cerebral paralysis will usually cause no diagnostic confusion to those who are acquainted with the well-marked contrast presented by these two affections ; and it is equally unnecessary, in my opinion, to detail, after the manner of many authors, the striking differentiæ between infantile paralysis and other forms of well-marked nervous disorder.

The practical difficulties of diagnosis occur where children, otherwise healthy, are more or less suddenly found to be suffering from inability to move a limb ; and I have several times seen this condition confidently diagnosed as infantile paralysis by medical observers when it was due to an accidental *muscular strain* or, occasionally, to *periostitis*, syphilitic or otherwise. Pain on pressure or on movement of the apparently paralysed limb does not, as we have seen, at first exclude the diagnosis of infantile paralysis ; when, however, it is marked and continued we need scarcely ever express the fears we may entertain, but may with considerable confidence give a good prognosis while enjoining perfect rest for the affected limb. Such strains as these are often caused by careless handling of a baby, and the leg or arm may hang almost as motionless and flaccid as in a typical case of infantile paralysis. *Hip-joint disease* need only be mentioned as causing frequent diagnostic mistakes which careful local examination should always preclude, even apart from the important symptom of the knee-jerk, which is absent in the paralytic disorder. *Peripheral paralysis*, affecting a large nerve from stretching, pressure or other causes, especially when one group of muscles, such as the peronei, are mostly or alone affected, and unaccompanied by much pain, may give rise to symptoms like those of the spinal disease; and in such cases, when we are unable to trace a distinct injury, the diagnosis should be postponed for a while. The palsies which date from birth and are due to intra-uterine causes or to obstetrical manipulation or instruments are usually referred with case to their true origin.

Prognosis depends on the amount of improvement made after the first week, when the paralysis has reached its fullest extent, until about the fourth or fifth month. The nature of the onset, whether accompanied or not by fever, gives no help to our forecast of the ultimate result. Limbs or groups of muscles which are limp and powerless at the sixth month from the onset will probably remain so for ever, and it is a rule almost without exception that muscles which quite fail to react to faradism applied to the nerve after six weeks from the attack may be regarded as permanently paralysed. Much wasting at any period is of very bad prognosis. For the first six weeks, however, after the attack, during which period the paralysis often remains nearly or quite stationary or its retrogression may be extremely slow, we must not give too grave a prognosis; for great and rapid improvement or, occasionally, even complete recovery may subsequently result.

Wasting is sometimes much obscured by a quantity of subcutaneous fat. In these cases especially the electrical tests both of failure of reaction to faradism and of too ready or altered reaction to galvanism are of great value. We must remember, however, that during the first six weeks we may find marked reaction of degeneration in muscles which subsequently improve or recover. After six months or so, when deformities appear owing to stretching of the paralysed muscles from the weight of the limb and to unbalanced action of the healthy ones, or when stunting of the limb is established, we know that little can be hoped for but some measure of mechanical or surgical relief. There are a few cases on record seemingly showing that improvement may sometimes occur with treatment, even after many months of an apparently stationary condition with no power of voluntary movement or faradic contractility.

Treatment in the earliest stage is rarely practicable for the reasons aforesaid; but when possible, or in cases where the disease is suspected, the patient should be kept absolutely still, and leeches or cupping-glasses may be applied along the spinal column, or blisters may be used in the same region. Later on, when the diagnosis is clear, the child should still be kept in bed, sedatives such as the bromides being administered in case of much restlessness. After six weeks, faradism, with a current just strong enough to produce contraction, may be used to the affected muscles which react at all, and the interrupted galvanic current to those which fail to respond to faradism. Ten minutes at a time, once or twice a day, is enough for either of these applications. It will be found, however, in most cases impossible to proceed with the galvanic current, owing to the great pain it so often causes in children, and even the faradic current has not seldom similar and other practical drawbacks. I believe that passive movements of the affected limbs, with thorough and daily repeated shampooing with the oiled hand, all care being taken at the same time to preserve continuous warmth to the limb by clothing and hot bottles in the bed, are not only more practicable but also more efficacious in every way than the electrical treatment. At any rate, though I have often ordered the manipulative method without the electrical, with, I think, beneficial results, I would never recommend the electrical alone. Whichever line be adopted, treatment should be persevered in for at least a year.

The general nutrition and hygiene of the child should be carefully attended to, and such medicinal "tonics" as iron, strychnia, arsenic or cod-liver oil may be required in some cases from time to time. Every encouragement to use the weakened limbs should be given, and much can be done in this direction by means of artificial supports, go-carts, wheeled chairs, and other mechanical devices.

Chronic Paralyses with Atrophy of Muscles.

Under this very general heading I include certain groups of cases of marked weakness with wasting of muscles, which begin almost always in childhood or early youth and are characterized by insidious onset and slow progress. In some, with apparently stationary periods of considerable duration, there is a tendency to death before adult age from general wearing-out or from some intercurrent disease; in others there is sometimes comparative or complete arrest. The best marked group with fairly distinctive clinical characters is known as "pseudo-hypertrophic" (or Duchenne's) paralysis; another, smaller and less clearly definable, may be classed generically as "progressive amyotrophy," and diversely specified as differing from the well-known type-form which begins, as a rule, in the small muscles of the hand. Of this I shall allude to two species known respectively as the "peroneal" and "juvenile" forms. This grouping is, however, exclusively clinical and confessedly imperfect, and by no means implies any conclusion or theory as to the pathology of these affections; for, although there may be reason for regarding some as due to primary disease of the muscles and others to disease of peripheral nerves or cord, it can scarcely be said that an exact pathological ætiology has been established in any; and I shall not enter here into any discussion of the differential diagnosis of myelopathies, neuropathies or myopathies.

In most of this probably heterogeneous class of affections there is usually believed to be a strong tendency to some kind of hereditary transmission.

Pseudo-Hypertrophic Paralysis.—This affection has long been recognised from its marked clinical characters, which are fully described in the text-books and in Gowers' well-known monograph. It begins almost exclusively in childhood, showing itself at first by a weakness of the lower extremities, which causes the child to stand with his back arched forwards, to straddle and sway in walking, to fall down readily, and to rise from the floor and mount stairs with difficulty; later on, by apparent enlargement of certain muscles and wasting of others, with a tendency to "pes equinus" from contraction of the calf-muscles or weakness of their opponents, and by marked aggravation of all the symptoms; and, finally, by inability to walk or stand, with extended wasting of muscles, including those which were previously enlarged. The paralysis is apparently proportionate to the wasting, and is not necessarily most marked in those muscles which are the seat of enlargement. The general health is usually good, and the symptoms may remain stationary for long; but the disease, once clearly established, is perhaps always progressive, and very few cases live to adult age. Death usually results from exhaustion or intercurrent disease; but our knowledge on this point is scanty, as most cases are lost sight of by their recorders long before the end, and, from loss of their characteristic symptoms, are probably unrecognisable by those in whose charge they die. One of my own well-marked cases, dismissed in an apparently stationary condition, was re-admitted under a colleague more than two years afterwards without a trace of muscular enlargement, quite incapable of any movement owing to excessive wasting of muscles, and with universal rigidity which suggested disease of the antero-lateral columns of the cord. Without the previous history, the nature of this case would entirely have escaped notice. The boy died from chest mischief, but a post-mortem was not permitted.

This disease may be suspected in the early stage from the straddling gait, the tendency to fall, and a certain difficulty in rising from the ground; the characteristic manœuvre of the patient's climbing up his legs, to assist the weakened extensors of the knee, hip and spine, being often postponed even until after some pseudo-hypertrophy has shown itself. In this early stage a marked hardness of the muscles which may afterwards become enlarged, and especially those of the calves and buttocks, may sometimes be observed and is an aid to diagnosis. But I must remark here, as bearing on the question whether this affection be ever recovered from or arrested in an early stage, that I have seen several cases, some of them after measles or other febrile diseases, where symptoms such as these, irresistibly reminding one of pseudo-hypertrophic paralysis, have lasted for weeks and completely disappeared, and others which have neither progressed nor shown any enlargement of muscles. I have seen, besides, a boy of six years old who had been slow and inactive for six weeks and unable to run for one week. There was marked "lordosis," a straddling gait, inability to rise from the ground without touching the knees, and notable enlargement of the calves and buttocks. The skin of the legs was mottled, and the knee-jerk was absent on the right and very slight on the left side. After six weeks, during which time I had not seen him, all these symptoms were much less, with the exception that both knee-jerks were now absent. A fortnight later he was quite well. I saw him finally six weeks after this, when he was still without symptoms, the knee-jerks were both present, and he walked and ran with ease.

In the established disease almost all the muscles of the body, including sometimes those of the face, may be more or less enlarged; but as a rule the most striking in this respect are the calf and buttock muscles, the spinati, and the deltoids. The pectorals, latissimi and other muscles of the upper part of the body often waste both early and rapidly. The skin is most often mottled over the affected parts, especially after prolonged exposure. It has been often stated that the surface temperature of the enlarged parts is higher than that of the rest of the body; but I am convinced, from repeated and careful observations in four well-marked cases, that this is a mistake. The simple precaution of eliminating the disturbing element of different periods of exposure by altering the order of observations sufficiently establishes the error of the prevalent statement on this point. The knee-jerks are said to be lost only in the later stages. I found them present in one case of seven months' standing, the boy having been quite active till $6\frac{1}{2}$ years old; absent in a boy of 9 who had difficulty in walking from the first, when he was 15 months old; and present, though slight, in a boy of 10 who had been quite well till 3 years old. Faradic reaction in the nerve-trunks is normal till very late, and then only diminished, owing probably to degeneration of the nerves; in the muscles it is always lessened from the resistance caused by the morbid fibrosis. Galvanism to the nerves does not show much change till late in the disease; but to the muscles produces a reaction diminished in proportion to the wasting of muscular fibre, and there may be a slight degree of the reaction of degeneration. In two of my cases I found that the faradic reactions in affected parts were altogether normal, one being of seven months', the other of seven years' duration; while in two more of several years' standing there was diminution of reaction to faradism both in the nerves and muscles of the legs, and scarcely any difference between the anodal and cathodal closure contractions with the galvanic current. All my cases, including several doubtful ones and the probable instance of recovery, were boys; and in none was there any ascertainable trace of heredity, direct or collateral. In many reported cases, however, it has been found that the disease occurs frequently in the same family, being apparently transmitted through unaffected mothers to their sons. Mental deficiency has been often noted. In one of my cases some obtuseness was reported from school, though the boy appeared quite intelligent in hospital; and in another speaking was delayed until the age of five. In the uncomplicated disease no sensory abnormality or loss of control over bladder or rectum finds place.

The clinical disease is certainly not congenital of necessity, but that there may be a latent congenital tendency is a hypothesis which cannot be disproved. In most cases no exciting cause can be found, and I regard as merely accidental the apparent development of one of my cases, aged 7 years, out of a definite attack, following on a fit and fever, of infantile paralysis of the left and, slightly, of the right leg five years previously. The right leg recovered from the paralytic attack completely, and the left to a great extent; but on admission there was marked lessening in length and bulk in the whole left lower extremity, where the surface temperature was persistently about five degrees lower than over the rest of the body. In this case there was marked wasting of the pectorals and deltoids, although the "spinate" and "triceps" muscles were bulky; while the left calf as well as the right increased by $\frac{3}{4}$ inch in girth during six months' stay in hospital. Two years afterwards the boy could not stand, his deltoids were powerless, and his buttocks wasted; but the triceps muscles and the calves were still very large.

The great weight of recent opinion on the *pathology* of this disease is on the side of its being a primary myopathy; although in some cases more or less wasting of the ganglionic cells in the anterior cornua of the cord and other morbid appearances have been found. In a case brought by Dr. Handford before the Pathological Society in 1889 there were the usual muscular changes, with degeneration of the diaphragm and heart-muscle, and marked degeneration and atrophy of the peripheral nerves of the affected parts, which, considering the absence of sensory symptoms, probably affected the motor nerve-fibres only and were secondary to the muscular degeneration. The cord was mainly healthy, although there was occasional slight degeneration of the ganglion-cells in the anterior cornua, and also an area of softening in the lumbar enlargement apparently due to hæmorrhage.

The muscular changes consist in great increase of connective tissue between both the bundles and the fibres themselves. There may be also an early overgrowth of fat. As the disease progresses much fat is deposited in the newly-formed connective tissue, and the muscular fibres are separated, atrophy, and disappear. Finally all muscular structure may be obliterated, and the newly-formed fibrous tissue degenerates, leaving nothing but fat and connective tissue; and the fat itself may at last disappear. The morbid process is essentially the same in the muscles which waste from the first and in those which have a stage of enlargement, and the pseudo-hypertrophy is mainly due to fibrosis.

The *diagnosis* of this disease is generally easy except at the beginning. It may occasionally be confounded with some of the rarer forms of progressive atrophy in children presently to be noticed, and, as I have often observed and as Ross points out, with various cases of retarded development and want of co-ordination, with or without signs of cerebral disease. I would especially remark here that the generalised forms of paralysis, not seldom seen after diphtheria and probably some other diseases, give rise to some of the symptoms which many thoughtlessly regard as pathognomonic of pseudo-hypertrophy. I have more than once seen the action of climbing up the legs, as well as the waddling gait, in clear cases of diphtheritic paralysis which perfectly recovered, where doubtless the same muscles were weakened as in the disease we are considering. A doubtful diagnosis may possibly be cleared up by extracting a small piece of muscle with one of the instruments invented for the purpose and examining it microscopically; but even when repeated, as it must always be, such a procedure is generally unsatisfactory in its result and not without risk, nor is it a worthy cause for administering the anæsthetic without which the operation is inadmissible.

The only hope for *treatment*, ignorant as we are at present of the cause of the disease, lies in the possibility of aiding any natural tendency there may be to arrest or recovery. I have seen cases improve considerably in hospital with good food and tonics, but one at least of these made subsequently steady progress downwards. Duchenne says that he arrested two early cases by faradism; but, in the light of the case quoted above of recovery under ordinary tonic treatment from symptoms which seem at least very suggestive if not identical with those of the recognised disease, such a statement is of little value. I would recommend in all early and all doubtful cases the best hygienic conditions, regulated exercise, shampooing of the affected limbs, and arsenic, iron, cod-liver oil, or a combination of these, as possibly helpful towards improvement or cure.

The Peroneal Type of Progressive Amyotrophy.—This affection has been clinically isolated of late years, and cases in point have been described by Leyden, Ormerod, Charcot and Marie, Tooth, Herringham and others. I showed at the June meeting of the Neurological Society in 1890 three cases in one family, recording them in *Brain* (vol. xiii. p. 456); and would refer the reader especially to Dr. Tooth's paper in the same periodical (vol. x. p. 243), and in the St. Bartholomew's Hospital Reports (vol. xxv.) for a full account of this class of cases. In established cases there is wasting and weakness of the legs, often with markedly diminished sensibility; and the feet are in the "varus" position. At least in the later stages there is absence of reaction both to faradism and galvanism. Subsequent to the leg affection there is wasting, with or without anæsthesia, and loss of electric contractility in the small muscles of the hands which ultimately assume the claw-like appearance of ordinary progressive muscular atrophy. In some early cases there are pains with tremors and unaltered electrical reactions, and in a subsequent stage there is the reaction of degeneration. Heredity, though often marked, is not always present. Dr. Ormerod's cases and my own were both groups of three, consisting of father, son and daughter; and the symptoms in all, save the father in the latter group, set in immediately after an attack of measles. The disease seems always to begin in childhood or early youth, and probably affects first the small muscles of the foot. Muscles besides those already mentioned may subsequently suffer; but there seems to be no doubt that many cases are arrested, for at least an indefinite time, after the legs and forearms are affected. In all my own cases the muscles of nearly all the rest of the body were normal in every respect, the girl, aged 17, affected since the age of 7, having been no worse for long, and the father, aged 42, first affected at 17, having been in a stationary condition for many years. The gait in all these cases was high-stepping (not spastic, as at first sight appeared) owing mainly to foot-drop from peroneal weakness; and the knee-jerks were present, with no ankle-clonus.

The *pathology* of this affection is not certainly known. On the whole, considering the at least occasional presence of fibrillary tremors in the early stages, the occurrence of pain in some cases, and the anæsthesia in many, a primary peripheral neuropathy seems perhaps to be indicated. Some regard the lesion as situated in the cord, and others as starting in the muscles. Too few necropsies have been published for the establishment of any anatomical point of ætiological importance. The immediate antecedence of measles in the cases above alluded to is to be remarked in connexion with other forms of paralysis apparently excited in the same way; but it can only be regarded as a favouring condition for the neuro-muscular breakdown which constitutes the disease. There is a general clinical likeness between this affection and pseudo-hypertrophic paralysis; but the pathology is even still more doubtful, and nothing of value can be said on the matter of treatment.

The "Juvenile" Type of Amyotrophy.—Of this form, described by Erb, I shall say but little, having only, with one exception, seen cases observed and published by others. Dr. Savill showed some very good examples at the Neurological Society in June 1890. The symptoms of wasting and weakness attack first, and are often for long limited to,

the shoulder and upper arm and the buttocks and thighs; there is lessening or loss of the knee-jerks, no fibrillary tremor, and no reaction of
degeneration, but only diminished electric contractility. The disease is
said to begin in later childhood or early youth. In some cases wasting
of the face and tongue has been observed, and in others almost complete
loss of the muscles which extend the trunk on the thighs, producing
an extreme degree of lordosis. As with the previously described cases,
there is nothing as yet approaching to certain ætiological knowledge of
this affection, which has only clinical claims to a separate position.

"Diphtheritic" Paralysis.

In a considerable proportion, probably amounting to one-fourth, of cases
of diphtheria which recover, various paralyses are apt to occur which are
of a sufficiently special character to merit the clinical title of diphtheritic
paralysis. Although, however, this group of cases is known to us chiefly
by its relation to diphtheria, it must be acknowledged that instances of
similar paralysis are found where there is no history or evidence whatever of diphtheria, or indeed of any previous disease ; and that some seem
to follow on other acute affections. I have seen sufficiently numerous
examples of paralysis of the so-called diphtheritic type presently to
be described, with no ascertainable connexion with even the slightest
throat-affection, to convince me that it is an unjustifiable closure of
inquiry to argue back to a hypothetical diphtheria as the necessary
cause of all such affections; and I hold this opinion in the face of the
fact that there are many slight cases of diphtheria with indefinite or
undiscovered signs in the throat. Except in the case of recognised
diphtheria such paralyses, in my experience, are not often sequelæ of
throat-affections, but either are apparently idiopathic or follow on
measles, enteric fever or other recognised or nondescript febrile attacks.
Those who regard this group of affections as always diphtheritic are
constrained by facts to admit that they must take place in a considerable number of cases after not only the mildest, but also completely
unrecognisable, diphtheria ; and there are some who maintain that the
frequency of the paralysis is in inverse proportion to the definiteness
and severity of the causal disease. It is certainly clear, when we consider the great and early fatality of diphtheria, that the paralysis under
notice can but rarely have a chance of existence in the worst cases; for,
although it may sometimes take place very early in the disease, it is far
most frequently first observed in the second or third week, and often
much later, after the subsidence of the acute attack. With this practical
and, as I believe, clinically important comment I shall shortly describe
the paralyses in question under the conventional term "diphtheritic ;"

for most of them, if not a large majority, occur in connexion with undoubted diphtheria.

The chief mark of these cases is a varying degree and extent of motor paralysis, most often symmetrical in distribution, slowly spreading, and often tending to attack fresh parts while others are recovering. Hyperæsthesia of the paralytic parts is sometimes noted at the onset, but rarely in children ; and some degree of numbness and anæsthesia is present in most cases. The knee-jerks are mostly, but not always, absent, faradic contractility is usually lessened or abolished, and in extensive cases of long standing the reaction of degeneration may be detected by the test of the galvanic current. In many instances that I have observed, the paralysis, which is generally incomplete, is accompanied by tremors or movement of the limbs indistinguishable from those of insular sclerosis. With certain exceptions, which are grave and often fatal, these paralyses tend almost always to complete recovery in the course of weeks or, it may be, of many months.

The **ætiology** of these paralyses is uncertain. Changes have been found post-mortem in several cases both in the nerve-centres and the nerves, but in others these structures are apparently normal. No definite cause as yet suggested covers all the clinical facts. The wisest utterance on the subject is, I think, that of Bristowe, who says that " on the whole it seems probable that a wave, so to speak, of slight inflammatory or other morbid process slowly traverses the medulla oblongata and the cord, and in some cases also the nerves in relation to the paralysed districts." This hypothesis leaves room for the possibility of some action on the nerves of the absorbed poison of diphtheria, analogous to that of other nerve-poisons which may modify or destroy functions while causing no demonstrable change.

Paralysis of the soft palate is by far the most frequent form of the affection which can be demonstrated, and is alone observed in a large number of cases; but a notable weakness and infrequency of pulse with enfeebled heart-sound, pointing to disturbed cardiac innervation, is a very common accompaniment. Sometimes, though not very often, there is more extensive paralysis of the pharynx, and possibly of the œsophagus, causing much dysphagia and the necessity of nasal feeding. In such cases the pharyngeal mucosa is insensitive, and, the upper part of the larynx being sometimes similarly affected, food may enter the glottis and occasion cough. The paralysis of the palate is evidenced by the motionless velum and pendulous uvula, nasal voice, snoring, and the return through the nose of swallowed fluids. Faulty accommodation from paralysis of the ciliary muscles, shown especially in loss of near vision, is almost as frequent as the palatal weakness, and can be established in a large number of cases where no impairment of sight is complained of. The pupillary light-reflex is maintained, but may be sluggish. These three phenomena,—the *palatal*, the *ciliary* and the *cardiac* weakness,—associated as a rule with *loss of the knee-jerks*, are the most characteristic marks of diphtheritic paralysis, and are in my experience mostly preceded by recognisable diphtheria. The further events of paralysis of the limbs, of the neck and trunk muscles, and of the respiratory apparatus, including both the intercostals and the diaphragm, are seen in a smaller number of cases; and it is in this more extensive class that strabismus of all kinds, but generally double and mostly divergent, and, though much less often, ptosis, facial paralysis, and affection of the abductors of the vocal cords, are apt to occur. Although any or almost all of the voluntary muscles may suffer in turn, or to some extent simultaneously, so that the patient may be unable to move or raise the head, the paraplegic distribution (one leg being always worse than the other), is the most frequent; and the arms are hardly ever alone affected. The skin over the weakened parts is usually more or less insensitive ; and in some cases symmetrical patches of anæsthesia may be observed over the body, especially on the extremities and at the tip of the nose. Wasting of the paralysed muscles is often marked, the gait may be ataxic as well as straddling, and the feet tend to catch the ground. Very occasionally there is impairment or even loss of control over the bladder. With regard to the knee-jerks it must be remembered that, although usually absent, they are sometimes present; and that, while their absence characterizes a large number of cases of diphtheria without discoverable paralysis, they may occur and even be brisk in some where extensive paralysis supervenes. This phenomenon is, therefore, of no great prognostic value. In two cases of paraplegia which recovered, one with left ptosis, and the other with double internal strabismus, but without any history of previous diphtheria, the knee-jerks were present throughout. It is in such cases as these, however, which are usually classed as diphtheritic, and also in others where, with generally absent knee-jerks, the paralysis is mostly or entirely confined to the limbs and trunk muscles, that a clear history or other evidence of diphtheria such as palatal, ciliary or cardiac paralysis is very often wanting. It is in these doubtful cases, too, that rhythmical tremors and ataxia are most frequently noted, and that weakness of the legs and back may give rise to a gait and mode of rising from the ground which reminds us of the phenomena of pseudo-hypertrophic paralysis. I have seen several cases of this kind, and one where the paralysis was of a remarkably shifting or metastatic character and repeatedly invaded both facials and the limbs on both sides. There were rhythmical tremors and variable anæsthesia, and the knee-jerks were present throughout, sometimes feeble and sometimes exaggerated. There was no history of any throat-affection. The first observed symptom was nasal speaking, and recovery did not take place for nine months. In most instances of unquestionable diphtheritic paralysis there is albuminuria, the presence of which may assist us in the diagnosis of some doubtful cases.

With the¹ exception of cases marked by great slowing or by much frequency and irregularity of the heart's action, and of those extensive paralyses which involve the intercostals or the diaphragm, ultimate recovery may almost always be expected, and the sooner the more slight and limited the paralysis. Especially should we be on our guard against the more serious complications when one part after another is successively attacked. Paralysis limited to the palate or to the eye mostly recovers in a few weeks; but, when the limbs or trunk are affected as well, months may pass before complete recovery of any part. With intercostal or diaphragmatic paralysis, the chief marks of which have been mentioned under the head of diphtheria, the danger is very great; some cases dying quickly, and most after no long time, with pulmonary complications. I have seen pronounced intercostal paralysis once, and diaphragmatic paralysis three times, end in almost sudden death after the child had taken a few short gasps for breath. In such cases the pulse is usually very frequent, and death is the immediate result of cardiac failure. It is well in practice never to pronounce a definitely good prognosis until all signs of paralytic nature have disappeared for at least a fortnight.

In the **treatment** of all cases of proved or suspected diphtheritic paralysis absolute rest in bed must be enjoined, and the greatest attention given to nutrition. When there is pharyngeal paralysis, and still more when the œsophagus and the larynx are involved, food must be given by the nasal or œsophageal tube. Alcoholic stimulants should be frequently taken, and, when there is much cardiac weakness, large quantities may be necessary. Strychnia is perhaps useful, but should of course be given cautiously, beginning with small doses. For the limb paralysis a daily application of the galvanic current of just sufficient strength to produce response may at times be ordered. Artificial respiration, three or four times a day for a quarter of an hour, is strongly recommended by Dr. Pasteur in all cases of respiratory paralysis, especially of the diaphragm, for the purpose of preventing or lessening the collapse and other pulmonary troubles which so frequently occur. Faradism of the chest has been also advocated by Duchenne and others. Iron and arsenic may be given systematically in all cases.

CHAPTER III.

ACUTE DISEASES OF THE BRAIN.

CERTAIN symptoms of disturbed brain-function, mainly consisting of spasm, paralysis, headache and affections of consciousness accompanied by varying degrees of fever, are referable to involvement of the surface of the brain in inflammation of the pia mater, such inflammation being due to various causes and either of primary or secondary origin. There is sometimes considerable difficulty at the outset in deciding on the true cause of these symptoms ; for results at first sight very similar may follow on the temporary brain disturbance from that modified blood supply, generally hyperæmic or toxæmic in character, which is part of many febrile diseases. Of these, pneumonia and enteric fever may be quoted as examples. In the case of young children this difficulty is frequent, and has an important bearing both on prognosis and treatment.

Great irritability, dislike of light, convulsion of varying degree, vomiting, and even temporary strabismus and contraction of the pupils may all be the results of but temporary brain disturbance without inflammation ; and, although it is generally true that the more suddenly such symptoms arise the less likely is their primary cerebral origin, I am convinced that it is often impossible to form, and rarely wise to express, a definite opinion at the outset of any given case. I have seen several instances of fatal meningitis, including some of tubercular origin, which, at least clinically, have begun suddenly with convulsions and high fever. There are certain symptoms now to be referred to which, occurring early, may lessen or remove doubt, but these are often not marked till later ; and, generally speaking, when we can make an absolutely positive diagnosis of brain-disease, the case is already too far advanced for pathology to permit of other than a grave prognosis. The symptoms pointing either to great probability or approximate certainty of acute inflammation, which, as arising most often from the membranes, is usually described as *meningitis*, are local spasms or paralyses, marked inequality or inactivity of the pupils, great and increasing drowsiness, unrhythmical breathing, especially when interrupted by sighs or spells of "Cheyne-Stokes" respiration, and distinct irregularity in the force, frequency and rhythm of the heart-beats. These and other important symptoms occur, some or all, sooner or later in the course of acute brain mischief; and their special significance will now be considered in relation to the various forms of meningitis. Of acute general *encephalitis* with red or white softening, examples of which, though rare, are occasionally met with, I need say nothing, as it has neither a definitely distinctive symptomatology nor any characteristics special to childhood. With regard to localised *cerebral abscess* I would only remark that it is not very common, and has no special marks in childhood. Besides resulting from bone-disease, ear-mischief and sometimes broken-down tumours, abscess in the brain is known to be occasionally attributable by way of embolic processes to suppuration in the lungs, bronchi or pleura, the morbid material entering the left heart by the pulmonary veins.

Acute Meningitis.

The symptoms of meningitis are those of involvement of the surface of the brain and are signified by motor, sensory and mental changes. In most cases in childhood this affection has its origin in the pia mater ; and is the immediate result either of tubercle starting in the membrane or in the brain beneath, or of some other less certainly differentiated inflammatory or infective processes, among which it is customary to mention " simple meningitis." In others definite primary lesions are found outside the pia mater, such as disease of the cranial bones and, especially, otitis media.

There are certain cases, too, in which some symptoms of meningitis or, more strictly speaking, of cerebral trouble are seen, and nothing is found in those who die beyond an excess of fluid in the cerebral ventricles. Such are, however, as far as we know, most probably referable to an inflammatory origin, and may be termed *subacute hydrocephalus.* For clinical purposes certainly they will best be considered in connexion with meningitis. I shall describe first the tubercular form of meningitis, not only as the most frequent, but also because it affords the widest field for a detailed study of symptoms from its generally more gradual onset and slower course.

In a practical work confined to disease in children it is unnecessary to give any pathological description of the various forms of meningitis, except so far as the pia mater may be concerned ; and I will only say, with regard to the clinical aspect of a class of cases which are due to extra-cerebral lesion, and where the inflammatory effusion is primarily in the dura-arachnoid cavity, that the symptoms are often at first those of rigidity, convulsion and pain, which may continue for some time before coma and paralysis set in as evidence of more profound involvement of the brain. Several cases in my note-books of purulent meningitis of the convexity of the brain, following on definite otitis, illustrate this statement, which, however, is far from being universally true.

Tubercular Meningitis.—I have no doubt from my experience both in children and adults that this affection, though it may occur at any age, is by far most frequent under twelve years. I have seen necropsies of several cases of one year and under, such as are often regarded as very rare, owing, probably, to the small number of hospitals admitting babies and to the difficulty of obtaining post-mortem examinations in private practice. A considerable minority of cases have no family history of phthisis or other tubercular disease, and several begin suddenly with cerebral symptoms in the midst of apparent health. The well-known premonitory symptoms, described at length in text-books and monographs, and probably referable to the process of tubercle elsewhere than in the meninges, doubtless obtain in most instances ; but, after a review of over one hundred carefully noted cases, besides many others, in my hospital books, I can say that the cerebral symptoms are quite as often seemingly primary in infants as in older children. Post-mortem examination shows too that tuberculosis elsewhere, and especially lung-involvement, is as much marked in older children as in infants. An hereditary history of tuberculosis and premonitory symptoms are however, I think, but seldom both lacking in the youngest patients, though I have seen a boy of fifteen months who, with a good family history, was positively stated to have been quite well till the day after a severe fall, and after a short course of the illness was found post-mortem to be the subject of extensive tuberculosis affecting many organs besides the brain.

I mention these points merely to discourage the undue weight often given to clinical averages in the diagnosis of individual cases, and would emphasise the fact that in nearly all cases of tubercular meningitis the presence of tubercle elsewhere is proved by post-mortem examination. In very many of my cases measles, so often apparently the determining cause of tuberculosis, preceded the onset of the meningitis by a few weeks, the children having ailed during the interval ; and in some the affection seemed to follow directly on otitis. We must not, in fact, when making our diagnosis, trouble ourselves much more about the origin of this special affection than about that of any tuberculosis at any age, although doubt may very frequently be lessened or perhaps removed by marked evidence of family phthisis, by a definite history or indication of previous brain mischief (which may be due to tubercular tumours), or by the discovery of tubercular processes in other organs. In none of its varieties, indeed, does tuberculosis play a more striking clinical part as a specific disease from apparently *ab extra* infection than in some cases of cerebral meningitis.

It follows that, in the absence of any observable local causes such as ear-disease, nose-disease, traumatism, and of any conditions such as, for instance, the epidemic or sporadic form of " cerebro-spinal meningitis,"

S

pneumonia, erysipelas, or other febrile affections out of which a non-tubercular meningitis is known to arise, we may always more or less strongly suspect tuberculosis as the cause of any given case presenting the appearance of meningitis. We have to fall back for a more definite diagnosis upon the fact of tubercle being by far the most frequent cause ; on the usually more sudden onset and rapid course either towards death or, sometimes, recovery in other forms of the disease hereafter to be discussed ; and on the prevalent predominance in the tubercular variety of signs of basic inflammation, such as marked irregularity of pulse and breathing, and local paralyses of the external muscles of the eye, of the face, or perhaps of a limb, or of one side of the body including the tongue. Meningitis, however, mostly vertical, with convulsions and excitement rapidly passing into coma, is not confined to the non-tubercular forms ; and non-tubercular basic meningitis is certainly sometimes observed. Therefore, however sure we may feel of the presence of meningitis, we are rarely quite justified in giving the almost hopeless prognosis which definite diagnosis of the tubercular variety entails ; and, further, since pathological science gives some reason to believe that even a tubercular case may at least temporarily recover, we should postpone our fatal forecast until well-marked coma and paralysis set in.

The first symptoms of tubercular meningitis, preceded mostly by definite or indefinite illness of varying duration with restlessness, irritability, and some wasting, are, generally, vomiting, headache, and hyperæsthesia with or without photophobia. The warning hereby given is strengthened by the presence of some fever, and by the absence, on careful and exhaustive examination, of any evidence of pneumonia or signs of acute local or general disease other than what might be referable to tubercular growths in brain, thorax, abdomen or elsewhere. More definite cerebral symptoms soon arise ; and, if we desire to describe the affection by stages, it may be said that convulsions, sometimes often repeated, an irregular and most often frequent, though sometimes very infrequent, pulse, irregular breathing accompanied by a tendency to sigh, somewhat contracted and often unequal pupils and great irritability to sound or movement may precede, for several days or a week or two, the signs of profound impairment of brain-function which mark the remaining course of the disease and increase until the end. Such are various paralyses, inharmonious movement of the eyes, drowsiness going on to coma, dilated and inactive pupils, and, perhaps, an infrequent pulse becoming generally much more frequent again before death. But the more I see of meningitis and re-read the notes of my many cases, the less practical value I place on any division of symptoms into orderly stages ; for, apart from the innumerable exceptions to even such a general succession of symptoms as above sketched, and the frequently early appearance of paralytic phenomena, we must take into account the remarkable and well-known fact, which my own cases amply illustrate, of the complete disappearance of many or, perhaps, most of the symptoms at even a very late period. Among other very striking examples of this I have notes of a child four years old who had had typical symptoms, and was for six days unconscious of her surroundings and irresponsive to all stimuli of touch, light and sound, but suddenly woke up for a period of four hours, moving her limbs and audibly asking to see her mother, and died in coma about two hours afterwards. I shall therefore refer seriatim to the various symptoms seen in tubercular meningitis, noting their diagnostic significance and usual clinical position, and remarking, by the way, that none absolutely differentiate it from non-tubercular affections.

Vomiting occurs at the onset in most cases, and may last for several days. It is rarely, if ever, absent in the ordinary form of basilar meningitis. When primary gastric disturbance can be with probability excluded, and there is no diarrhœa, or still more when there is constipation, this should always be regarded as a possibly grave symptom. *Headache*, with some photophobia as a rule, either complained of or inferred from the child's appearance or actions, is probably of almost universal occurrence ; when persistent, it adds much to the probability of the cerebral significance of vomiting, and, in cases where its existence is certain, it is one of the most important facts in early diagnosis. Of these two symptoms, vomiting usually ceases early in the disease, although it sometimes persists ; while headache most often seems to last until signs of coma appear. There is nothing very characteristic in the nature of the vomiting itself, which may indeed take place, as I have frequently noticed, only after swallowing food or drink. The headache is often frontal, but more frequently quite general. The oft-quoted distinction between cerebral and gastric vomiting on the ground of the former being sudden or "projectile" and unattended by nausea is, in my opinion, of no great value, especially in young infants. When the time has passed for the appearance of characteristic eruptions in the exanthemata, and pneumonia has been excluded as far as possible, any suspicion of enteric fever, which often begins with severe headache and vomiting, is much lessened by the presence of one or more of the following symptoms which are more or less distinctive.

Irregularity of the pulse and breathing, and especially of the latter, are valuable diagnostic signs. Irregular breathing, accompanied by marked sighing, or by periods of more or less marked Cheyne-Stokes respiration, persists as a rule to the end, the pulse, however, varying much both as regards irregularity and frequency, and always becoming regular in proportion to increase in rate. In the early stages of tubercular meningitis, especially in infants, these signs may be absent or at least occur only in spells. Infrequency of pulse, often a valuable indication of primary brain mischief in febrile conditions of adults, is, in my experience, but rarely observed early in the meningitis of young children, the majority of my cases showing either acceleration throughout, or only a late slowing, followed usually by much increased rate before death. *Constipation* is the general rule, and is mostly accompanied by a *retraction of the abdominal walls* which, after a while, may become extreme, and give rise to the well-known "boat-shaped" belly. In a considerable minority of cases, however, there are loose motions or diarrhœa, the abdomen being then almost always either of normal appearance or distended. *Inelasticity of the skin* of the abdomen, with a distinct *feeling of doughiness* on palpation, is unquestionably very frequent, especially in the later periods of the disease; but it is in no way confined to tubercular or even to cerebral cases, being fairly often seen in an extreme degree in alimentary disorder with much wasting, and especially in enteric fever. I have long believed, from clinical and post-mortem experience, that all abdominal signs and symptoms other than the prevalent constipation and retraction are mainly dependent on the abdominal tuberculosis which so frequently precedes and increases with the meningitis. Tubercular meningitis is extremely often the last event of general tuberculosis; and it is mainly in the non-cerebral symptoms and in the history and course of individual cases that we must search for the diagnostic points which distinguish the tubercular from other forms of meningitis. *Cervical opisthotonos* is not infrequently seen, but it is usually somewhat late in appearance and referable either to considerable ventricular effusion or to meningitis of the cervical cord. As an early symptom it rather points to a non-tubercular meningitis, and especially to the cerebro-spinal form. *Nystagmus* is very frequent in the later stages of the disease, but has no special diagnostic significance; and the same may be said of *grinding of the teeth* which often occurs quite early. *Convulsion* is by no means rare even as an initial symptom, but is not distinctive even of cerebral disease unless definitely one-sided, and not always then. When frequently repeated, as it not seldom is, until coma sets in, it often points to an involvement of the convexity. *Rigidity of limbs* of varying degree and duration is frequently observed. *Paralyses* of the cranial motor nerves, especially of the sixth and less often of the seventh, are valuable signs of basic meningitis, and, therefore, most often of the tubercular form; as also is definite hemiplegia, which may be due to softening of the brain ganglia so often seen post-mortem in connexion with softening of the commissures and intra-ventricular effusion. The paralyses in tubercular meningitis are often varying and evanescent in character. More or less *rhythmical* or *repeated similar movements* of arms or legs are of very frequent occurrence in meningitis, occurring generally in the later stages with increasing drowsiness. These movements are not of convulsive character, but are, in appearance at least, voluntary or the result of some feeling of irritation, and may continue for hours together. *Bright and long-continued flushing of the face* and other parts of the body, or "tache cérébrale," either on movement only, or following light touches with the finger, or sometimes permanent with fluctuations in degree, is often seen, and is usually a sign of advanced disease, as also is the mucoid *film* on the conjunctivæ so frequently seen to a marked extent shortly before death. Even a slight amount of this flushing at the outset or quite early in the disease may increase already-formed suspicions of meningitis, but I have many times seen it well-marked in fevers, and cannot regard it as such an important aid to diagnosis as I once was inclined to do.

The temperature is raised more or less throughout in most cases, ranging as a rule between 100° and 102° or 103°; it is subject to great variations in the same and different cases; but, generally speaking, it ranges higher in infancy than in later childhood. Although it may sometimes be normal or even markedly subnormal before death, my experience is that a great rise at the end, often as high as 106° or 107° and sometimes higher, is very much more frequent, even in cases which have previously had a low febrile register.

The state of the pupils in the early stage is often not characteristic. They are, perhaps, mostly unequal in some degree, often contracted when the patient is irritable or in the early convulsive stage, and almost always dilated and immobile later on. Marked inequality of pupils is doubtless a valuable positive symptom of brain mischief, but I have several times observed continuous equality even in severe cases.

Hyperæsthesia, including photophobia, is a very marked symptom in most cases even in the earliest stages, and may last for many days. Movement is resisted, often with screaming, and in a few cases and most often at night the so-called "hydrocephalic cry" is suddenly heard. I am not sure whether this may not be referable to terror as well as pain, night terrors with hallucinations and very often with screaming being frequent in the notes of the older cases, and occurring quite early in the disease. Sudden crying out, too, is sometimes referred by the sufferer to pain in the epigastrium, but the pain is, I think, more frequent in the cerebro-spinal form of meningitis presently to be mentioned, and may then perhaps be referred to involvement of the roots of the spinal nerves. *Anæsthesia*, on the other hand, increasing in degree is the rule in the later stages, and may sometimes be observed quite early when the disease is rapidly advancing. The conjunctiva may be comparatively insensible before the power of speech and spontaneous movement has ceased.

Delirium of any extent or duration is certainly not often seen in the meningitis of children, and is perhaps only compatible with an inchoate lesion. The mental as well as the physical symptoms of meningitis are those of progressive impairment and loss of function.

I have often seen profuse and continued *sweating* in tubercular meningitis; and in a very few cases, but in none with constipation or abdominal retraction, rose-spots, indistinguishable from those of enteric fever, appear.

The longest *duration* noted among my cases with necropsies is a little over four weeks from the first observed cerebral symptoms, but the disease may sometimes last considerably longer. One case which ran a course of fifty-three days, with otherwise typical symptoms and a history of previously failing health, was found post-mortem to be non-tubercular. The average duration of the cases which appear clinically to be most uncomplicated is probably about three weeks. There is nearly always marked wasting. Convulsions very often immediately precede death, whether or no they may have been observed earlier. The younger the patient the shorter, as a rule, is the course of the disease. The meningitic symptoms may be greatly masked or almost entirely obscured by those of more general disease, especially in cases of abdominal or pulmonary tuberculosis, and when there has been much wasting and exhaustion. In some indeed I have seen no symptoms, besides irritability and drowsiness, which could with any reason be referred to cerebral mischief. A remarkable case in illustration of this is reported by Dr. Sturges in vol. i. of the Westminster Hospital Reports. Occasionally there is a temporary initial stage of wild excitement or acute mania which may last for some days. This is seen mostly in older children and adults; and may obscure the diagnosis to even a practised observer.

Tubercular meningitis supervening on tubercular tumours in the brain, or, more often, on peritonitis or extensive tubercular disease in the lungs, usually runs a very rapid course, as exemplified by many of my cases.

The varying and special symptoms of tubercular meningitis are in some degree connected with the amount of intra-ventricular effusion, and with the extent and locality of the deposit of tubercle and its accompanying meningitis. In some cases unconsciousness is observed almost from the first, and the earlier symptoms above enumerated may be altogether wanting. The early appearance and prominence of general paralysis and coma are sometimes due to large and rapid ventricular effusion.

The best reasons for diagnosing tubercular meningitis during life are a history of family predisposition to tubercle, the discovery of tubercular disease elsewhere, a period of premonitory symptoms, local evidence of disturbance at the base of the brain, a gradual course of the disease, and

an absence of any more demonstrable origin. There is one sign however—the presence of miliary tubercle in the choroid—which, were it of
more frequent occurrence and less often mistaken, even by experts, would
be diagnostically very valuable ; but it is not often discovered, except in
cases of clearly general tuberculosis, and has not seldom been suspected
during life and disproved by examination of the fundi after death.
Moreover ophthalmoscopic examination is very difficult with meningitic
children before the paralytic stage of the disease, when the diagnosis is
usually certain. Neuro-retinitis in varying degrees is most often present.

The *morbid anatomy* of tubercular meningitis does not show constant
variations explanatory of the many different clinical forms of the disease.
We may, however, often correctly diagnose the somewhat exceptional
occurrence of much lymph on the surface of the hemispheres from a
rapid clinical course, with convulsions and rigidity passing into coma,
in cases apparently uncomplicated with much disease elsewhere. In a
large majority of instances different degrees of effusion of lymph or pus
at the base of the brain are found, with miliary tuberculosis of the pia
mater of very various extent. The tubercles usually follow the course
of vessels and are most especially abundant in the Sylvian fissures, in
the choroid plexuses, and often on the inner surface of the hemispheres.
We may find much lymph with but few tubercles, as well as numerous
tubercles, especially on the convexity of the brain, with little or no
lymph. Masses of yellow or caseous tubercle are not seldom seen, and
may be, especially when just under the pia mater, the starting-point
of the meningitis. Initial convulsions, especially when unattended by
marked paralytic symptoms or subsequently impaired consciousness, and
when of unilateral distribution, are certainly sometimes referable to such
tumours localised in the cortex. I have either seen or obtained the
history of such convulsive attacks in several cases which at different
periods subsequently developed meningitis.

There is nearly always some, and often very large, effusion into the
ventricles ; and the surrounding brain structures, including the basic
ganglia, are very frequently much softened or quite diffluent. In a few
cases I have seen the signs of basilar meningitis with no visible tubercles
except in organs other than the brain. Such cases are rightly regarded
as tubercular, and it is a fact that without microscopical observation
slight miliary tuberculosis of the pia mater is often overlooked. It may
indeed sometimes be felt, as granular matter, where it cannot be seen
by the naked eye. The pia mater in the upper part of the cord is not
very rarely found to be the seat of inflammatory effusion and of tubercle.
In some few cases with this post-mortem phenomenon there had been
marked and rigid retraction of the head. Purulent collections in both
middle ears are often found, as also in cases of non-tubercular meningitis,

without any disease of the bone or of the dura mater. Tuberculosis, both miliary and caseous, of the bronchial, mesenteric and other glands, is of exceedingly frequent occurrence, and we very often find these processes in the lungs, pleuræ, peritoneum, liver, spleen, kidneys and other organs. In some cases extensive tubercular peritonitis or ulceration of the intestinal mucosa explains many symptoms observed in life. A careful examination will almost always discover at least some small amount of caseation in gland or other organ. I have never seen a case myself, although instances undoubtedly occur, of tuberculosis strictly limited to the pia mater.

Of the ultimate *ætiology* of this disease there is but little to say, the modern theory of bacillary infection of a presumably ready subject covering meningeal as well as other tuberculosis. It may, however, at least be suggested that in the frequent apparently primary forms of meningitis, with but little disease elsewhere, the rapidly developing brain and richly vascular pia mater of the growing child are especially obnoxious to the tubercular poison. I am sure that measles in some way greatly encourages tuberculosis, whether the caseous changes so frequently found after this disease are themselves tubercular in essence, or only the nidus of a subsequent infective material. It is with some hesitation that I express an opinion, not unsupported by authority but resting on no demonstrable basis, that in some instances constant and excessive mental application, leading to hyperæmia of brain, may be the determining or at least an important contributory cause of tubercular meningitis. The hypothesis is, at the worst, reasonable, and from certain cases I have seen I cannot but consider it probable. Severe falls and blows on the head may also be regarded as possibly exciting causes.

I shall reserve the subject of treatment until I have spoken of the remaining forms of meningitis in childhood, and can add nothing to what I have incidentally said on the matter of prognosis except the statement that tubercular meningitis is and must be almost always fatal, not because of the mere presence of tubercle, but because tuberculosis, being essentially progressive, has likewise progressive effects.

Meningitis unconnected with tubercle, but with symptoms often indistinguishable from those of the tubercular form, not seldom occurs in children. It is sometimes apparently idiopathic; sometimes closely connected with various morbid processes such as, for instance, pneumonia, erysipelas, ulcerative endocarditis, the exanthemata, "cerebro-spinal fever," exposure to heat, syphilis, and perhaps rheumatism; and sometimes clearly secondary to certain cranial lesions, morbid or traumatic, of which purulent otitis is the chief. After reviewing all my many noted cases of meningitis in children which were, either most probably or certainly, unconnected with tuberculosis, I cannot find any grounds, either clinical or anatomical, for making any clear or practical division between them, and shall therefore but shortly mention the main groups. Generally speaking, non-tubercular meningitis begins more suddenly, runs a shorter though sometimes a very much longer course, tends oftener to at least partial recovery, is much more seldom only basic in position, more frequently involves the convexity of the brain, and is less rarely purulent in character than the tubercular form already considered. With the exception, perhaps, of those cases which are either traumatic in origin or clearly secondary to disease of the petrous bone resulting from otitis, we must recognise in all these (including probably many where purulent otitis media is found without carious bone) a distinct vulnerability and proneness to variously caused inflammation of the highly vascular pia mater which invests the rapidly developing brain of early childhood. It is remarkable that an enormous majority of the subjects of the epidemic form of cerebro-spinal meningitis, so little known or at least so seldom described in England, are young children ; and it is not improbable that very many cases of so-called idiopathic meningitis, as well as those which occur in connexion with pneumonia and other acute diseases, are due to specific infection especially attacking the cerebral membrane. Practically, in a large number of cases of meningitis in infants where there is absolutely no symptom or discovery of tuberculosis anywhere in the body, we are obliged to forego a causal diagnosis, although several may be with great probability referred to one or other of the sources above-mentioned, examples of many of which I have seen myself.

Of **apparently idiopathic meningitis**, clinically uncomplicated and without any other post-mortem lesion, I have seen several instances ; and some cases seemingly of this nature have completely or partially recovered. That partial recovery, for a time at least, can take place is fairly evidenced ; first, by cases with marked symptoms which get well and, dying after some time from a similar or quite different affection, show post-mortem evidence of old meningeal inflammation ; and, second, by the profound mental deterioration, indicative of serious brain lesion, so often observed after recovery from most of the physical symptoms of acute meningitis. In these partially recovered cases the special senses, such as sight and hearing, are often much impaired or destroyed. It remains probably true, however, that only a very limited and short-lived meningitis can miss a fatal result ; and many reported cases of recovery are due to the still wide-spread error of lightly diagnosing meningitis from almost any marked cerebral symptoms accompanied by fever. Many indeed of the early symptoms in all cases of meningitis are doubtless due to the initial and acute hyperæmia alone.

Cases symptomatically identical with many of those known as epidemic cerebro-spinal meningitis or cerebro-spinal fever, especially of the non-eruptive form, are unquestionably of no rare occurrence, and, like the epidemic cases, not seldom recover. It may possibly be better to regard these as sporadic examples of the epidemic type-disease than to refer them to the unsatisfactory class of "idiopathic cases," but their ætiology is at the best hypothetical.[1] In addition to the symptoms of cerebral meningitis, usually of the base, there is marked cervical and often dorsal opisthotonos, and much pain in the body and limbs, especially increased by any kind of movement. Post-mortem there is found cerebral meningitis, often purulent and mostly at the base, and much involvement of the pia mater of the cord. The onset of the illness is more or less sudden, with vomiting, headache and often varying degrees of delirium, and the temperature averages somewhat higher than in tubercular meningitis. The duration of the affection is very variable, sometimes of many weeks. The cases which recover perfectly are usually, however, of shorter febrile course, though convalescence may be protracted. Most of the cases which we meet with are parallel to those of the epidemic form, where the brain is more affected than the cord, and tonic contractions of the extremities, though frequently present, are not very prominent. Lung consolidations occur sometimes, and in one well-marked case I observed a copious eruption of small red spots, especially marked on the chest and back, appearing after admission and lasting until death. Of the epidemic disease, well-known in America and Ireland and described in the text-books, I shall say nothing, for want of personal knowledge; but it has been undoubtedly recognised from time to time in Great Britain.

Purulent meningitis is but an extreme form of what we have considered under other headings, and may occur in any form of the disease. The term is mostly used, however, to signify the cases which are secondary to disease of the petrous bone or some other lesion in the cranial cavity. It is found both at the base alone and also extending over the convexity of the brain. I refer hereafter to otitis and its clinical bearings, and would only remark here that it is at least probable, from some cases that are apparently meningitic, that a limited meningitis of a non-suppurative character may be set up by otitis. We should, then, examine the ears and also the nose in every case where meningitis is suspected.

All these forms of meningitis may occur, as the tubercular variety does sometimes, with very slightly marked symptoms or perhaps none at all. This symptomatic latency however is almost entirely confined

---

[1] Since writing the above I have read an article on Cerebro-spinal Meningitis in *Brain*, Part LVII. 1892, by Dr. E. F. Trevelyan, which strongly corroborates this view and much amplifies its grounds.

to children who are the subjects of previous severe disease or are markedly depressed and wasted.

Simple Acute Hydrocephalus.—Under this unsatisfactory and probably imperfect title I include a group of cases which, though not frequent, are by no means rare. They are marked by symptoms of cerebral disturbance, probably sometimes recover, and are] shown, in those who die, to be connected with an effusion into the ventricles with no other evidence whatever of the process or products of inflammation. The symptoms, briefly speaking, are some fever with sickness and often constipation, followed soon by pain in the head which is not necessarily severe nor accompanied by photophobia; the child may be irritable, but is more often drowsy though easily roused, and may continue to feed well and even talk spontaneously. This condition may last for days, without any flushing, disturbance of pulse or breathing, convulsion, pupillary signs, or other symptoms characteristic of more than passing disorder of brain. But drowsiness most often increases to coma, and before death there are sometimes strabismus, convulsion, irregular breathing, and other symptoms usually seen in the final stage of meningitis. The complete absence for even many days of any localising signs, and especially those of the ordinary basic effusions, the undisturbed intellect, and the undetermined and often mild character of the symptoms, which rightly lead us to exclude meningitis as generally understood, may also mislead us to regard the cerebral phenomena as temporary, and symptomatic of some general or local disorder which may have eluded observation. I have seen such cases referred to dietetic causes or to "gastric catarrh," and, in those which recover before any marked signs of intra-cranial pressure have set in, it is as impossible to disprove such a fanciful diagnosis as to establish the presence of ventricular effusion. But many, if not most, of the cases I allude to end in death, and I know of no certain sign by which we can avoid a great risk of error in giving a favourable prognosis. If we are fairly certain of the absence of other disease in a previously healthy child, and the symptoms of listlessness continue for some days with any headache or fever, however slight, and, still more, if a yet open fontanelle distinctly bulges, we must be well on our guard, and increasing lethargy will soon justify a very grave prognosis. Many more mistakes are made on the optimistic than on the pessimistic side in cases such as these, even by practised observers who are well aware of their pathological entity. Children under two years of age are most often the subjects of this affection, but I have seen cases over three. It is more than likely that many cases of the familiar chronic hydrocephalus are the results of the acute affection above described.

At the close of this review of acute affections of the cerebral membranes and closely allied conditions, it is well, for practical reasons of diagnosis, to refer very briefly to that set of cerebral symptoms which follows on exhausting disease, especially in infants, and has been described under the name of "hydrocephaloid." I have frequently seen this state mistaken for primary brain-disease. The child, who is most often the subject of diarrhœa, lies in an apathetic condition with eyes half open, is with difficulty roused, may breathe irregularly, and often sighs. The pupils are sluggish and often unequal, the pulse frequent, small, and sometimes intermittent, the complexion pale, and the skin cool. Probably all these symptoms may be explained by imperfect blood supply from venous stasis. In some cases there is said to be an early stage of restlessness and screaming, with great sensitiveness to light and noise, frequent pulse, hot skin and contracted pupils; but such a stage is certainly very exceptional. If not properly treated, and especially if actively treated by "antiphlogistic" measures, the patient is in imminent danger and may die in coma. Alcoholic stimulation, artificial heat and good nourishment will very often quickly relieve or cure. In these cases there is usually no fever, except sometimes from such adventitious causes as continuing acute diarrhœa or a concomitant lung affection; and the fontanelle is usually much sunken. With due care and inquiry into the history of the case mistakes in diagnosis should scarcely ever occur. The ventricular effusion and œdema of the pia mater which are found in some of these cases that die are due to venous stasis from retarded circulation following on exhaustion. Sometimes also there is thrombosis of the cerebral sinuses.

The urine should be examined for albumen in every case of suspected meningitis or acute brain mischief, for many cerebral symptoms are common to these affections and to acute or chronic nephritis.

Treatment.—A careful consideration of what has been said regarding the diagnosis and course of meningitis in its various forms will make it apparent that little more success can be expected than is actually attained from attempts at curative treatment, when there is good evidence that a discoverable meningitis, whether tubercular or not, has been established. Yet we can rarely or never know for certain, from symptoms, when the line between hyperæmia and actual exudation has been crossed; and we are aware, not only that grave cerebral symptoms in connexion with disturbed blood supply very frequently pass away, but also that actual meningitis does sometimes spontaneously recover.

It is only in a small minority of cases that, either in hospital or private practice, the child is seen in the earliest stage of cerebral disorder. This is the time when good may possibly be done by purging a few times in cases with constipation, by the abstraction of blood by means of leeches, and by the administration of antimony, opium or such other drugs as may be believed to have anti-inflammatory action. I must, however, state that, having used the two former methods very often and the latter not seldom, I have but little reason to place faith in them; but the earlier the case, the more I would advise this procedure, in the ascertained absence of the "hydrocephaloid" condition or of enteric fever. But we must remember that at this stage enteric fever cannot always be definitely diagnosed or excluded. I always give bromide and sometimes iodide of potassium in cases which come under notice in an early stage; the former is at least soothing in action, and the latter, besides covering cases of possibly syphilitic origin, is believed by many good and not credulous authorities to be of use in meningitis other than tubercular. In all early cases suspected to be meningitis, whether tubercular or not, the hypophosphite of lime or soda may be tried. They are believed by some to be of use, and are certainly harmless. An ice-bag to the head is occasionally very quieting in cases with much headache, but its application is often fruitless owing to the very restlessness it is intended to relieve. Mercury, from much experience of it in past years, I am sure is useless. I never give it now, believing it to be decidedly objectionable in cases which might spontaneously recover. There is no hindrance to the use of opium or chloral in certain cases marked by a prolonged period of great pain and restlessness; and opium may be given in early cases, even when these symptoms are not marked.

At the outset the ears should always be carefully examined, and if, from the presence or a previous history of otitis or from a bulging of the tympanic membrane, there be a suspicion of retained pus, free evacuation should be secured by puncture, and the meatus stuffed with antiseptic wool.

Every suspected case must be kept cool and quiet, and as much in the dark as possible.

In advanced stages, or when increasing drowsiness or any paralysis is observed, the best that we can do, while we need not omit the iodide or hypophosphite which may have been given, is to attend to the patient's nourishment with the faint hope of the remarkable recovery occasionally seen.

Prophylaxis can only apply to the tubercular form of meningitis, which is often preceded by symptoms due, either to the deposit of tubercle in the meninges causing hyperæmia without actual exudation, or to disease elsewhere. In suspected or actually tubercular cases, therefore, all sources of cerebral excitement should be carefully guarded against; and, when possible, children of markedly tubercular families should be similarly protected. We may be well advised in discouraging much mental labour in cases of delicate children with close phthisical relationships.

CHAPTER IV.

CHRONIC DISEASES OF THE BRAIN.

THE brain of the young child may be affected by many of the pathological conditions familiar to us in adults, and in addition by extensive effusion into the ventricles, usually described as chronic hydrocephalus. After dealing with the last-named affection I shall give a short account of cerebral hæmorrhage, embolism, thrombosis, tumours, and sclerosis, as symptomatically indicated or found post-mortem in childhood.

Chronic Hydrocephalus.

This is a name generally given to a well-known class of cases where there is increasing enlargement of the head, in connexion with an effusion of fluid usually ventricular, but sometimes outside the pia mater. Neither on unsuspected ventricular effusion found only post-mortem and occurring in tubercular and other wasting diseases, nor on that other important form without much or any head enlargement, which, as we have seen, [is so common in association with marked meningitis, is it necessary to dwell here. I would nevertheless emphasise the fact that we occasionally meet with cases in young children of intermittent pyrexia with much irritability, accompanied by paroxysms of more or less severe pain in the head without localised paralysis or other symptoms, where practically nothing is found post-mortem beyond extensive ventricular effusion. Among a few instances of this, one case, which I saw in the practice of a colleague at Westminster Hospital, is especially prominent. The child was two years old, and the fontanelles were quite closed. Symptoms, as above-mentioned, endured for months; a high temperature, sometimes reaching 104°, occurring almost daily. There was much effusion in the ventricles but no post-mortem sign of inflammation, with the exception of one or two very minute spots of past meningitis.

In the **diagnosis** of hydrocephalus we must always remember the large head of rickets, where the fontanelles are alike open and ossification is retarded. Observation of the more globular shape in hydrocephalus, and the more or less flattened vertex in rickets, will prevent confusion in many cases; but it must be remembered that a considerable proportion of hydrocephalic infants are also rickety. A word of warning is necessary against regarding a tense or pulsating fontanelle as a sign of hydrocephalus. Such a phenomenon may, on the one hand, be an accompaniment of acute meningitis, and, on the other, a merely temporary occurrence in the course of various febrile conditions.

The skull-bones in hydrocephalus are nearly always very thin, although rarely they may be abnormally thickened; and sometimes there is an abundance of "ossa triquetra," which may become displaced and strikingly manifest if the fluid be withdrawn by aspiration. In extreme cases there are wide intervals between the bones, and sometimes the whole cranial vault is made up of mere osseous islets in the surrounding membrane. The surface veins are large, and the hair scanty. The eyeballs are often directed more or less downwards, leaving much of the whites exposed; and there may be weakness of the levatores palpebrarum or of any of the muscles of the globes. Frequently there is optic atrophy and enlargement of the fundal veins, and the eyesight may be impaired in any degree. Imperfect vision or blindness may, however, exist with no ophthalmoscopic abnormality, and may then be due to pressure on the cortical centres. The sense of hearing is likewise very often defective.

Mental impairment of any grade may occur; although in some cases, and especially in those where the affection appears to have set in late, but little or no deficiency is observed.

In many instances there is weakness or loss of motor power in the legs, and sometimes ataxia of the upper extremities. Rigidity of limbs and trunk, and tremors on movement may occur, probably connected with pressure on the cerebral motor-tract; and convulsions, including glottic spasm, are frequent. Often there is general retardation of bodily development. The general organic health, however, may remain good for long, although in time wasting generally sets in. In marked cases death happens mostly early, with emaciation and often with convulsions; but sometimes the patients last for several years, and a very few outlive childhood. There is probably a considerably greater number of cases of hydrocephalus arrested before the affection has become extreme than can be easily demonstrated. All are familiar with some instances of this event, where, even after notable enlargement of the head for a short time, arrest takes place, and recovery is good with no mental or physical impairment.

The **origin** of chronic hydrocephalus is, as a rule, obscure. There is, practically, no clinical differentiation possible between the so-called "congenital" and "acquired," or between the inflammatory and non-inflammatory cases. Post-mortem examination often shows *thickening and roughness of the lining membrane of the ventricles and the choroid plexuses,* and sometimes lymph is found obstructing the ventricular communications. In other instances absolutely no traces of inflammation are seen. In the congenital cases there is often further evidence of faulty development, cerebral or otherwise. *Tumours* of the cerebellum or pons, though often occasioning much effusion into the ventricles, are not frequent causes of the clinical class of cases under notice, comparatively seldom giving rise to very notable enlargement of the head. The undoubted inflammatory origin of many cases is only sometimes referable to a distinct *meningitis*, but it may be believed that such an affection may have existed and have undergone arrest. Various opinions are recorded as to the connexion between hydrocephalus and *syphilis*. In my experience the association is decidedly frequent, and Fournier and others have reported instances tending to support the view that this association is a causal one.

In a large majority of cases the enlargement of the head is observed in the first few months; in a very few, examples of which I have seen, the pressure of the effused fluid has been first noticed owing to the reopening of already closed sutures. In a case of a child, aged five years, where the sutures re-opened, there was found a slight basic meningitis which had closed up the openings of the fourth ventricle.

It is possible that non-inflammatory ventricular effusion arising in connexion with malnutrition, and an imperfectly developed cranium leading to insufficient antagonism to the intra-vascular pressure, may be much encouraged by the occurrence of additional strain on the cerebral vessels from such a cause as violent coughing. Hydrocephalus certainly seems sometimes to arise clinically out of *whooping-cough*.

Hydrocephalic fluid is sometimes clear and sometimes turbid with lymph or even pus, with or without some trace of blood. In the clear fluid of apparently non-inflammatory cases there is but a trace of albumen, while in those marked by signs of inflammation albumen may be in considerable quantity.

The symptomatic **treatment** of this affection is very unsatisfactory, nor does it appear to me that there have been any direct artificial cures. We know that some, and believe that many cases are naturally arrested with improved nutrition, and also that but few live long, however treated, when the enlargement of the head has been great and rapid. We must before all things attend to diet and hygiene, giving cod-liver oil and the mineral tonics, and appropriately treat any concurrent disease, not forgetting the possibility of syphilis. Following such methods we may sometimes be agreeably surprised at the result. I have patiently tried in many cases the effect of mercurials, both internally and externally, of iodine and the iodides, and of other medicines; I have bound the head with elastic and other bandages; and on three occasions I have aspirated the fluid. I am convinced that all these plans are useless, and that aspiration is more likely to bring discredit on the physician than even temporary relief to the patient.

Cerebral Hæmorrhage, Thrombosis and Embolism.

Some of the clinical aspects of circulatory obstruction in the brain are treated of under the heading of cerebral paralysis. From a practical point of view there is but little to add concerning the above-named conditions, which have much that is common in their symptomatology.

Hæmorrhage into the meninges is frequently met with in the new-born, and is mainly due to injury during protracted labour or, less often, to instrumental delivery. Sudden increase of intra-cranial pressure during arrested breathing, and notably in a paroxysm of whooping-cough, may perhaps determine hæmorrhage even with healthy vessels; and the vascular changes which may take place in the course of such acute infectious diseases as scarlatina, enteric fever, measles, diphtheria, or smallpox are doubtless predisposing causes to hæmorrhage variously excited. Hereditary syphilis, too, must be ranked as a known though infrequent cause of vascular disease. Hæmorrhage into the substance of the brain is an acknowledged occurrence, and takes place most often in the cortex.

The *symptoms* of meningeal hæmorrhage in the new-born are convulsions, coma or paralysis observed at birth or soon afterwards. Monoplegia, hemiplegia, or bilateral paralysis is a frequent result in cases which do not soon die, and complete recovery probably takes place in some cases. In cerebral hæmorrhage in older children the symptoms are similar to those in adults, depending on the locality of the lesion and the extent of the hæmorrhage. With slight effusion there may be rapid recovery.

Thrombosis of both arteries and veins, and especially of the sinuses, is well-known in childhood and of various causation. *Syphilis* accounts for some cases, and instances have been reported in the course of *diphtheria* and other poison-diseases. Sinus-thrombosis occurs in exhausting affections, and it has been suggested in this context by Dr. Gowers that thrombosis of the superficial cerebral veins may explain divers cases of hemiplegia in childhood. A common cause of thrombosis is *disease in the neighbourhood of the sinus*, such as meningeal inflammation, tubercular or otherwise, injury and disease of the skull-bones, internal ear mischief, or erysipelas and suppurative processes outside the skull. The paralytic symptoms of thrombosis are usually gradual in onset, preceded by headache, and of long duration ; varying, of course, in nature and gravity with the extent and locality of the lesion.

Embolism may occur, in children as in adults, in those conditions which involve disease of the heart-valves or favour clotting in the left side of the heart, especially the auricle. Rheumatism supplies most of

T

the examples of this; scarlatina and other infectious diseases a few. In young children the subject of cerebral embolism is not extensive. The symptoms tend to be sudden, and may be accompanied with convulsions, especially when the vessels of the cortex are implicated.

Tumours of the Brain.

Tumours, both cerebral and cerebellar, often occur in childhood, and are not very rare even in early infancy. Tubercular tumours are by far the most frequent, and form a very large proportion of all cases below the age of seven, becoming gradually rarer in later years. Next in frequency are glioma and sarcoma, or a mixed form of these growths. Others, such as syphilitic gumma, myxoma and carcinoma, are but very seldom seen; while parasitic cysts, though apparently not rare in Germany, are scarcely met with here.

Tubercular tumours are of all shapes and sizes, varying from small aggregations of miliary tubercles to large caseous masses; they may be single or multiple, encapsulated or widely diffused. Although occurring in any part, the most favourite seat, especially of the larger ones, is in the cerebellum or the pons and its neighbourhood. They very often occupy the cortex, and frequently originate in the pia mater. Almost always they are found post-mortem to be associated with tubercle elsewhere; and particularly affect scrofulous or tubercular children who have often, besides, a family history of such diseases. The symptoms vary much according to the size and position of the tumours, which may be of very different duration. In some the symptoms may endure, with remissions, for months or even years; in others, which remain latent, the first symptoms noticed may be those of the meningitis which so often forms the fatal chapter in their clinical history. Small tubercular tumours, with or without symptoms, occasionally become obsolete through cretification; and large caseous ones sometimes form abscesses which may be the foci of more generalised tuberculosis.

Gliomata may occur anywhere in the substance of the brain, most often, on the whole, in the white matter; they do not involve the pia mater. The pons is a very favourite seat. They may be hardish, or very soft; fairly well-defined, or with no limitation from the surrounding brain substance; and often attain a great size. Their growth is sometimes very slow. In one case, aged thirteen, where definite symptoms of brain-disease had lasted off and on for nearly six years, I found the whole of one temporo-sphenoidal lobe involved in a glioma, the rest of the body being quite healthy. The harder and more definite tumours may soften and break down, and there is in all a liability to internal or circumferential haemorrhage owing to their vascularity.

Sarcomata may be found in any part of the cerebrum or cerebellum, or may start in the membranes or from the bones. They are usually round- or spindle-celled, for the most part grow very rapidly, and are seldom multiple. They are mainly seen in early childhood, and are unassociated with similar disease elsewhere. They have undoubtedly a close clinical and anatomical relationship with gliomata.

Concerning the general **ætiology** of brain tumours I can but endorse from my own experience the usual confession of ignorance and the perhaps equally usual belief that in many cases, difficult as proof may be, severe blows and falls on the head are the real starting-point. It is at least, I think, very probable that such accidents may occasion the growth of tumours of all the kinds above mentioned, though of course in many instances tubercular tumours, especially when small and multiple, must be regarded as merely expressions of a general tuberculosis.

The **symptoms** indicating tumour within the cranium are usually described in two classes, general and local; the first, such as headache, vomiting, optic neuritis, and, we may add, bilateral convulsions with affection of consciousness, being as a rule common, some or all of them, to all brain tumours in infancy of whatever position ; the second, such as unilateral or local spasms and paralyses, in-coördination and disturbance of equilibrium, and affections of sensibility, speech, swallowing or respiration, being referable to the position of the lesion in the brain, of whatever kind it may be, and often more or less exactly indicating that position. From a practical point of view, however, and especially in relation to these affections in childhood, such a division seems to be of little worth. Localising symptoms are often absent altogether when the tumour occupies certain parts of the brain, a fact which, especially in infants and young children who cannot express themselves, may render the diagnosis obscure for long; and the convulsions so very frequently associated with brain lesions in infancy, being often general, are devoid of localising value, and indistinguishable from those which are independent of coarse disease and classed as idiopathic convulsions or epilepsy. I shall indeed pass over with but short notice the matter of localising symptoms in cases of brain tumours, as neither special to our subject nor sufficiently illustrated in my own experience by post-mortem examinations in children. Several of my best cases were taken out of hospital before death, and in several others the necropsy was forbidden.

An early and enduring symptom of tumour in the brain is *headache,* which is often severe and may be either constant or recurrent. In cases of growing tumour there is probably always some headache with frequent exacerbations. The seat of pain is but rarely an indication of the seat of the disease as proved post-mortem; but my experience is certainly in accord with that of many observers who note frequent concomitance of constant occipital pain with tumours of the cerebellum or bulb. I have, however, seen several cases where all other symptoms of tumour were well-marked, both in infants and children who could speak; and some where a cerebellar tumour was found post-mortem, in which there was for long but little evidence of headache, and the patients, with intervals of slight paroxysmal pain, were quite cheerful.

Convulsions of general distribution, with impairment or loss of consciousness, are very frequent, and are recorded in many histories as the first noticed phenomena. In several cases, where the symptoms were those of cerebellar or of pontine growth, general and repeated convulsions, slight or severe, occurred very early. It is only when a convulsion, limited to one side or to one muscular group, is recurrent, and especially when it is not accompanied by marked evidence of defect of consciousness, that it becomes a valuable evidence of a lesion in or close to the motor cortex of the opposite side of the brain.

Vomiting at intervals, often unexcited by any discoverable cause, occurs in most cases, wherever the tumour may be : and in connexion with constant headache is a very serious warning. *Optic neuritis*, which in some degree is present in a very large majority of cases, is of value as a sign which may help us at once to distinguish some severe cases of frequently recurrent migraine from cerebral tumour. It may exist for long without producing any complaint of visual trouble, is usually double, and, though it accompanies tumours of any kind or seat, is especially frequent with those of the cerebellum and the base of the brain and then often causes early blindness. Optic neuritis is however by itself, as is well known, no absolute proof of brain-disease, even when other symptoms seem to indicate such mischief. Messrs. Ashby and Wright record a case of otitis without cerebral disease in which optic neuritis was observed. *Vertigo* is a feeling complained of in most cases from time to time, although it is most constant in the case of cerebellar tumours where there is also evidently faulty equilibrium. *Nystagmus* is frequently observed, especially with tumours in the cerebellum or pons. *Drowsiness, irritability,* and all kinds of *mental changes,* from apathy to delirium and acute maniacal excitement, may frequently be noted, very few cases being unmarked by any kind of psychical disturbance. When the pain is great and continuous there may be persistent *insomnia.* Most growing tumours are accompanied by *wasting,* which is especially marked in long-standing cerebellar cases. Instances are common where children continue steadily wasting for weeks or months after vomiting may have ceased and pain is no longer evident, dying at last from gradual failure of the nervous centres and sometimes with convulsions. In one chronic case with the symptoms of brain tumour, but not examined post-mortem, there was a sudden arrest of breathing and discoverable heart-action, the pulse, however, returning with artificial respiration performed at intervals, and remaining palpable for more than an hour after all signs of respiration had ceased. In cases of tumour at the base, especially of the cerebellum and pons, pressure on the veins very frequently causes effusion into the lateral ventricles, and various degrees of hydrocephalus may be found. At the same time great enlargement of the head, even in infants, is not so frequent as we might be led to expect. Death in cerebral tumours may take place from pressure on the vital parts in the medulla, is often coincident with convulsions, and very frequently the result of a meningitis set up by the growth.

The presence of several or all of the above-mentioned symptoms is a strong indication of brain tumour; and if, in addition, there are definite local spasms or paralyses or other signs of limited lesion, such as altered gait, impairment of speech, breathing, swallowing or the like, the diagnosis is tolerably certain. In the absence, moreover, or indistinct presence of many of the so-called general symptoms, these local manifestations are of the greatest importance.

Passing by the whole subject of the *differential diagnosis* of the various sites of tumours of the cerebrum proper, which is a matter of the knowledge of cerebral topography, I shall but shortly deal with those of the **cerebellum**, which are by far more frequent in childhood than in later life; touching also, incidentally, on the almost equally frequent tumours of the **pons**.

The chief characteristic of cerebellar as distinguished from other tumours is *disturbance of equilibrium*, especially shown by a more or less general ataxia when the body is unsupported, without any true paralyses. Thus the body swings in standing and walking, and also, though in early cases to a less extent, in sitting. Some uncertainty is also, I think, observable sometimes in the large movements of the arms, and there is often difficulty in holding them erect; while the child can use its hands well when the body is at rest with the arms supported. Thus it may be in defect of equilibrium, not only of the body generally but also of the limbs in certain positions, that the so-called cerebellar ataxia is evidenced.

But apart from the great fact of defective equilibration, seen in both station and locomotion, which is undoubtedly a sign of failure of cerebellar action and is connected with disease of the middle lobe of that organ, symptoms of cerebellar disease may be inextricably confused by those of the involvement of other parts, and especially of the pons and medulla which are so often coincidentally affected both directly and indirectly.

The motor tract in these bodies is exceedingly often pressed upon by tumours of the cerebellum, while, on the other hand, the cerebellum or its peduncles may be interfered with by tumours growing from the pons or its neighbourhood; and the rigidity and tremors which are so frequently seen in these cases, as also the exaggerated knee-jerks, are in all probability the result of disturbance in this tract. It is said by many that the knee-jerks are either exaggerated or at least not abolished in disease of the cerebellum itself. I have, however, observed that not only spasms and rigidity, as Sharkey has, I think, satisfactorily demonstrated, but also exaggerated knee-jerks are apparently confined to cases where the cerebral motor tract is affected, either by the growth or by indirect pressure. On the other hand, in two cases regarded during life as cerebellar tumour, where the middle lobe of the cerebellum was found post-mortem to be largely occupied by a tubercular and a sarcomatous tumour respectively, with no apparent pressure on the cerebral motor-tract, there was no rigidity or fine tremor on movement, and the knee-jerks were entirely absent throughout. I mention this only as an indication, wanting, of course, much more observation for its establishment, that the presence of knee-jerks may probably imply some degree of cerebellar activity; that destructive cerebellar disease probably causes diminution or loss of the knee-jerks; and, therefore, that this phenomenon cannot be regarded as a diagnostic point between the so-called cerebellar and spinal or peripheral ataxia.[1] The diagnosis of the locality of brain tumours is very often obscured by multiplicity and indefiniteness of lesion; and we but very seldom meet clinically with the pure symptoms of cerebellar disease alone.

General *convulsions* occur in the history of cases of cerebellar tumour quite as often as in that of others, being frequently among the first symptoms noticed. In my experience the opisthotonic seizures, described as peculiarly cerebellar in origin, are very rare. Among many cases of cerebellar and pontine tumours, either demonstrated or suspected, I have but rarely seen any fits during their long residence in hospital until meningitic symptoms set in and death was imminent.

Tumours of the pons may be suspected or diagnosed from motor paralysis, especially when it is bilateral or crossed, the cranial nerves being affected on one side, the limbs on the other; and involvement of the medulla is often indicated by altered breathing or heart-rhythm, or by impaired deglutition. I have, however, occasionally been surprised

---

[1] Since writing the above, I have read the report of a case by Dr. Handford in *Brain* (Part LIX. and LX., 1892, p. 458), of "cerebellar tumour with loss of the knee-jerks," which, with Dr. Handford's comment, corroborates this view. In this instance the whole of the middle lobe of the cerebellum was destroyed by a sarcomatous growth, and the medulla was somewhat flattened thereby; but the knee-jerks were throughout entirely absent.

at finding tumours of considerable size occupying, and apparently growing from, the floor of the fourth ventricle, where there were no bulbar symptoms until quite the end of the case.

In making the general diagnosis of brain tumours we should remember that the symptoms may be more or less latent, and may sometimes very closely simulate those of chronic meningitis, of cerebral abscess, or, when the effusion and head-enlargement secondary to pressure by the tumour are exceptionally great, of chronic hydrocephalus ; and that in some patients approaching the period of puberty hysterical symptoms are sometimes easily mistaken for those of tumour, or the two affections may co-exist, the hysterical phenomena being, as it were, grafted on the more serious nervous disorder and greatly obscuring it.

Although some cerebral tumours may be latent for long and, even when symptomatically evidenced, may still run a protracted course, the prognosis must almost always be that of a fatal result. In children there is scarcely ever the possible chance, present in some adult cases of syphilitic tumour, of arrest by specific remedies. Such medication, however, is often tried as a forlorn hope. When therefore the locality of the tumour has been diagnosed with the best approach to accuracy, according to modern knowledge of cerebral topography, it is clearly justifiable to advise recourse to trephining and possible extirpation, however little ultimate success may have as yet been attained by these methods.

Sclerosis of Nervous Centres.

With regard to this condition, whether diffused or disseminated or limited to a few small patches, I can but say, with Dr. Goodhart and others, that, although occasional examples of it in the brain are undoubtedly found post-mortem, our clinical knowledge of any symptoms connected therewith is very scanty. I have seen cases, where small patches of sclerosis have been unexpectedly found in the brain, which were marked during life by no salient nervous symptoms ; and, on the other hand, I have sometimes met with groups of symptoms, more or less similar to those referred to disseminated sclerosis as best known in adults, which either diminished, or disappeared, or remained without further development. I have never had an opportunity of seeing the brain in any such case. With extensive sclerosis as the occasional result of a chronic meningo-cerebritis there is marked mental failure or idiocy ; with small and localised patches due to some inflammation or softening there may be, as has been said, no symptoms at all.

Cases have been described by several observers as " multiple " or " disseminated " or " cerebro-spinal " sclerosis occurring in children, the symptoms being more or less the same as in the adult disease, but more often beginning suddenly with tremors after a convulsion. For a comprehensive résumé and comment on the literature of this subject I would refer to Dr. Pritchard's article in Keating's *Cyclopædia of the Diseases of Children*, where, however, the symptomatic and anatomical disquisition is chiefly based on the recognised disease as known in adults. A few of the cases quoted seem, from the clinical point of view, to have full claim to be classed in this category; but the author states that in a much larger number the diagnostic data are far less decisive. Of the latter variety I think I have seen several examples. They were, however, cases which I am as yet unable to classify. The chief symptoms were tremor, increased or occurring only on movement, varying degrees of ataxia, and sometimes nystagmus. I have elsewhere referred to this when speaking of paralysis after acute disease, some examples having occurred in such a connexion. Without more post mortem observation, it is, I think, impossible to dogmatize about this affection in childhood; the symptoms of the recognised type in adults are confessedly variable, from the very nature of the disease; and it seems unadvisable to attempt a systematic account of "disseminated sclerosis" in childhood which must be based almost entirely on clinical conjecture. All that may be said, considering the extremely scanty anatomical knowledge concerning this disease in children, is that its occurrence in some of the cases referred to is at least highly probable. Closely allied, symptomatically, to many of the instances described as multiple sclerosis are the rare disease known as *Friedreich's ataxia* or hereditary tabes (for a description of which I refer to the text-books), some cases of brain tumour, and some of chronic meningitis. In cases of this ill-defined class we may practically regard the presence of nystagmus, optic atrophy, definite local paralyses either sensory or motor, rigidity, recurrent convulsions, and marked headache as, each and all, more or less strongly indicative of organic disease, and consequently as of bad prognosis; while we should hesitate long before condemning those cases which are marked mainly by ataxia and tremors, whatever be the condition of the knee-jerks. When, however, marked knee-jerks with ankle-clonus are demonstrable in childhood and continue indefinitely, the symptoms are of grave prognostic import as to ultimate recovery.

CHAPTER V.

CHOREA.

IN treating of this well-known but much discussed disorder I shall but scantily describe it and make no attempt at reviewing the numerous and conflicting theories of its ætiology. I have chosen rather to give, as briefly as practicable, the results of my own observations, based on a study of much larger numbers of out- and in-patients than are included in the series to be referred to in some detail, and on the consideration of much that has been weightily and diversely written on the subject. In illustration of various points I have analysed 162 cases under fourteen years old observed in my wards at Shadwell Hospital, a series which has many advantages over those which include out-patients only; and have also made use, in some respects, of the registrars' abstracts of 178 cases under the care of my colleagues and myself at Westminster Hospital. The histories of these were taken by many different hospital residents from numerous medical schools in London or elsewhere, and the records as a whole are probably freer than some from the drawbacks of bias so difficult to eliminate from the notes of a single observer.

Chorea is described by Sturges[1] as an "exaggerated fidgetiness, an extravagant exaltation of that continual unrest which is the natural characteristic of childhood." These words give both the true pattern of the disease and the best key to its explanation. The typical picture of chorea, as well as its varying grades of severity, I assume to be familiar to all of any practical experience. It is essentially a disorder of childhood, affecting the female sex nearly three times as often as the male, and tending almost always to recovery; and, though it may recur many times or even persist indefinitely, its onset in later years is rare and almost always associated with some kind of psychical disorder. In its ordinary form its age limit is mostly from six to fifteen. The youngest patients observed by myself were two girls of three. Out of 310 cases under fifteen years there were two boys and two girls under five, 150 between five and ten, of whom 65 per cent. were girls, and 156 between ten and fifteen, of whom 78 per cent. were girls. After puberty the proportion of females affected is overwhelming.

The characteristic *movements* begin either gradually, as an almost imperceptible increase of fidgetiness; or suddenly, with or without ascertained exciting causes. They usually affect the hands first, and tend

[1] See "Chorea," Smith & Elder, 1881.

to become more or less general in a large majority of cases, the upper part of the body almost always suffering most. In the older children facial grimace is sometimes the first observed symptom. For a full consideration of the place of origin and of the distribution of choreic movements I would refer to the work of Sturges, who shows that the more specialised muscles, or those most in want of control, are usually the first to suffer and are always most affected throughout the disorder. In 61 of my own cases, where special inquiry was made on this point, the movements were first noticed in both hands in 6, the right hand in 6, the right hand, arm and leg in 6 (several of all these having the face affected as well, either at once, or very soon), the right arm and leg in 1, the right arm and face in 3, the face in 3, both arms in 2, both legs in 2, one leg in 1, the left arm and leg in 10, the left arm in 4, the left hand in 1 ; in the rest the movements were said to be general from the first, with an almost universal preponderance in the hands and arms. In all my cases I find only three possible instances of movements confined to one side throughout ; but the notes of these are but scanty, and I am well convinced, from observation of numerous alleged cases of "hemi-chorea," that strictly unilateral choreic movements of prolonged duration are of remarkable rarity in childhood, if indeed they ever occur.

The face, especially in the older children, rarely escapes altogether ; the eye-movements are frequently affected; and the muscles of respiration, of speech, and pre-eminently the tongue, mostly suffer in some degree. In some severe cases chewing and swallowing may be difficult or impossible, and the patients may need to be fed by the nasal tube. Sometimes the movements are exceedingly violent, the patients injuring themselves against the bedstead and even jerking themselves out of it if not properly protected and watched.

Some amount of weakness (or "paresis") of the affected limbs always obtains, and marked paralysis after the movements have subsided is not nearly so uncommon as is often taught. I have seen numerous instances of this, including four of complete paralysis of all the voluntary muscles, with faltering heart, irregular breathing and great difficulty of swallowing, lasting from one to three weeks. Such patients are apt to become fatuous, to waste excessively and to suffer from bed-sores. Three of the four mentioned made complete recovery and one died.

The movements almost always cease during sleep and, in most cases, are increased by attention called to them or by attempts at coercion ; but in a considerable minority, especially in older children, they can be greatly controlled by voluntary effort. I have notes of many patients who were always quieter when being talked with or when endeavouring to exert control. In properly chosen cases this fact, often contradicted or overlooked, gives useful indications for treatment.

Emotional disturbance is marked in very many cases even in little children, but especially in those over ten years; and may precede the motor symptoms for some time. *Analgesia* of varying extent, especially in the parts most affected by movements, is not seldom met with, but is rare in early childhood. I have, however, seen one well-marked instance in a child of three. Very various *hysterical* phenomena often, and definite *mental* disturbance sometimes, occur in the older cases; but I am of opinion that the intellectual powers of most choreic children are not below the average, and that psychical disorder is mostly shown in the sphere of feeling.

Dilatation with or without sluggishness *of the pupils* is a very frequent phenomenon, most often disappearing on recovery; and *incontinence of urine*, apart from definite cases of choreal paralysis where it is the rule, is not so rare as I once thought.

The *heart's action* is almost always accelerated, often irregular in force and rhythm—a fact surprisingly often denied—and in a large number of cases, besides those with the rheumatic endocarditis presently to be noticed, there are either distinct murmurs, or reduplication or alteration of the sounds. With respect to the cardiac irregularity which seems, as Sturges teaches, to be more marked in younger subjects, we must remember that in a slight degree it is physiological in early childhood. The special cardiac signs of chorea I shall discuss separately, but it may be said here that "hæmic" murmurs, either confined to the base or ventricular as well, with loud humming sounds in the cervical vessels, are very frequent, and sometimes occasion difficulty in the observation or discussion of heart-affection in chorea.

The *temperature* of the body is notably subnormal throughout in a large number of cases, being often as low as 97° or 96°, as is abundantly shown by routine observation during many years at Shadwell Hospital. I have but rarely noticed a rise of temperature at the height of severe attacks, and believe from the cases I have seen (a few of them being strongly marked in older children and adults, and attended by mental disturbance or even mania) that it is probably due to causes not proper to chorea.

Chorea has a great tendency to *recur* at varying intervals, about one-third of my cases being admitted in at least their second attack and several in their fourth or fifth. Further salient points in the clinical history of chorea, and bearing importantly on its ætiology, are its close association with other *nervous* phenomena and definite hereditary neuroses; its very frequent excitation by special nerve-disturbances which are evidenced by various emotions, especially *fright*, immediately preceding the attacks; and the notable proportion of cases which are the subjects of *rheumatism* or have a distinctly rheumatic family history.

I shall now notice briefly the relationships of chorea to rheumatism
and to heart-affections respectively, postponing the matter of the nervous
excitants and associations of the disorder until I remark on its ætiology.·

The relationship between **Rheumatism** and Chorea is sufficiently frequent to necessitate its consideration with regard to ætiology.   In a small
number of cases chorea occurs in the course, or as the immediate sequel,
of an attack of acute rheumatism ; and a considerably larger number
of choreic patients suffer from rheumatism at some time or other before
or after the chorea.   It is also stated by many that there is an unduly
great proportion of rheumatism in the families of choreics ; and, further,
there are those who so exalt and magnify the part played by rheumatism
in the production or encouragement of chorea as almost to incur the
logical necessity of admitting no other mode of causation.   It is clear
that to be accurate in this matter we must be definite in our use of the
term "rheumatism" and have an approximate notion of its usual incidence on the population.   It must be borne in mind, however, that acute
rheumatism occurs often in childhood with but little or even no demonstrable arthritis, and I have consequently been careful to include in
the "rheumatic" class of cases of chorea all those with pericarditis, or
with endocarditis marked by symptoms or signs other than a mere apex
murmur, as well as many others which might be very doubtfully
called rheumatic, and some with limb-pains and especially wrist-pains
which were probably not rheumatic at all.   Even with this very liberal
construction of rheumatism I cannot class more than 35 per cent. of
my chorea cases as in any sense rheumatic.   Inquiry as to family
rheumatism is difficult and often unsatisfactory, especially among the
working-classes on whom chorea is chiefly incident ; but, after many
investigations on this point, including over 90 cases of chorea where
the histories were fairly ascertainable, I find that rheumatism occurred
in about 30 per cent. of the immediate families of choreics, and that
rather more than half of these rheumatic families belonged to children
who had themselves suffered from rheumatism.   The ordinary incidence
of rheumatism on those who frequent London hospitals from all causes
being probably not less than 20 per cent., it would appear that rheumatic heredity has a certain but not a high degree of importance in its
association with chorea *per se*.   The connexion of rheumatism with
chorea may well be more extensive than statistics can show, for undoubtedly rheumatism may follow long after chorea, and early rheumatism may escape notice ; but, from the above-given and subsequent
reasons, the causative *rôle* of rheumatism itself seems to be strictly
subordinate to certain nervous conditions essential to the production of
chorea.   There is no clinical difference between the chorea of rheumatic
and non-rheumatic subjects other than the permanent heart-disease

which so often marks the rheumatic class. Hence it is clear that there
can be no ultimate explanation of chorea in the fact of its frequent
alliance with rheumatism; and it must be remembered that, while some
choreics are rheumatic, a very small proportion of rheumatics become
choreic. Doubtless rheumatism is not infrequently, though in less than
10 per cent. of my own cases, the immediate excitant or antecedent
of chorea; and it may be that an hereditary proclivity to rheumatism
may favour the development of the disorder. An enormous majority
of the cases of definite heart-disease with chorea are found in rheumatic
subjects and are mostly subsequent to unquestionable attacks of rheumatism ; and it seems to me at least probable that the valvulitis often,
though by no means always, found in cases dying with chorea is of
rheumatic origin albeit undiscovered. How rheumatism acts as favouring or exciting the nervous disorder which is chorea we do not know.
Sturges has suggested, and I have seen some cases in point, that the
pain of rheumatism may be the immediate excitant of a chorea which
follows directly on acute attacks, but this hypothesis will not go far.
We know, however, that, although the pathology of rheumatism is still
obscure, its concurrence or association with chorea is by no means its
only alliance with nervous disorder; and we may reasonably believe that
the true nexus between rheumatism and chorea, which is plainly observable in but a small minority of choreics, is probably to be found in some
ancestral nervous disorder common to both complaints.

In most cases of chorea there is at some period **Heart-affection** of
some kind, consisting of either *cardiac murmurs*, mostly systolic and
apical, *altered or reduplicated sounds*, or *irregularity in force or rhythm*
of the heart-beats ; and in a certain number there is clinical evidence of
permanent heart-disease. In 111 out of 156 cases, personally observed
by myself and by others as well, and all available for reference on these
points, I found one or more of the above-mentioned phenomena. It
is especially the question of the permanency of cardiac murmurs in
chorea and of their relation to endocarditis and rheumatism that is
here important. Apex murmurs, generally systolic only, but sometimes
" præsystolic " or both, and in a few cases accompanied by aortic diastolic
murmurs, were observed in 80 cases ; these systolic murmurs were often
very soft, variable and evanescent, altered by position, or accompanied
by a basic murmur, and disappeared before the patient's discharge. In
34 out of these 80 cases rheumatism was noted as having occurred, some
few of them, however, being very doubtfully rheumatic. The murmur
was persistent on discharge in 25 of the 34, and disappeared in 4, the
notes of the remaining 5 being deficient on this point. All the diastolic
murmurs are included in these 25 persistent cases.

Of the remaining 46 cases of apical murmur without history or sign of rheumatism, 26 were discharged without any abnormal heart-signs, 12 with both murmur and movements, 1 died with a structurally healthy mitral valve as demonstrated post-mortem, in 2 the notes are silent as to murmur on discharge, while in 5 the murmur remained when the patients left hospital free from all movements. In 2 of these 5 cases there was a præsystolic murmur as well as a systolic, indicating a probability of valvulitis ; but in none was there any further sign or symptom.

As an important supplement to these observations, showing the very great preponderance of rheumatism in the history of cases discharged well with murmurs, I will shortly recount the results of an examination, by Dr. Hastings and myself, of 44 patients who had chorea in my wards at periods varying from two to twelve years previously. Although several of these were not among the number of my fully reported cases, having been imperfectly noted or, in a few instances, altogether omitted from the case-books, the lesson taught is sufficiently striking. Of the 44 patients thus examined 18 had murmurs, mostly apical alone, and 13 of these had had rheumatism. In 7 of these rheumatic cases the heart-disease had been noted on discharge from hospital, in 2 others the acute rheumatism had taken place after discharge, and in the rest a murmur on discharge was not mentioned but, being previously noted, was probably present. In the remaining 26 the heart was normal both as to size and sounds. Of these, three had had rheumatism certainly, and two doubtfully ; and ten more were found noted in the books as having had systolic apex murmurs or irregular heart-action. Thus out of 16, or possibly 18, cases with rheumatism there was persistent murmur in 13, accompanied in nearly all by well-marked signs of heart-disease, while in 26 cases without any evidence of rheumatism there were but 5 with murmurs, two of which, though apical in position, were accompanied by loud venous humming and marked anæmia. In the third, aged 11, examined two years after a single attack, the murmur was systolic and apical, and unaccompanied by any further sign or symptom of heart-disease ; in the fourth, the murmur heard at the lower part of the sternum was diastolic in time (the aortic second sound being still heard at the base), and, being accompanied by a loud humming and arrested by pressure at the root of the neck, was deemed to be of venous origin ; and in the fifth case, aged 18, examined five years after a third attack of chorea, there was a double mitral murmur with some cardiac enlargement. A careful study of these apparently non-rheumatic cases, 10 in all out of the two sets of observations, will show that not more, and possibly several less, than 7 seem to be referable to organic heart-disease ; and the whole series of cases abundantly indicates that permanent heart-disease is rarely, even in appearance, the result of chorea *per se*.

Subtracting from my list of 156 cases above-mentioned the 80 with apical murmurs, there remain 45 with normal hearts and 31 with re-duplicated or altered sounds, irregular action, basic murmur or venous humming, about two-thirds of these last leaving hospital without abnor-mality. Both of these latter classes contain several cases with a history of previous rheumatism. I would add, lastly, that in the 178 West-minster cases, summarised by the registrars, there were 29 of mitral murmurs, constant while under observation, of which 16 had suffered from definite acute rheumatism; while in 77 there were varying or occa-sional apex murmurs quite disappearing before the patient's discharge.

Seeing, then, that permanent heart-disease in choreic patients is in far the largest number of cases unquestionably rheumatic, and bearing in mind that rheumatism or other disease sometimes escapes observation in childhood, it is at least very probable that such heart-disease is either rheumatic in origin or due to some other cause of valvulitis apart from the chorea. But here a certain difficulty meets us in the occasional presence of small vegetations on the edges of the mitral valve in patients dying with chorea, without history of rheumatic symptoms of heart-disease or even observed murmurs, of which one case occurred in my wards at Westminster in May 1891, and has been quoted by Dr. Sturges[1] in connexion with this subject. But it seems as reasonable to refer this valvular affection, found in some of those exceptional cases which die, either to the well-accredited rheumatic cause (albeit undiscovered) or to some other morbid process, as to contend that chorea has an endocarditis of its own apart from rheumatism. Chorea, it must be remembered, nearly always recovers; and, as we have seen, permanent heart-disease very rarely follows on any but definitely rheumatic cases. It is also true, although frequently forgotten, that healthy valves are not seldom found post-mortem after death with chorea. Of this I have seen four examples, in three of which well-marked systolic murmurs were more or less per-sistent until death. I do not discuss the possible or, it may be, probable theory that the fine valvular vegetations found in presumably non-rheumatic chorea are caused by the mechanical deposit of fibrin, for such an hypothesis can neither be demonstrated nor disproved. We certainly know that rheumatic endocarditis may occur without murmur, and there is good reason to believe that ordinary valvulitis may occasionally dis-appear; but it remains true that such a frequent recovery as must be involved in the theory of the constantly endocarditic origin of choreic murmurs is not in accord with clinical and pathological knowledge. I would further urge, not only that irregularity of the heart-beats is much more marked in simple chorea than in at least the early stages of organic mitral disease in children, but also that the invariable absence of the signs of aortic regurgitation in non-rheumatic cases, however the heart

[1] See International Journal of the Medical Sciences, Dec. 1891.

may be otherwise affected, tells considerably against the endocarditic theory of the ordinary mitral trouble in chorea.

Many of the murmurs in non-rheumatic chorea are probably of dynamic origin, and connected with faulty innervation of the nervo-muscular apparatus of the heart and valves ; while some may be referable, in part, to temporary cardiac dilatation.

The **morbid anatomy** of cases dying with chorea throws but little light on pathogeny, as might be expected from the prevalent tendency of the disease to recovery ; and it is clear that no permanent lesion can be the true cause of the phenomena to be explained. Death with or from chorea below the age of puberty is very uncommon, and in boys extremely rare. Great mental excitement or other impairment is the rule in cases dying with otherwise apparently simple chorea, and acute rheumatism figures largely in the fatal cases. I saw a case, however, many years ago in a child of five where the chorea was neither severe nor rheumatic, but where there was an apex murmur. Death followed on a short attack of vomiting, with violent palpitation and epigastric pain ; but *post mortem* the heart and all other organs appeared perfectly normal.

After consideration of much that I have read and something that I have seen regarding the post-mortem appearance of the nerve-centres, I believe that morbid anatomy has contributed little of importance towards ætiology, excepting a considerable amount of evidence of either recent or long-standing hyperæmia of the brain and, perhaps, of the cord ; and that we may reject the theories of neurologists generally on this point, including one of the latest which suggests sclerosis of the nerve-centres as explanatory of choreic phenomena. The fact of changes in the medulla and other parts, due to hyperæmia, as shown in some cases by the enlargement of vessels and a marked cellular accumulation in the perivascular spaces, may indeed be referred to vaso-motor origin ; but it is clear from the whole natural history of the disorder that such appearances are secondary, and that a wider cause must be sought for the explanation of both the vaso-motor paralysis and all the varied phenomena of the disease.[1] Of the heart-changes found after death with chorea I have already spoken. They do not appear to have any connexion with the

---

[1] Since writing the above I have seen Dr. Charlewood Turner's valuable paper (in the *Trans. Path. Soc. of London*, 1892), describing lesions of some of the large pyramidal cells in the deeper layers of the cerebral cortex in the Rolandic region. The lesions, observed by him in five cases dying with chorea, consisted in marked œdematous swelling. Dr. Turner argues that such lesions, being presumably recoverable unless in their highest grade, are in accordance both with the clinical fact of the uncontrolled voluntary movements of chorea (whether temporary or indefinitely persistent), and also, as indicative of nutritive defect, with the pathogenic conditions with which the occurrence of chorea is associated. Dr. Turner admits that these lesions in his cases were probably in part due to the exhaustion of the patients in the

mortality of the disease and may be absent in fatal cases of otherwise typical character.

The **origin of chorea** is probably to some extent explicable by the unstable condition of the developing motor nerve centres in childhood, and the natural history of the disease shows that it makes its usual appearance at the very time and in the very manner that we might expect from the various stresses and disturbances incident on these centres before they are duly organised for controlling muscular movements. The disorder affects first and especially those parts of the body which mostly lack due control, at a time when that control is more and more in requisition; and its subjects are mainly girls and little boys in large and crowded towns, whose nervous systems are unavoidably exposed to the many rude buffets which are the heritage of poverty with all its negative and positive evils. This enormously preponderating incidence of chorea on the children of the poor, with its bearing on both ætiology and treatment, must never be lost sight of.

It has been fully shown by Sturges that the uncontrolled movements of many fidgety children are of the same pattern as those of established chorea; and it may be said that an exaggeration of what may be called a physiological neurosis of childhood is the true neurosis which makes chorea possible. For the production of the marked clinical disorder such a basis is necessary, as well as some exciting cause for its special display, such exciting cause being definite and powerful in inverse proportion to the grade of the neurosis. Children, indeed, are all potentially choreic, even as all adults are potentially hysterical.

There is abundant evidence of various *nervous disorders* in the persons and families of choreic children, and the older the child the more prominent this relationship becomes. The frequent emotional symptoms already mentioned illustrate this point, and well-marked migraine and less special forms of headache are exceedingly common. Out of 162 cases, excluding for the moment 42 of them presently to be quoted in connexion with the excitants of chorea, I find at least 39 who were spontaneously stated by the parents to be always exceptionally "nervous," "excitable" or "irritable;" several subject to night terrors; and many hysterical. In 30 histories from this list of choreics I find a definite note of epilepsy, insanity, chorea or marked hysteria in one or more members of the family, and in all but four instances (of grandfather, aunt and uncles) the affected relatives were

---

moribund state; but a careful consideration of his observations and reasoning seems to establish a great likelihood that future research in this direction will prove that such lesions as he describes are necessary for the production of those choreic movement-symptoms which are referable to disturbance in the Rolandic (or so-called "motor") region of the cerebral cortex.

father, mother, brother or sister. Chorea was noted in 13 families (thrice in the mother and once in the father), epilepsy in 8, and insanity in 5. In many of my cases the family histories were taken without any reference to nervous disorder. It may be concluded, therefore, that nervous heredity probably plays a much more important part among the antecedents of chorea than is shown by these statistics, striking though they are.

As a definitely exciting cause of choreic attacks, nervous disturbance, evidenced by great emotion or distinct fright preceding chorea at periods varying from a few hours to two days, is noted in detail in 42 out of 162 cases. I have particularised 26 of these (belonging to a series then numbering 105) in vol. i. of the Westminster Hospital Reports (1886), and cannot dwell longer here on this important point. Besides these I have several more recent cases, and some which illustrate Sturges' observations of the origin of movement in the limb or limbs which have been the subjects of fatigue, injury, or some other impression causing fright: as, for instance, in a leg after treading on a cat, in a limb which has been struck, or in hands overworked with writing or sewing.

It is the custom of many authorities on this subject, and especially of some who hold the almost exclusively rheumatic pathogeny of chorea, to neglect, deride or exclude all evidence or allegations of nervous excitants as vague and unappreciable ; but I would urge from a study of my own cases, apart from the experienced statements of Sturges and other weighty authorities, that there is a positive and undeniable proof of definite nervous disorder, both personal and hereditary, as showing predisposition, and of definite nervous disturbance as an excitant to chorea, in a far greater number of cases than there is evidence of either family or immediate rheumatism. It must, moreover, be clear to any one who reflects on this matter that many hereditary cases, especially of insanity and epilepsy, may be concealed ; and that, from the nature of things, and notably among the poor, large numbers of efficient causes of fright and other nervous disturbance in childhood must necessarily escape notice. Rheumatism, as we have seen, enhances the neurosis of chorea, and certainly excites the attacks in many instances ; and the two affections may have neural relationships of old ancestral date. It must, however, be remembered that very often a definite nervous shock immediately precedes chorea in cases which at some previous time have had an attack of rheumatism.

In closing these remarks on ætiology I must refer to an important instance of the varying topical distribution of chorea, which was given by Dr. Ranke of Munich at the International Medical Congress held in London in 1881. Out of 40,723 children, seen by him during fourteen years, there were only 19 cases of chorea. Dr. Ranke then informed me that there was no lack of rheumatism among his patients. It would thus appear that it is in other and more complex conditions that an explanation must be sought of the frequency or rarity of chorea in different places.

Of **prognosis** in chorea something has already been said incidentally. The most suddenly beginning and severest cases which soon reach their height are by no means the most protracted. When properly treated, indeed, they are in my experience considerably below the average in duration. I have repeatedly seen patients, who soon required padded beds and very careful attention, recovering completely in three weeks or less. The more chronic the case and the less simply motor it appears to be, the more doubtful is the prognosis as to duration or ultimately perfect recovery; and the history and nervous concomitants of each individual case are of great weight in respect to the probability of recurrence. The average duration of chorea is not a question of much importance, for in many cases its onset is gradual and its date unnoted. Excluding some exceptionally chronic cases, however, I am inclined to set down the average period from the first noticed movements to their complete cessation as about eleven weeks; but with proper treatment from the beginning this might possibly be much less. Very chronic cases are usually slight; and in those which last for years the movements are generally limited, and may frequently be confined to the hand or face alone or to a single group of muscles. I have seen two cases, as the remnants of chorea, of very obstinate but ultimately curable spasmodic movements of the sterno-mastoid or other neck-muscles. One of these had lasted nearly five years. Control over the tongue, as shown by the power of retracting it steadily and slowly, is usually regained before facial and hand movements cease, and is generally a herald of good recovery; but there are many slight cases where the tongue is but little or not at all involved.

Chorea is much influenced by **treatment,** the right direction of which is already indicated by much that I have said. The most complete rest, both of body and mind, must be enjoined or encouraged, and the patients should in all cases be kept in bed until the movements have nearly ceased. In severe cases which show no signs of improvement, and especially when complicated or attended by much mental disturbance, absolute quiet and comparative darkness are, I believe, strongly advisable; and the child should be visited by none but its attendants. In ordinary cases, however, the quiet routine of treatment in a ward for children only seems very beneficial, affording occasions for both companionship and amusement. From a comparison of cases treated over many years at Shadwell with those at Westminster I find that nearly two-thirds of the former leave hospital perfectly recovered, but scarcely more than one-third of the latter. This is not to be accounted for by difference of period of stay in hospital or of medical treatment, but is probably due to all of the Westminster cases having been treated, until recently, in wards with adult patients, and thus more variously disturbed than at Shadwell.

Choreic cases should not be placed within sight of one another; for, although I have never seen an instance of chorea arising from imitation, I am sure, from experience of some cases of recurrence after such juxtaposition and of many more of strikingly rapid improvement on removal from a choreic neighbour, that the nervous effects of chorea upon chorea are bad. In all severe cases injury should be prevented by padding the bed; but coercion of the limbs should only be practised, as a rule, in cases where the child might otherwise hurt itself. As complete freedom as possible, physical as well as mental, is indeed ever to be aimed at in the treatment of chorea. In a few cases, however, Dr. Sturges has obtained a good result from wrapping the child in a sheet, the sufferers expressing a feeling of relief in such confinement of their limbs. I have myself tried this method with success in one or two instances; but, as Dr. Sturges insists, the child must always be consulted before the movements are thus restrained. Nasal feeding must be employed when there is dysphagia, and all care taken to keep up nutrition whenever it seems to fail. Constipation, which is even more frequent in choreics than in most patients confined to bed, should be corrected by occasional enemas or aperients; and disturbed sleep or sleeplessness must be combated. If warmth to the feet, feeding at frequent intervals, quiet and darkness fail, I give wine or brandy in varying doses, opium, a combination of chloral and ammonium bromide, or sulphonal. This last drug is very useful; but I can say from experience that it should be given with caution, always under observation, and never pushed so far as to produce continuous sleep. One dose in the twenty-four hours of about 10 to 15 grains for a child of ten is probably enough. The drug takes long to dissolve in the body, and is apparently cumulative in action. When the nightly sleep is good, narcotic and paralysing drugs should, I think, never be given. Formerly I have often tried these methods, but never with good, and sometimes with evil, results; and, having not seldom given conium in doses which produced their physiological effects, I am convinced that it is much worse than useless.

In most cases the child's attention should be directed away from its movements; and as soon as any improvement is shown, but not before, spontaneous attempts at definite use of the hands should be encouraged. When voluntary control lessens the movements, regulated use, as in writing, knitting or needlework, may be practised for a short time daily. In some mild cases, where improvement lingers during rest in bed, I have found it well to allow the children to sit up or even walk about. The effect of this change must, however, be carefully watched.

As soon as it is deemed advisable to get the patient up, a change of scene is often of great benefit. The child should be sent into the country to lead an unexciting life under the influences of as much fresh air and sunlight as possible, with gentle and regulated exercise.

Although many cases are apparently in good health, many more are in need of both natural and medicinal nervine tonics ; and it is in these that drug-treatment is often of great service. I have had no doubt for many years now, having previously persevered with many medicines indiscriminately, that the numerous cures by drugs and especially by arsenic, reported from time to time in the journals, are to be fully explained in this way ; and I am wholly in accord with Sturges when he says that arsenic and other medicinal tonics are indicated not by the chorea itself but by its many associated weaknesses. Among these anæmia is prominent ; and, believing in the very great value of arsenic in many cases of this affection, I frequently give it in chorea when it is thus or otherwise indicated. At the same time I am well convinced that in a very large number of cases of chorea there is neither promise nor potency of cure by any known drug. I have tried arsenic, as the best accredited medicine, in many obstinate cases, and in large and rapidly increased doses ; but have never found improvement except when other conditions were changed as well. As regards the use of arsenic, I believe that larger doses than from 3 to 5 minims of the " liquor," thrice daily, need never be given ; and that in children, whether choreic or not, there is quite as much susceptibility to poisoning when the drug is pushed as in the case of adults. Cod-liver oil and iron are also most useful tonics. In some cases, especially of long standing, indigestion may require treatment by regulated diet and appropriate medicines.

In the paralytic cases the greatest care should be observed in moving the patients, for the heart is often markedly feeble ; unremitting attention must be given to feeding, the nasal tube being used if need be ; and frequent small doses of alcohol may be indispensable.

CHAPTER VI.

HYSTERIA AND FUNCTIONAL NERVOUS DISORDER.

Although many systematic, and more especially foreign, writers on nervous disorders have fully recognised the frequency of hysterical phenomena in childhood, this subject has been but slightly glanced at or altogether ignored by the authors of most works on disease in children. The admirable lecture, however, on hysteria in Professor Henoch's work, and the comprehensive article, richly illustrated by cases and quotations from many sources, by Dr. C. K. Mills in Keating's *Cyclopædia of the Diseases of Children* leave little to be desired by the clinical student. In this country Dr. Wilks, in his lectures on Diseases of the Nervous System, long ago called attention to hysteria in young boys as well as girls ; and interesting cases have from time to time been published by different observers. A compendious but very practical account is given by Messrs. Ashby and Wright in their text-book of the Diseases of Children.

I shall mainly devote this chapter to a few cases, culled from large numbers in my note-books, in illustration of the chief varieties of hysterical and other phenomena in children under the age of fourteen ; omitting, perforce, the consideration of multiform nervous disorders incident on the periods of puberty and adolescence. Being precluded by limits of space from attempting to discuss or define hysteria, I would refer the reader for my own views on the subject to the article under this heading in Tuke's *Dictionary of Psychological Medicine ;* and would only state here that I construe the term widely as signifying a *neurosis largely due to hereditary constitution,* regarding the various concrete expressions of the disorder as the result of multiform excitants acting on this vulnerable nervous material. In the case of children *fear* and *pain* rank high among the exciting causes of hysterical display. It must be remembered that hysteria is a psychosis as well as a neurosis : that some degree of mental disorder, evinced in the sphere of feeling rather than of intellect, colours and underlies all its phenomena, predominantly physical in expression though they often are. We thus both expect and find two characteristics in the hysteria of early childhood. Owing to the still imperfect and at the same time rapid development of the higher nervous centres, whose action is associated with the control of the emotions and of sensory and motor action as well, less obvious external stresses upset control and produce hysterical phenomena in the child than in the adult. By reason, again, of the less perfect development of the higher cerebral functions, the less complex are the relations of the child with the external world; and the less, in consequence, is the sphere, and the fewer are the varieties, of possible aberration. Although, as we shall see, hysterical phenomena, comparable with many that are seen in adults, whether prominently mental, sensory, or motor in expression, are frequently met with in childhood, yet their character is less protean than in later life, being limited by the child's stage of development. As soon, however, as cerebral development has advanced so far as to render mental action evident, the time of possible hysteria has arrived. I would mention here once for all, with respect to the neurotic relationships of hysteria, that many cases show a large proportion of family histories of insanity, epilepsy, chorea and hysteria itself; and that in hysterics there is not seldom a history of infantile convulsions. A considerable number of choreics are markedly hysterical, and a history of definite epileptic fits is, in my experience, very common with hysterical children.

As in adults, so in children, hysteria may show itself mainly in psychical aberration; or its most prominent features may be motor disturbance, either spasmodic or paralytic, sensory disturbance in the form of either anæsthesia, hyperæsthesia or pain, or some other disorders of function. In most these various elements are, some or all of them, inextricably blended; and the psychical factor, however latent it may be, must be thought of as always present. For practical purposes, however, I shall dwell shortly on the mental aspect of hysteria before giving illustrations of the more prominently physical expressions of the disorder.

The cardinal fact in the **psychopathy** of hysteria is an exaggerated self-consciousness dependent on feeling uncontrolled by intellect; and we know that even in the normal child there is an ample supply of this material. Besides this, evinced in many vagaries of conduct, there is often some intellectual disturbance proper as well; but the chief mental abnormality is evidenced in the sphere of *feeling*, and mainly by excessive impressionability and tumultuous emotion on slight excitation. In close association with this are phenomena which are the direct results of simulation or with difficulty distinguished therefrom. In hysterical adults of both sexes we are all familiar with the frequent coincidence or alternation of clearly involuntary disorder with equally certain malingering, and outrageous, or practically criminal, conduct. This connexion is not seldom seen in childhood; but continuous and elaborate hysterical display, whether mingled or not with imposture, is but rarely met with in young children, their mental development being generally inadequate for its production. The following case, exceptional in so young a child, will serve to illustrate this and other points.

A girl of ten was admitted complaining of headache and giddiness, stiffness of the left leg, and great difficulty in walking, from which she had been suffering for several months. She had come from another hospital, with the provisional diagnosis of spastic paralysis. Two years before these complaints set in she had been subject to occasional fits which were seemingly epileptic. Her left leg was apparently shorter, certainly somewhat smaller, and of lower temperature than the right; the knee-jerk was excessive, and there was marked ankle-clonus. On the right side the knee-jerk was brisk, and slight ankle-clonus was obtainable. The gait soon became much worse, the left leg seeming nearly an inch shorter than the right. On examination, this apparent shortening was found by measurement to be due to tilting of the pelvis, and at the same time a large phantom tumour suddenly appeared in the abdomen. The left lower extremity was very rigid. It was then decided, in her presence, to examine her under chloroform. A week later this was done, upon which all rigidity and apparent shortening disappeared, and spine, pelvis and limbs were found to be quite normal. It was discovered, while she was still insensible, that her nostrils were stuffed with cotton-wool; this device to avoid the effect of the chloroform having been carried out by herself shortly before my visit, as she afterwards confessed. All the symptoms reappeared before she fully regained consciousness; but, having been told that her leg was cured, she rapidly improved and after a fortnight's absolute neglect was discharged perfectly well in every respect. She was cured partly by the detection of her trick and partly by her belief in the treatment.

In the especially mental class of hysterical cases, which usually implies a markedly bad neurotic heredity, we must place numerous vagaries of conduct, and excessive and apparently causeless emotional display, arising more or less clearly out of extreme self-consciousness. Frequent results of this are acts of destructiveness, such as the smashing of glass and furniture; setting fire to bedclothes, curtains &c.; self-injury, especially in the form of scratching the skin; and other conduct which may perhaps be differentiated from insanity only by greater amenability to the moral control of others, by the temporary nature of the outbreaks, and by frequent associations with sensory or motor disorders of the hysterical type. Many of these cases, indeed, are on or within the border-line of insanity; for we can scarcely class otherwise those actions of arson, murder, suicide and the like which we read of from time to time as committed by boys and girls.

Typically acute *maniacal* attacks, usually of short duration, are sometimes seen in children of markedly hysterical temperament, and, with proper care, may disappear without recurrence. I have also seen some cases, with a history of a recent fit or fits, which were marked by eccentric conduct of various kinds, such as is often observed in the post-epileptic conditions of adults. A not uncommon phenomenon is the imitation of the noises and habits of animals : for instance, barking like a dog and biting at bystanders. This is sometimes seen in combination with spasmodic phenomena, such as coughing with a whoop, or "croupy" and rapid breathing; and I have seen several cases of paroxysmal attacks of grunting expiration, lasting many hours, in children with marked mental evidence of hysteria, although this phenomenon often occurs as the most prominent symptom. It is rare to find well-marked examples of this pre-eminently mental form of hysterical display in young children ; and it is doubtless during a year or two before puberty that the worst cases of this kind occur. In such cases, too, I am of opinion that a markedly bad family history of the graver neuroses is excessively frequent; and several that I have met with have suffered notably from infantile convulsions. The hysteria of childhood well illustrates the truth that the less obvious the exciting cause the profounder is the fundamental neurosis ; many of the attacks which we are considering being almost, if not quite, inexplicable by their immediate conditions. The following case is a good example of this form of hysterical outbreak.

A boy of nine, begotten by an actually intoxicated father who had been a drunkard from his youth, was violently passionate from his infancy, and soon developed somnambulism. His schoolmaster reported him as tractable and intelligent but very restless. Some months before admission he had several attacks of furious and apparently causeless passion, foaming at the mouth with fixed jaws and rapid breathing. On more than one occasion he said that he would kill his brother. In one of these attacks he was brought to the hospital. During his stay there of some weeks he was quiet and docile, and seemed perfectly intelligent.

The next case, of an older boy nearly fourteen, open, according to some, to the interpretation of malingering, exemplifies at least the difficulty of accurate diagnosis. He had no bad family or personal history. About a year before admission to Shadwell Hospital he complained of severe headache, often sleeping almost continuously for a week, with much somniloquence, and appearing quite rational in the intervals. He soon became subject to attacks of "wildness," grinding his teeth, swearing, and smashing furniture, and was afterwards apparently unconscious of what he had done. A little before admission prolonged drowsiness was broken by these outbreaks alone. He feared solitude, fancied he saw rats and mice, and frequently screamed with terror. In his sleep his limbs were often seen to twitch. He never hurt himself nor any one else. On admission he appeared healthy, and nothing abnormal was found on a searching examination. While in hospital for a fortnight he had no symptoms. Two years and a half afterwards he came to me as out-patient at Westminster Hospital, with a string of complaints which I at once deemed imaginary. I did not recognise him until he subsequently told me that he had been at Shadwell.

Night terrors with trembling or screaming, and with or without definite hallucinations described by the patient, are very common in hysterical children. They are, however, by no means confined to this class, for they may occur occasionally to almost any child as the result of undue excitement and various kinds of nervous disturbance. *Somnambulism* and *somniloquence* are also frequent, and *nocturnal enuresis* is exceedingly common.

The **motor manifestations of hysteria** in children, in the direction of either spasm or paralysis, are various and frequent. They are characterised by an absence of evidence of all recognised organic causes; but their ultimate test is usually to be found in concomitant manifestations, however slight, of mental disturbance. "Fits" of various kinds are very common in the subjects of these disorders, either with or without apparent loss of consciousness. Sometimes the whole attack consists of partial or generalised spasm, or of tonic contraction of certain groups of muscles, without any discoverable affection of consciousness. When evidence of defect or loss of consciousness is established, the diagnosis of epilepsy cannot be easily rejected; and such diagnosis must always be made, at least provisionally, in those not infrequent cases of falling with giddiness and apparent temporary losses of consciousness without spasm which are by no means uncommon in young subjects of many kinds of hysterical display. The diagnosis of hysteria as the sole cause is easy only when the spasms occur with no loss of consciousness and are accompanied by well-recognised hysterical phenomena. Again, attacks of convulsions, with evidence of complete loss of consciousness, often happen in near association in time with typical hysterical attacks, or alternate with them at varying intervals. Such attacks must be regarded as true *epilepsy*. In a girl of twelve years old I repeatedly observed both kinds of attacks. From the one, consisting of opisthotonos, violent throwing about of the limbs, clenched fists and screaming, she could at once be aroused by a faradic current which soon induced her to give rational answers to questions; while upon the other, where the spasms were more of the regular epileptic order and there was no screaming, no stimulation whatever had any effect. In all the attacks of either kind the eyes were open. Of the typical *hystero-epileptic* attack, as rendered classical by the description of Charcot, I have had no experience in children; and it is the rule that hysterical attacks of general convulsions approaching to this description are seen only in girls or boys nearing puberty. I have observed several attacks of opisthotonic seizures, apparent hallucinations, uttering of various noises, screaming, talking, rolling out of bed and other phenomena, with certainly much indifference or complete want of response to painful stimuli, in boys between twelve and fourteen years old.

The following case of hysterical disorder is fairly typical of several that I have met with in both sexes between the ages of eight and fourteen. A girl of eleven began to have frequent attacks of screaming and then "fainting" (*i.e.* falling suddenly, with apparent loss of consciousness sometimes for half an hour or more) soon after being roughly handled and much frightened by a man, when she was out in the street. Sometimes she would "bark like a dog," and sometimes "laugh idiotically" on coming to. For two months before admission she had spasmodic seizures, kept her bed, and was said to be unable to stand. The attacks observed in hospital were marked by opisthotonos, the fingers being clenched, but the arms thrown wildly about. When I first saw her she would neither speak nor act as she was bid. Being placed on the ground and told to rise, she cried and barked alternately, and then had a convulsion which lasted five minutes. A strong faradic current was applied to her legs until she gradually rose and danced about as in extreme rage. Two days afterwards the child was playing about, talking and walking naturally. She left hospital in a fortnight quite well, having had no relapse other than an easily mastered reluctance to walk by herself, after she had been up three days. This girl had been subject to infantile convulsions, and had an insane aunt. A very similar case occurred in a girl of seven with no ascertainable bad heredity ; but she had had measles, scarlatina, whooping-cough and chicken-pox at short intervals, not long before the nervous troubles set in with "pains all over her, especially in the legs."

Localised spasms of groups of muscles, especially of a tonic character, are not seldom met with. Besides those of legs or arms, I have seen a few instances of rigidity of the muscles of the back and neck, and one of long-continued contraction of the muscles of one shoulder, which was always kept elevated except in sleep. There were but few other hysterical phenomena to mark the case ; but the symptoms ultimately yielded to neglect after making no response to faradism or any active treatment. In this connection I will but just allude to frequent instances in children of laryngeal spasm, of the well-known short "hysterical" cough, and of attacks of rapid breathing (sometimes like paroxysms of asthma), in conjunction or alternation with various hysterical phenomena ; and perhaps "hysterical vomiting," of which I have seen several well-marked examples, may be mentioned here, as possibly due to spasm of the stomach.

Minor degrees of *catalepsy*, consisting in a dazed condition with the state of "flexibilitas cerea" of the limbs, or in the latter state alone without any observable defect of consciousness other than lessened sensibility of the skin, are, I think, not uncommon in hysterical and "nervous" children. I have several times been able to induce this condition readily, either by merely placing the limbs in awkward positions, by previously closing the eyes, or by other simple "hypnotic" methods. It may be said incidentally here that some degree of hypnotism is readily induced in many hysterical children of both sexes, as has often been shown in the wards at Shadwell. I would, however, as a rule deprecate the frequent repetition of this practice in individual cases; except perhaps, in certain instances, with therapeutic intent. Spontaneous examples of cataleptic stiffness of limbs may occasionally be observed in hysterical children of any age. The youngest I have seen was in a girl of three, who was sent to me as an instance of infantile paralysis because she had never walked. I found her surprisingly emotional, readily crying and laughing, with the manner and kind of self-consciousness of a much older girl; and it was observed before long that she indulged in both manual and femoral friction of the vulva. After a few days in hospital she was induced to stand and walk a little with support. She often remained sitting in one position for long; and I soon ascertained that her limbs could be placed and retained for a considerable time in the most uncomfortable attitudes. For more than a quarter of an hour on several occasions she sat with her thighs elevated at an angle of $45°$ to the seat of the chair, her legs extended, and her arms held vertically; occasionally whimpering a little, but making no effort to change the position in which her limbs had been placed. This condition gradually disappeared, but during the whole of her stay in the hospital she was highly emotional.

Other disorders of movement, occurring in the hysterical, are *rhythmical tremors* supervening on falls or other kinds of shock. Such cases, however, are not confined to the hysterical; and some have seemed to me to be referable to the category of organic mischief. In two cases of this kind, aged eight and nine respectively,—one with the "deep reflexes" in all extremities much increased, the other normal in this respect, nystagmus being absent in both,—a few weeks' rest in bed was followed by complete recovery. In a few others, beginning in later childhood and making no improvement, although accompanied by hysterical symptoms and showing no evidence of structural change anywhere, there seemed grave reason to fear organic disease such as disseminated sclerosis. I may here remark that, in children as well as in adults, whatever their symptoms, we should never omit to search as carefully for organic disease when hysterical phenomena are prominent as in those where they are absent. Organic disease and many structural lesions concur with hysterical display, and are often among its exciting causes.

Repeated and larger movements, such as head-nodding and head-rotation, bowing of the body &c., already mentioned under the heading of "local spasms," occur in the hysterical, and sometimes are equally evanescent with the other symptoms; but such movements, again, are by no means always of hysterical nature. In this connexion I would mention one case I have seen, in a little child, of frequent attacks of rotation while in the sitting position on the bed or floor; the child seeming dazed at the time and afterwards exhausted. In close alliance with this are the cases described by many writers as "chorea magna," a combination of attacks of running, jumping, and various co-ordinated movements with all kinds of psychical and sensory affections. Many striking instances of this are detailed by Professor Henoch.

Of *impairment or loss of motor power* hysteria supplies many instances. Simple ataxia of the limbs is not common in my experience ; but I have seen some cases in young and older children, in association with obvious psychical hysteria, where there seemed to be no evidence of further disease. Some became rapidly well with ordinary routine treatment, such as hospital life affords ; while in others, which persisted, both diagnosis and prognosis appeared obscure.

Motor paralysis of whole limbs, as often seen in hysterical adults, is not prominent among my cases and is rare in young children. I have seen a few instances of temporary paralysis of an arm or leg in children under five years old, which from the whole circumstances of the case appeared to be clearly hysterical ; and a case of eighteen months old, reported by Gillette, is quoted by Dr. C. K. Mills in his above-mentioned article. Much more common are local paralyses of the eye-muscles, such as ptosis ; and *aphonia*, caused by paralysis of the vocal cords. I have seen a number of cases of the latter affection in boys and girls between seven and fourteen years old, some of them supervening on chest-colds or definite laryngeal catarrh ; while others were not referable to any previous local disorder. Hysterical aphonia in children is often eminently curable by the infliction of pain which causes crying. I have thus cured many cases of long standing by the use of a strong faradic current, one or both electrodes being always placed on the front of the neck in the case of children old enough to appreciate this direction of local treatment. I am well convinced, from numerous and various trials, that the cures of hysterical aphonia which are reported from time to time as the result of "laryngeal faradisation," whether external or internal, are entirely of psychical nature, and in no way referable to any direct effect on the faultily acting cords. It is always well to induce the child to count or repeat sentences as a condition of the cessation of the painful application. When aphonia is the chief or last lingering symptom of hysterical dis- order, a single use of the battery is often enough to abolish the affection permanently. In many cases, however, the aphonia returns, or is replaced by some other hysterical phenomenon. It may be taken as certain that, as long as functional aphonia is treated locally and chronically, there will be no improvement. The cases which yield to the treatment above mentioned yield at once; those which do not must be treated not locally, but by change of surroundings and other general measures. One case, which was quickly cured by faradism, was in a boy of eleven who had been subject to epileptic convulsions and to frequent pains in the head until he was six years old. Four months before admission fits of apparent unconsciousness recurred without convulsions, and he became completely aphonic. On admission he looked miserable and only whispered; but after one application of faradism all his symptoms disappeared, and he left in three weeks quite well and lively. He had four elder sisters, all subject to hysterical attacks.

Hysterical failure of the peroneal muscles on one or both sides, causing talipes with or without contraction, is very common in hysterical children between ten and fourteen; and sometimes we meet with other forms of deformity from a similar cause. Such symptoms are apt to begin suddenly; some however, as in aphonia, supervening on pain or traumatism. I have seen one or two typical cases growing directly out of true rheumatism; but pain and tenderness over the feet and ankles, apparently of purely hysterical kind, without any trace or suspicion of further disease, is a fairly common antecedent. In some instances the feet preserve the normal position when the patient is lying down unobserved. These cases are often perpetuated by orthopœdic instruments and operations, of which I have seen some disastrous examples, as well as of others whose disability was indefinitely prolonged by confinement to bed owing to mistaken diagnosis. A striking case is that of a highly emotional and precocious girl of twelve years old who suffered from double functional talipes varus of long standing, accompanied by great tenderness of the feet and alleged inability to stand. The diagnosis of chronic rheumatic arthritis had been made, apparently from the pre-existence of some slight swelling; and the prognosis of a life in bed had been pronounced and accepted. With a little firm "moral" treatment out of bed and systematic neglect the child rapidly improved, and after a few weeks was running and riding about.

Prolonged and marked hysterical paralysis, as I have said, is not frequent; is, I think, seen only in children approaching the age of puberty; and, is perhaps always accompanied by other marked signs of the hysterical neurosis. A girl of thirteen was admitted into hospital about a month after she had lost the use, first of the right leg, and then of the left, soon after a fit which was, judging from the history, almost certainly epileptic. She had had several convulsive fits up to two years old; but had since then been healthy and of good intelligence, though very excitable. There was a history of much consumption in the family of her father, who had suffered from several fits during infancy and childhood. She was apparently unable to move her lower extremities at all; passed urine and fæces under her; and had much though not universally diminished sensibility to touch, heat, pricking and faradism. She was plump, with a somewhat silly neurotic expression; and behaved generally like a much younger child. When her legs were lifted up as she lay, and let go, they dropped heavily and were apparently toneless; but they were drawn up slightly, and never dragged, when she was removed from bed and held up by the armpits. Insensibility to all forms of cutaneous stimulation over the whole of the lower extremities appeared almost absolute. The knee-jerks were normal, there was no ankle-clonus, and all reactions to both kinds of current were everywhere natural. After ten days, with encouragement and repeated faradism, she was able to stand a little with support; but this improvement soon declined. The cutaneous anæsthesia so increased in extent and character that she became seemingly insensible to almost everything, the strongest faradism being only felt on the face and ears. She was then isolated, attended for a while almost wholly by nurses, and treated by occasional cold douches and once or twice by the actual cautery to her legs; but as a rule was observantly neglected. After a month I found that she could walk with but little support; and very soon, on being promised return to the general ward when she could walk alone, she got about by herself and was quite cleanly. At this period the temperature rose several times to between 101° and 104° F., with extreme flushing of the face; and three weeks after the improvement had been noticed she had two severe epileptic fits in one day, with intense flushing of the whole surface of the body. She continued to improve otherwise; but after another fortnight had twelve more fits near together which had all the signs of epilepsy. For a month subsequently, until discharge, she seemed perfectly well in mind and body, occasionally however passing water involuntarily. This case in several points differed from one of pure hysteria, mainly hysterical though it clearly was.

Among **sensory disturbances** *anæsthesia and analgesia* of varying distribution are, in my experience, by no means rare in children, either as the leading symptoms or as subordinate to other hysterical phenomena. I have alluded to this under the head of chorea, and have seen several cases in little children comparable to an interesting series published by Dr. T. Barlow in the *British Medical Journal* of December 5, 1881. The following case which I reported at length in *Brain*, Part XXIV., is striking enough for short quotation.

A boy of thirteen, with an epileptic father, and a history of headaches,

occasional falling with "faintness," and a "fit" in the night shortly
before admission, came into hospital complaining of pain and tenderness
over the outer surface of the right thigh, of complete anæsthesia in his
right thumb, and of sometimes "seeing everything red." There was
no evidence over the whole of the right thumb of feeling either touch,
pricking, burning or faradism; and a needle was several times thrust
suddenly and deeply under the nail when he was off his guard or carefully blindfolded. The boy could walk well, and stood steady with his
eyes shut. Some weeks later he complained of pain in his legs. It
was then found that there was no response to any of the above-mentioned stimuli applied to his lower extremities up to an inch and a half
above the upper border of the patellæ. Blisters and strong faradism
and other severe applications to the affected parts were repeatedly and
ineffectively tried. After about five months, during part of which time
he had been at home in the same condition, he was, at my request,
admitted into the London Hospital under Dr. Hughlings Jackson, where
after strong faradism, all his symptoms disappeared in a few days.
While he was at Shadwell there was no psychical aberration observable
in the boy, who was reported to be fond both of his school and his home.
If this case be regarded, as it was by some, as malingering, I would
submit that even the absence of discoverable motive is far less remarkable than the complete control shown over the expression of what must
have been severe suffering. It is scarcely possible but that this striking
case of anæsthesia was perfectly genuine.

Complete hemi-anæsthesia with hemiplegia I have not yet seen below
the age of puberty. Dr. Goodhart, however, quotes two cases in boys of
eleven and twelve, giving the details of one which appeared to be typically
hysterical in its character and history. Neither have affections of the
senses of sight, hearing or smell been at all frequent in my experience,
although I have seen some examples of apparently unilateral amaurosis
and loss of smell, and more of seeming impairment of taste. The
investigation, however, of alleged anæsthesia, and notably of the special
senses, is very difficult in children, with the exception of those cases
where complete analgesia can almost certainly be established by severe
tests.

Cutaneous *hyperæsthesia* and complaints of great *pain* in joints, limbs,
abdomen, head and many other parts are of very frequent occurrence
in children, and are sometimes, though not often, distinguished with
difficulty from the results of organic disease. Numerous examples might
be given of the hysterical joint so familiar in patients beyond the age
of puberty. Sometimes there is slight, and occasionally considerable,
swelling; but, as a rule, this affection can be established as functional by
the absence of objective evidence of arthritis, by free movement under

force or chloroform, and by some of the usual psychical accompaniments of hysteria. More obscure, for a time at least, are cases marked by complaint of severe pain and tenderness over the abdomen, which, with rigidity of the abdominal walls and sometimes much abdominal distension, closely simulate peritonitis. Such cases are also apt to be marked by some pyrexia; and I may here observe that I have many times established the fact of an association between severe abdominal pain and at least temporary rises of temperature at all ages, especially in childhood and youth, with or without suspicion of hysteria. Among many and various cases of hysterical pain and tenderness I may mention an instance of a boy of eight, who had lain in bed for five months with the complaint of acute abdominal tenderness, and was completely cured by one application of strong faradism; and another of a girl of eleven, with similar suffering and frequent attacks of paraplegia off and on for four years, who recovered after three faradisations, and remained well to my knowledge for three years.

Of *nervous pyrexia* I have spoken in the section concerning fevers. I will here but refer to one remarkable case out of others that were perhaps more certainly of hysterical origin, selecting this on account of its peculiarity. This case was under my observation for two months, and closely simulated tertian ague. There was, however, absolutely no other evidence of this disease, and both quinine and arsenic were wholly ineffectual. In spite too of antipyrin and other measures taken to reduce temperature the attacks of pyrexia continued, but at last gradually disappeared. In the intervals of these attacks the child seemed usually very well; but she had marked psychical evidence of hysteria and often had fits of screaming and apparently causeless vomiting in the apyretic intervals. There was no abnormality of the discs, nor any other evidence whatever of organic disease.

To conclude this sketch of the symptomatology of hysteria I would mention but two more cases. One was that of a girl under my observation at intervals for some years between the ages of ten and fourteen. She had at first chorea, most marked in the left arm, rigidity of the left leg, and complete anæsthesia of the right forearm with rhythmical movements like those of disseminated sclerosis. There were frequent recurrences of these and other symptoms in various combinations, and throughout she showed a markedly hysterical character. Her last admission was for a first attack of acute rheumatism some months after the subsidence of her hysterical symptoms. She had danced several times in public before she was ten years old. The next case illustrates the trance-like conditions often seen at or after the age of puberty, but very rarely, I think, at an earlier age; as well as those equally rare disturbances of nutrition with which we are familiar in older patients under the name of "anorexia nervosa" or "apepsia hysterica"—a class of cases which often yield to the fashionable and expensive massage treatment, but, much more often than is generally believed, relapse thereafter, or fail in some other but equally or more deplorable direction. A girl of eleven, with deeply neurotic parents, herself the idolised centre of the family, had suffered for many months from headache of the nature of clavus, with intervals of complete apathy ; lying with wide open and un-winking eyes for several days together. Occasionally she had attacks of general convulsions, but could then always be roused by vigorous measures. She wasted rapidly, and, although she made attempts to eat, was apparently unable to take more than an occasional biscuit, vomiting up nearly everything else she attempted. She had had "massage" at home by a trained nurse for many weeks, but all trials of forced feeding were failures. Removed from home in an extreme stage of emaciation, and isolated for a short time with no active treatment, she improved slightly in flesh, and solid food placed and left by her was eaten and not vomited. Her parents insisted on her return after less than a fortnight, and I was informed several weeks afterwards that she was being "massaged" and was still "almost in a dying condition."

Of the **diagnosis, prognosis** and **treatment** of hysterical affections in children much has been said incidentally in the foregoing remarks. The differential *diagnosis* of the numerous hysterical affections which more or less closely simulate organic disease can only be made by means of careful observation and reflection, and a sound knowledge of the signs and symptoms of such disease and of the various methods of investigation. The previous and family history of the case and the psychical condition are most important factors in diagnosis when the most salient symptoms complained of are physical, such as convulsions, or paralysis, or are prominently those of pain or loss of feeling. At the same time I must repeat that in children, as well as in adults, both slight and serious organic diseases are often apt to excite and to be concealed by phenomena of clearly hysterical nature.

Generally speaking the *prognosis* of hysterical affections in children below puberty is good, provided that the treatment be judicious. It is more or less grave as a rule, especially as regards recurrence in one or other form, when there is a bad family history of neurotic disorder. In the well-to-do classes the ultimate forecast is perhaps on the whole rather worse than among the poor; for in many cases in the latter category external excitants, such as fright and pain and traumatism, are more numerous, and often play a larger proportionate part than the constitutional neurosis in the production of hysterical disorder. In other words, though great stresses may require a less deeply neurotic constitution for the production of hysterical phenomena, the disorder will so much the more readily disappear on the removal of those stresses and the supply of fresh surroundings. On the other hand, a deeply neurotic constitution may almost spontaneously or, at least, with the slightest excuse breed hysterical display ; and the ultimate cure of such cases, often met with among the well-to-do, is difficult or doubtful. However this may be, this much is certain, that, taking all kinds of hospital cases together and comparing them, in point of response to treatment, with those occurring in the well-to-do and, especially, the wealthy classes, the former group have a decided advantage over the latter. Much of this difference is of course to be accounted for by the great aptness of hospital treatment to most cases of hysteria ; for in the general ward of a hospital the child is in a much less self-important position than it can be even in any private institution, although quite separated from its own home and relatives. In a hospital, again, more than anywhere else can the system of observant neglect, so essential to the cure of many cases, be efficiently carried out.

It is of course all-important in the *treatment* of hysterical children to minimise or remove all conditions, both physical and psychical, which tend to emphasise their neurosis or to occasion its display. All matters of hygiene and nutrition should be carefully attended to, anæmia or any other coincident malady energetically combated, and an outdoor life insisted on as much as possible. The patients must always be removed from the care of nervous or hysterical relatives ; and bad cases, as a rule, are better treated away from home. But few subjects of marked hysteria are of physically robust constitution, unless the exciting causes of the display are very plain ; and such drugs as cod-liver oil, arsenic and iron are often of the highest value. I have already indicated the cases in which local treatment of affected parts may be of use. In many instances, however, local treatment of any kind perpetuates the mischief. Each case must be treated on its own merits, and more or less success will generally be the reward of good judgment and management. The child's attention must be directed to external things, and active sympathy must be withheld ; but, all the same, systematically harsh treatment, and even the least degree of the vindictive attitude, should be carefully avoided. It is the great difficulty of steering a course between the petting and scolding of hysterical children that renders most mothers and relatives the worst attendants possible. But very few among even intelligent people can appreciate the nature of hysteria ; and those who are once convinced, as some may be, that the hysterical child is not the subject of what they understand as serious disease, conclude as a rule that the case is one of shamming and viciousness, and treat it according to the dictates of their own ethical or religious creed.

Being of opinion that hysteria in the young is far more often either overlooked, mistaken or disastrously mismanaged and maltreated at both

lay and professional hands than in adults, and desiring to give prominence
to what is called the moral method of cure, I shall say nothing here
of the minor details of drug or other treatment which individual cases
very often necessitate. I would only warn the reader to avoid as much
as possible the use of any sedative or narcotic medicines, including the
bromides; and rarely, if ever, to continue them.

From certain experiments in hypnotism with "suggestion," conducted
by Dr. E. E. Ware, the resident medical officer at Shadwell Hospital,
I am somewhat inclined to believe that benefit may result in some cases
in childhood from the use of this method. In two instances children
who had long and daily suffered from excessively frequent fits, of a nature
difficult to distinguish from epilepsy, remained well for several weeks
after a few applications of hypnotism, during which they had been told
that their seizures would not recur in the future. In one case there
were no fits for several months; after which time, however, the patient
made no further appearances at the hospital.

CHAPTER VII.

HEADACHE.

As in adults, so in children, headache is a symptom of various morbid
conditions; but it is not until after the age of five or six years that
it becomes recognisably prominent as the mark of a more or less independent neurosis. Before the speaking age headache may be evidenced
by facial expression, with knitting of the brows; and by great irritability, restlessness and frequent rolling of the head. In quite young
children we must think of ear-disease; meningitis, especially tubercular;
brain tumours and abscess; syphilis; and the onset of any of the acute
febrile diseases, especially pneumonia and enteric fever: making our
clinical search accordingly for concomitant symptoms of these several
affections. In older children, besides these causes, faulty ocular accommodation, and notably hypermetropia, must always be remembered as a frequent source of headache which is mainly frontal in site, sometimes accompanied by squinting, and usually remittent or altogether absent when
the eyes are not used for reading or with other fixed purpose. Anæmia
alone, from whatever cause arising, is frequently associated with headache,
especially in cases beyond early childhood, and is often apparently causal;
but I have over and over again failed to find any evidence of pain in the
numerous cases I have seen of profound anæmia in young children in

connexion with splenic enlargement. Rheumatism is frequently accompanied by headache, and we should look for evidence of this affection in cases not otherwise explicable; bearing in mind that it is often inconspicuous in childhood, especially as to its arthritic manifestations, and that its diagnosis is not seldom aided by the discovery of a rheumatic family history. Syphilis should always be remembered as a possible cause of headache in children at any age; and such headache is not necessarily accompanied by more definite symptoms of syphilis. Violent or repeated coughing is a common cause of headache both in children and adults; and in some cases of chronic cough, which are neglected or regarded as incurable, the symptom of headache alone may be complained of and unsuccessfully treated in ignorance of its real causation. Hysterical headache, of which I have seen several examples both in boys and girls, can often be clearly discovered by concomitant symptoms and the success of appropriate treatment; but purely neuralgic headache, of the type so common in adults, is, like other neuralgiæ, of uncommon occurrence before puberty. Gastric disturbance, in popular, and, sometimes, in medical parlance, covers a multitude of headaches; but, apart from acute attacks of gastric catarrh, which are as a rule demonstrably dependent on injurious ingesta and accompanied by vomiting, the stomach in my opinion is but rarely accused with justice of causing prominent and recurrent headaches in childhood. I am convinced by long experience that in children, no less than in adults, an immense number of cases of headache, although often apparently induced by dietetic causes or associated with vomiting, and then commonly called " bilious attacks," are due to the underlying and practically primary neurosis presently to be referred to under the name of " migraine." In some cases, however, headache is undoubtedly part of the symptoms due to what is known as lithæmia. Of the headaches which accompany valvular heart-disease, renal disorder, fevers, and many other maladies, presenting nothing peculiar to childhood, I need not speak. The foregoing sketch of the clinical conditions of headache refers only to this affection as an apparently isolated or, at least, pre-eminent complaint.

Migraine, by which we are to understand a neurosis, expressed by paroxysms of headache and disturbed vision, accompanied often by vomiting, and tending to recur through a great part of life, is a very frequent disorder of childhood, making its first appearance in most cases before the age of ten years, and, not seldom, much earlier. Like other neuroses, such as hysteria and epilepsy, it may be latent, and thus roused into expression by strong stimuli alone. This is evidenced by its occurrence, in some few persons, only after prolonged fatigue, mental excitement, or the nervous depression caused by severe illness; but we rarely meet with those who can number only one or two attacks. My own experience is in accord with the well-known dictum that migraine is often allied with a tendency to gout.

For full description of the clinical symptoms of migraine, and discussion of its pathogeny, I must refer to larger and systematic works; and would only state here that my experience of what I regard as this affection in childhood justifies the now prevalent conception of migraine as a primary neurosis, and therefore as not ultimately attributable to erroneous dieting or to faulty processes in the stomach, liver or other organs. The hypothesis of "lithæmia" as the cause of this malady, scantily supported by facts at the best, and failing to cover either the symptoms or the clinical concomitants of migraine, is, in my opinion, in no way supported by the study of the disorder as it undoubtedly occurs in childhood.

In the earliest cases attacks of vomiting are very prominent, although always preceded by either the evidence or the definite complaint of headache; while the existence of the subjective ocular symptoms, but rarely quite absent in the adult, can of course be far less often established in childhood. The popular term "sick-headache," which is often applied to this affection, is especially appropriate to its manifestations in childhood; and in many cases we find that, as years go on, the sickness diminishes or disappears, leaving only the headache with or without the visual disturbance. In many of the past and current writings on migraine in childhood the symptomatic description is taken from the typical accounts of the disease as best known in adults, instead of being based on direct clinical study from young subjects; and therefore its frequent confusion with gastric disorder or "bilious attacks" is ignored or slighted. In a modern article on this matter I find, as an example of this error, the statement that the chief difficulties in diagnosis are caused by organic cerebral disease and petit-mal. Now the prominent vomiting in the migraine of childhood, and the not infrequent sequence of the attack on either a surfeit or an impropriety of diet, cause a vast number of cases to be attributed to diet or disorder of stomach or liver; and lead to much laboriously unsuccessful treatment by strict dieting with lessening of nitrogenous food, by purges, mercurial and otherwise, and by chemical remedies directed to modify the gastric secretions. The avoidance of certain articles of diet, which seem, in some few cases, to determine an attack, is clearly therapeutic up to a certain point; but my experience has amply taught me that not only do numerous cases of paroxysmal migraine occur when the child has been uniformly well dieted and has had good health in the intervals, but also that, in almost every case I have inquired into where the attacks have been attributed by the parents to indigestion or biliousness, there was an acknowledged absence of any such exciting cause in many or most of the paroxysms.

In proportion, indeed, to the intelligence of the parents I have usually found that their diagnosis of "biliousness" is admitted by them to be merely an inference from the fact of sickness, and not from the observation of any dietetic cause, at the absence of which they very frequently express their wonder. I can further say that, in the immense majority of recurrent sick-headaches which I have treated in children who were the alleged subjects of bilious attacks and who had been strictly dieted under medical orders, I have never seen increase, but usually decrease, of the symptoms after the discontinuance of all strict dieting and of drugs directed to disorder of the alimentary system. In the sick-headaches of children, as in those of adults, we find an hereditary history of migraine and other definite neuroses, such as neuralgia, epilepsy, hysteria and insanity, in many cases; in children, indeed, according to my experience such a history obtains in a considerable majority. Further, whether there be such an hereditary neurotic history or not, the nervous temperament is very marked in most cases, and the immediate occasion of an attack is often found in shock, excitement or overwork of the mind. In older children a tendency to asthma is sometimes observed. The affection is much more common in city than in country children, and in those with generally unhygienic surroundings. Apart from the prevalent prominence of the gastric symptoms, the migraine of childhood has generally less distinctive characters than in the adult; owing, partly, in all probability to the common subjective symptoms, such as visual derangement, noise in the ears, giddiness, chilliness, and numbness or tingling of the extremities, being less often expressed. The headache, too, is in my experience, judging from the cases where the patient is old enough to describe it, less often one sided in onset than in the adult; but the frequency, even in the adult, of a diffused frontal headache is certainly great enough to render the term "migraine" or "hemicrania" of ques tionable propriety. Light and noise are as a rule avoided by the patients, who usually desire to lie down. The appetite is generally lost. All the varieties of migraine, with its varying vaso-motor symptoms of pallor or flushing and dilatation or contraction of the pupils, may be sometimes observed in children. I have but seldom noted or suspected the occurrence of pyrexia in the paroxysms, as mentioned by Gowers. Mere pain may certainly be accompanied by a rise of temperature in young children. We must always be on the watch for *otitis* in cases of apparently one-sided headache in young children, especially when it is accompanied by pyrexia and vomiting.

The transient aphasia which occasionally occurs in adults is rare in childhood. I have never observed it; but in two cases of well-marked migraine, in children that I have seen, the frequent occurrence of aphasia at the outset has been clearly described to me.

In making the **diagnosis** of migraine in children we must be careful to exclude other causes of headache by instituting a thorough clinical examination and inquiry ; and, in those cases where the headache is very frequent, as well as the very few in which it is almost constant, we must search for ophthalmoscopic changes or any localising or other symptoms of brain disease before definitely pronouncing the headache to be of non-organic origin.   Migraine is essentially paroxysmal ; and usually there are several days, or more often several weeks, between the attacks.   The paroxysms do not as a rule last so long in childhood as in adult life, being rarely of more than a day's and often but of a few hours' duration.

The **treatment** of headache will of course depend on the cause, and that of the various symptomatic forms is often clearly indicated and successful.   We must examine for faulty visual accommodation or astigmatism, and remedy any discovered defect by appropriate glasses. In any case of apparently idiopathic headache, whether migrainous or purely neuralgic in nature, we may always try the effect of antipyrin in two doses of three or four grains, with a four hours' interval, for a child of five years old.   This drug is capable of notably alleviating attacks of migraine in numerous instances, and of completely arresting those of pure neuralgia in many more.   It is of more importance, it seems to me, in the case of children, to aim at antagonizing the tendency to frequent headaches, than to cure the individual attacks ; for the longer this tendency lasts the more intractable it becomes.   With this object the child should be encouraged to be out of doors as much as possible, and subjected to the best hygienic influences.   I always, and often successfully, prescribe arsenic and iron, with or without strychnine, and very often with cod-liver oil ; and forbid all continued or excessive mental strain.   The systematic administration of quinine is occasionally of very good service ; and, according to some, the same may be said of iodide of potassium.   I have never observed any benefit from the bromides, except as sometimes tending to lessen the severity of an attack when given in a large dose.   Dr. Eustace Smith strongly recommends the persistent use of strychnine and ergot.   For the symptomatic treatment of the attacks, different drugs are unquestionably efficacious in different cases ; and we may be reduced to an empirical trial of one remedy after another, frequently with good success at last.   Indian hemp, chloral hydrate, chloride of ammonium, gelsemium, alcohol, guaranine, caffeine, and strong coffee or tea, have all been found useful in some cases.   My own experience, both with children and adults, is quite in accord with that of the late Dr. Fagge, who, in spite of the acknowledged tendency of this disease to cling more or less to a patient through the greater part of life, says that migraine "if systematically taken in hand" (and, I would add, with due regard to the circumstances of each individual case) is very amenable to treatment.

CHAPTER VIII.

OTITIS.

OTITIS of various kinds, especially of the middle ear, is common in infancy and childhood; and, from some of its less generally recognised manifestations, seems to require separate consideration which may find place here. In many instances otitis causes much general febrile disturbance, or, it may be, symptoms closely simulating meningitis, without any prominent sign of local trouble. Many cases of marked and enduring pyrexia in infants, mostly of a remittent but sometimes of a continued form, and causing great diagnostic difficulty, are due to otitis. Unless the ears be examined with the speculum, when bulging of one or both membranes may sometimes be found, or, failing this evidence, unless the membranes be punctured and the pus evacuated, these cases may be for long undiscovered and wrongly diagnosed as enteric fever, tuberculosis, or some other febrile condition. Such mistakes, several times made in the wards at Shadwell, have led us of late years to make a careful examination of the ears, or puncture of the membranes, in many cases of pyrexia in young children which were not otherwise explained; with the result of finding purulent otitis media in many instances, including some where nothing abnormal was observed with the otoscope, and where there was no complaint or evidence of ear-ache. I have been much instructed by the frequent detection of these cases, among both in- and out-patients, by Dr. E. B. Hastings, recently resident medical officer at Shadwell.

Cases of both catarrhal and purulent otitis media seem to be frequently set up by extension of naso-pharyngeal inflammation along the Eustachian tube, or by tonsillitis or post-nasal growths; and are encouraged by any blocking of the nares which facilitates the forcible entry of air or liquids into the tympanic cavity during coughing, swallowing, or vomiting. In children unable to speak a catarrhal inflammation of the tympanum may occur, of sufficient importance to cause considerable and prolonged febrile disturbance, without definite evidence of local disease, and with no subsequent discharge of pus. In severe cases, however, pus is often discharged after a while, with subsidence of symptoms. Otitis, therefore, should always be thought of in otherwise inexplicable attacks of fever; and the proof of its occurrence is of course very strong when there is constant crying, or evidence of local pain or deafness, which can usually be discovered in older children. These latter cases, of plain nature, are best treated at first with the continuous application of dry heat to the ear by means of bran-poultices, which will frequently relieve the pain in a short time. The tympanum may also be inflated by the Politzer method. If, however, persistence of symptoms, and more especially rigors and bulging of the membrane, point to a purulent inflammation, the membrane should be at once incised. In the cases where only fever is present, and careful observation has excluded other causes as far as possible, the membrane or membranes should certainly be punctured if there be any oozing or bulging; and, even in the absence of these signs, this operation should not be long delayed in otherwise inexplicable cases where the fever is persistent and high. Properly performed it does little or no harm, even when its result disproves the suspicion of ear-mischief. Whether the opening is spontaneous or surgical, insufflation of the powder of boric acid or iodoform should be practised, and the meatus carefully cleaned and stuffed with antiseptic wool.

Purulent otitis is not very often the result of simple catarrh, but is exceedingly frequent after many of the exanthemata, especially scarlatina or measles, and often begins insidiously. It occurs also after diphtheria, sometimes after enteric fever, and in association with any form of cerebral meningitis. Especially with meningitis the inflammation seems to begin in both internal ears, and may or may not affect the tympanum. It is certain that purulent double otitis media often occurs in connexion with meningitis, tubercular or otherwise, without any disease of the petrous bone, and is then probably to be regarded not as causal but as concomitant with the wider affection. *Otitis interna* may be evidenced by vomiting and cerebral symptoms, especially giddiness and unsteady gait; and results in complete deafness. The deafness which is observed after recovery from apparent meningitis is probably often due to this affection.

It must be remembered, in connexion with the clinical importance of otitis, that, among several other micro-organisms found in association with the lesion, the "diplococcus pneumoniæ" has been observed. Some cases of this disease may thus be probably regarded as instances of independent specific infection.

Disease of the petrous bone is well known as a sequel of otitis, and often causes purulent meningitis, cerebral or cerebellar abscess, thrombosis of the lateral sinus, or facial paralysis.

In all cases of chronic otorrhœa sedulous attempts should be made to check the affection by antiseptic and astringent injections. For this purpose a lotion of sulphate of zinc or borax, or both, of the strength of five grains to the ounce of water, is useful; and, in obstinate cases, a similar or somewhat weaker solution of silver nitrate.

CHAPTER IX.

TETANUS.

ALTHOUGH rare, or at least rarely coming under medical observation, in England, tetanus in infants is a disease of such importance and fatality as to necessitate a short notice. I consider it in connexion with nervous disorders because of its prominent symptoms. Present knowledge as to its ætiology almost conclusively shows that it is strictly an infectious disease. Among the large number of young babies admitted into the Shadwell Hospital during the last eighteen years or more I have seen but very few examples, apart from occasional traumatic cases under surgical care. It would appear from statistics that traumatic tetanus is on the whole comparatively not rare in children beyond infancy, and that it is most frequent during early youth. Nothing need be said concerning this form of the affection, which is described by all systematic writers, except that it occurs after punctured, lacerated and contused, rather than incised, wounds; that dirt in the wound is a strongly favouring factor; and that modern research has, practically, proved that the essential cause is the operation of the specific bacillus of Nicolaier, introduced at the seat of the lesion. This bacillus, several excellent specimens of which were shown at the London International Congress of Hygiene in 1891, is marked by one knob-shaped extremity caused by the development of ovoid spore-formation. Of the so-called *idiopathic* or non-traumatic tetanus, which may occur at any age and numbers more recoveries than the commoner traumatic form, it can only be said that it is probably also specific in its origin ; the germ being introduced in other ways, possibly by the mouth, and its poison acting under less favourable conditions. The high temperature that marks some cases is almost surely due to affection of the nerve-centres which are concerned with the regulation of the body-heat.

"Tetanus neonatorum" is almost certainly, considering both its similar symptomatology and its favouring conditions, of the same pathology as the tetanus of later age. Tetanus in animals has been produced by Beumer by inoculation with inflammatory material from the umbilicus of a fatal case of tetanus neonatorum. It occurs mostly in association with dirt and neglect, with injury or inflammation of the umbilical cord, and with lesions during birth ; and is apparently favoured by damp and cold. The symptoms usually appear between four and eight days after birth ; and it has been remarked by Niemeyer that its limits of appearance are between the first and fifth day after the separation of the remains of the umbilical cord. The disease is fatal in about 90 per cent. of all cases; and death usually takes place between four or five hours and three days after the onset of symptoms.

In the **diagnosis** of tetanus, other than that of the new-born, *strychnia poisoning* must of course be thought of; and perhaps especially now, when strychnia is so often given to children. The spasms of strychnia poisoning begin as a rule suddenly, not gradually, as in tetanus; the arms are involved; the jaws are not affected until late in the paroxysm instead of at the beginning; and the seizures are usually separated by periods of complete general flaccidity of all muscles. From *tetany* the diagnosis is mostly of no difficulty; the hands and feet only being tonically contracted in this disease of infancy, which, as we have seen, is closely allied to the convulsive condition.

The earlier the symptoms appear in traumatic cases, the worse is the **prognosis**. The non-traumatic instances are much more hopeful than others. In "tetanus neonatorum," which is in all respects comparable to traumatic tetanus, the only slight element of hope is the duration of the symptoms beyond two or three days. Cases of tetanus with wounds should be **treated** in all respects, both as regards surgical dressings and subsequent disinfection of bed and bedding, on the principles applicable to infectious diseases. The wound should be freely incised and thoroughly cleansed with an antiseptic. In non-traumatic cases it has been suggested, and it is probably advisable, to give an initial purge and such an antiseptic as salicylic acid. Forced feeding by a tube will be necessary, and alcohol should be given from time to time, as well as bromide of potassium with chloral in as full doses as can be safely borne. This can be judged of only by its effect, and therefore small doses should be administered at first. For the same sedative purpose inhalation of chloroform, or morphia injections, may be tried with great caution. Calabar bean has been extolled by some and rejected by others. It was useless in the few cases of tetanus neonatorum where I tried it, as it certainly is in traumatic tetanus. In one case of so-called idiopathic tetanus, in a boy of ten years old, recovery took place while calabar bean was being taken; but improvement had already set in before the drug was prescribed. The best mode of administration is by hypodermic injection of eserine, beginning with $\frac{1}{100}$th of a grain.

SECTION V.

DISEASES OF THE RESPIRATORY SYSTEM.

SECTION V.—DISEASES OF THE RESPIRATORY SYSTEM.

I HAVE thought it well, for practical purposes, to depart in some particulars from a strictly systematic or anatomical classification of the disorders to be treated of in this section. The chief symptoms pointing to disorder of the respiratory passages are frequent or laboured breathing, impairment or loss of voice, cough, and expectoration. Some of these symptoms may occur singly or in marked prominence, or in varying proportions and combinations. Affections of the *larynx* are marked by alteration or absence of voice or by dyspnœa in all degrees; persistent cough becoming prominent in relation to the amount of involvement of the trachea and larger bronchi. The special symptom of affection of the *trachea and larger bronchi* is cough; dyspnœa being marked when the larynx or smaller bronchi are much engaged. The leading characteristic of disease of the *smaller bronchi and the air-cells* is dyspnœa or hurried breathing; cough being sometimes marked, but often insignificant, and occasionally absent. Proceeding, then, partly on a clinical and partly on a regional basis of classification, I shall treat first of diseases of the larynx and upper air-passages; next of the affections of the trachea and larger bronchi, including chronic bronchitis; and lastly of those of the smaller bronchi and the lungs. The subjects included under these headings will frequently overlap one another; but no one strict principle of classification, be it anatomical, ætiological or clinical, is free from these objections, and the arrangement I have chosen has the advantage of calling attention to what appear to me the main clinical groups of respiratory disorders.

In all cases where the symptoms point to affections of this class it is desirable to *examine the whole body* of the child as completely as possible at the outset. By this practice many simultaneous observations may be made, and many blunders prevented. The nature of the breathing, the retraction, if any, of the soft parts of the thorax and neck, the state of the abdomen, the appearance of the skin, and the amount of general nutrition may thus, among other points, be rapidly noted by eye or hand, and we may at once learn that we have to do with something more than disease localised in the thorax. Examination of the fauces should never be omitted.

335

Considering the frequency of general diseases, infectious or otherwise, being prominently evidenced by, or complicated or merely coincident with, the symptoms of respiratory trouble, the value of this caution as to exhaustive examination is manifest. I have frequently known neglect of it, both in my own practice and that of others, either prevent or postpone the discovery of exanthematous or other disorders. Enteric fever has thus been missed, and marked deformities, especially those of rickets, passed over. More than once have I seen diphtheria ignored, with deplorable results, by omission to examine the throat and nose in cases, not only of general febrile illness, but also of coincident and even chronic respiratory disease. It must then be insisted on that the most careful clinical examination be made of the thorax and the whole body as soon as possible in every case. It may frequently, indeed, be made then, once for all.

In examining a young child's chest there are certain points to be remembered that may escape those who are inexperienced. It is not my intention to dwell long on the methods of handling or speaking to children, for I believe that his own tact and a few trials will enable any teachable man to adapt the mode of his examination to the age and idiosyncrasy of any of his patients. Although in some cases much can be done as regards auscultation with the ear alone, the stethoscope should generally be used for the sake of accuracy; and, notwithstanding the child's possible crying and restlessness, good results will soon be attained by practice. It is well to percuss when the child is quiet; for noise and movement interfere far more with percussion than auscultation. It is consequently sometimes better to percuss last in the order of examination, when the child may have become reconciled to the procedure. I believe it is always better to percuss lightly with one finger on another, considering the ready awakening of the general chest resonance that follows on more vigorous strokes. More definite and brilliant results of merely local diagnosis may occasionally be attained by the use of plessor and plessimeter; but these, I am sure, can always be practically dispensed with, especially in the case of children. The transmission of morbid sounds from the affected to the other side is certainly much more marked in children than adults; but care and a good ear will usually prevent erroneous conclusions. " *Puerile* " breathing, which is so familiar in name to the student, is very frequently forgotten in practice; and the normally harsh and sometimes almost grating sound of inspiration, as well as the frequently somewhat prolonged expiration in young children, is not seldom mistaken for a sign of disease. Several times have I known slight pleural effusions give rise to the diagnosis of disease in the opposite lung. It is essential, considering the great frequency of *bronchial* breathing in young children without any chest affection, to be especially careful in the comparative examination of both sides of the chest before pronouncing on the existence of disease. Another difficulty is constantly presenting itself to the beginner in percussing the chests of infants, owing to the great difference in note frequently observed on the two sides when there is nothing the matter. This is due to the varying amount of air entering the lungs from the still incompletely regulated action of the nervous apparatus concerned in respiration. We must also remember here that the respiratory rhythm in infancy is often very irregular from the same cause. It is only at a later time that this irregularity becomes a definite sign of disease. In percussing the chest we must not rely on comparative observations made during inspiration on one side and expiration on the other, for such a procedure may lead to surprisingly false results. It is quite common to note prolonged inspiratory holding of the breath by infants during examination, and also long pauses after expiration. Much may be learned from listening to the long inspiration which follows the pause. The type of infantile breathing is diaphragmatic and mainly nasal, oral breathing being a later acquisition; and the frequency of the respiratory act is great, gradually decreasing till it reaches the normal adult standard about the age of two years. In every case the results of percussion and of auscultation must reciprocally check one another; and it will often be well to postpone our diagnosis until we can repeat our examination.

CHAPTER I.

AFFECTIONS OF THE NOSE.

THE nasal passages are subject to various diseases which more or less interfere with respiration.

Nasal catarrh, whether due to chill or infection, may be observed at all ages, but is not common in very early infancy. It may be the herald of a laryngeal, tracheal, or bronchial and broncho-pneumonic catarrh, and is sometimes a very early indication of measles. In diphtheria, too, it may be an early symptom, and, when marked, is of bad omen both in that disease and in scarlatina. In every case of nasal catarrh the child should be stripped, and the throat and chest carefully examined.

The chronic form of nasal catarrh is often seen in *scrofulous* children, and is generally accompanied by other symptoms, such as dermatitis or mucous inflammations, especially about the eye and ear, and swelling of the cervical glands. It may go on to ulceration of the mucosa and disease of the bone, giving rise to ozœna with purulent or sanious

Y

discharge and all degrees of fœtor.   One of the commonest forms of
chronic nasal mischief is the *syphilitic;* and persistent nasal discharge
in infants, whether purulent or not, should always be at least suspected
to be of this nature, even if it be not accompanied by the characteristic
ulcerations or cracks between the nostrils and upper lip.

Owing to the predominantly nasal character of infantile breathing, any
blocking of the nares gives rise to more or less dyspnœa and impedes the
act of sucking.

Acute nasal catarrh should be treated by confinement to bed and
warmth ; and in marked cases, in view of the possible sequel of laryngitis,
severe bronchitis, or broncho-pneumonia, a steam-tent should inclose the
bed.   In chronic cases anti-syphilitic treatment should usually be tried.
I can confirm the statement of Goodhart that a course of grey powders
seems sometimes to be of use in chronic "snuffles," even when there is
no further evidence of syphilis.   It is advisable, in all cases of chronic
nasal catarrh, to keep the nostrils clean by local treatment, painting them
with the glycerine of borax or of tannic acid, or syringing them with a
weak solution of zinc sulphate (gr. iij. to the ounce) or of silver nitrate
(gr. i. to the ounce).   Any detected or suspected disorder underlying the
symptom should be treated by appropriate dietetic and hygienic remedies.

Over-growth of the glandular tissue at the upper part of the pharynx
and in the posterior nares has been for some time recognised as connected
with a definite set of symptoms, and is of frequent occurrence.   It is
generally known under the name of "**nasal adenoids**" or adenoid
vegetations.   Often met with in conjunction with tonsillar enlargement,
it may nevertheless be found alone, and causes snoring during sleep,
nasal voice, and a vacant expression mainly due to the habitually open
mouth.   There is frequently a discharge of mucus and blood into the
mouth, and more or less constant headache.

These growths, of various size and shape, sessile or pedunculated, can
be felt by the finger passed behind the palatal arch ; and their presence
may often be guessed from the open mouth and facial expression above
mentioned.   This affection is often connected with the scrofulous condition, and seems to follow on repeated catarrh ; but in many cases it
appears in very early life without discoverable exciting cause, and no
ætiological generalisation can be made.

The best treatment is the careful removal of the growths by the
finger-nail or an instrument invented for the purpose ; and subsequent
cauterisation of the part with nitrate of silver is probably advisable.
Strikingly good results are thus obtained.   The operation should not
be delayed ; for, among other untoward consequences of this affection,
chronic catarrh of the middle ear and purulent discharge from a ruptured
tympanum may follow on long neglect.

CHAPTER II.

LARYNGEAL AND LARYNGO-TRACHEAL AFFECTIONS.

LARYNGEAL affections are marked clinically by altered voice with more or less cough, by dyspnœa, or by a combination of these symptoms. In very early life their pre-eminent characteristic is dyspnœa, owing to the narrowness of the infantile glottis, and to a liability to general spasm which especially affects the laryngeal muscles. A comparatively slight inflammation of the laryngeal mucosa may thus materially impede the breathing, inspissated secretion alone being able to cause considerable obstruction ; and severe dyspnœa may occur in cases where the spasmodic habit is marked, even apart from the exciting cause of demonstrable affection of the laryngeal structures.

Excluding certain sources of laryngeal trouble, presently to be glanced at, and mostly with no marked peculiarity in early life, we shall find that the laryngo-tracheal affections, which, from their special incidence on child-hood, must chiefly engage our attention, fall clinically into three groups : the *first* being characterised by pure spasm ; the *second* by spasm arising from local affection of the larynx, however slight ; and the *third* by direct obstruction of the upper air-passages from acute and chronic inflammatory changes, membranous or otherwise.

Before discussing these groups, which, as will at once be seen, have in practice no absolute line of demarcation between them, I briefly touch upon the following causes of greater or less stenosis of the larynx and trachea, which must be borne in mind for the sake of diagnosis. **Warty growths** (papillomata), though not very common in childhood, are probably more so than is generally taught, judging from the results of post-mortem examinations. They may occur in very young infants ; and the symptoms may begin either suddenly or insidiously, cases of the first kind being most liable to mistaken diagnosis. The voice is impaired, and all degrees of remittent or continuous dyspnœa may be observed. Laryngoscopy, the only means of making a positive diagnosis, is always difficult and mostly impossible in quite young children ; and here emphatically so from the usually concomitant pharyngeal catarrh and irritability. This condition may be suspected from the absence of fever, the long continuance of the symptoms, and the evidence of the obstruction being solely laryngeal. When demonstrated, the growths should be removed if possible ; and, when the symptoms are severe and obstinate, tracheotomy should certainly be performed while the diagnosis is yet incomplete. The growths may then perhaps be at least partially removed, or thyrotomy may subsequently be performed.

Foreign bodies may obstruct the windpipe, causing paroxysmal dyspnœa and, unless soon otherwise expelled, indicating tracheotomy even before the development of urgent symptoms, when the nature of the case is clear. In some cases the body becomes lodged in one bronchus, giving rise to signs of bronchitis and of deficient entry of air on one side.

Laryngeal œdema, from inflammation or ulceration of the neighbouring pharyngeal structures, may seriously impede breathing, as also may **paralysis of one or both abductors.** Such incidents may occur in the course of tuberculosis, syphilis, measles, scarlet fever, diphtheria, small-pox, and enteric or other fevers.

Pharyngeal abscesses or **enlarged glands,** such as thyroid, thymus, or bronchial, may narrow the laryngo-tracheal tube; and even the most careful examination and study of these cases may not seldom fail us when seeking for an accurate diagnosis. A pharyngeal abscess should of course be rarely, if ever, missed; but the following case, which happened at the East London Hospital for Children, shows how difficult it may be to distinguish the effects of a caseous gland from that of a foreign body in the trachea. Sudden dyspnœa occurred in a child of one year old in apparently perfect health, who was at once brought to the hospital. Tracheotomy was performed by Dr. Hastings, giving slight temporary relief; but death followed in an hour and a half from the first symptoms of the attack. At the post-mortem a caseous gland, which had ulcerated into the trachea, was found just above the bifurcation, entirely occluding one bronchus.

Generally speaking, for a correct diagnosis of the causes of laryngo-tracheal obstruction, the history of the case must be carefully studied; and any concomitant symptoms, such as fever or other constitutional disturbance, must be taken into account. The diagnosis of obstruction of the upper air-passages from that caused by extensive involvement of the bronchial tree is not always to be made at first sight, especially when clearness of voice excludes marked laryngeal mischief; for extreme retraction of the soft parts of the thorax may be seen in both affections. Stridor, however, points to tracheal obstruction; and, especially when the larynx is involved, as in membranous inflammation, it is expiratory as well as inspiratory. Loss of voice and hoarse or whispering cough are pathognomonic of laryngeal trouble.

Laryngismus Stridulus.

Under this title I would include all laryngeal attacks in children which are strictly paroxysmal, the voice and breathing being unaffected in the intervals. A very large majority of these cases are undoubtedly referable to instability of the nervous mechanism, and are, further, usually associated with rickets. Some cases, unconnected with rickets or evident convulsive tendency, are probably due to a recurvation of the epiglottis, causing a close approximation of the ary-epiglottic folds, as shown by Dr. Lees in a case which died from some other cause; and still others may possibly be due, as was once widely believed, to reflex irritation from the pressure on the vagus by enlarged mediastinal glands. It must, however, be borne in mind that the essentially paroxysmal nature of laryngismus is not well accounted for by this last supposed exciting cause, and that the symptom is but rarely accompanied by either clinical or anatomical evidence of this condition. The special nervous proclivity which confessedly underlies most cases of laryngismus is probably a necessary element in the production of the spasmodic attacks which have been referred to such peripheral excitation.

True laryngismus is mainly to be studied in rickety children under two years old, although the tendency may sometimes be observed at a somewhat later time. It is, strictly speaking, a nervous affection—a disturbance of the respiratory rhythm. In its simplest form it is marked by glottic spasm, evidenced by a suspension of breathing and pallor or blueness of the face, followed, after a while, by a more or less pronounced crowing inspiration. The attack is frequently observed on the child's awakening from sleep. Excitement indeed, such as a fit of crying, often produces it; but in most cases, as in many allied spasms, the immediate occasion of the discharge of the unstable nerve-centre is not discoverable. In many cases, perhaps in most, the careful observer will note some evidence of further spasm, such as twitchings of the muscles of the face; and in the severe forms general convulsions are marked. Here the child throws itself back in evident distress, the neck and back are arched, the chest and abdomen stiff, the eyes turned up, and the limbs are tonically contracted with thumbs doubled, fingers extended, wrists flexed, legs thrust out, soles turned in, and toes stretched apart. There is often, too, discharge from rectum and bladder. After a few seconds the opening of the glottis is accompanied by the crowing breath. The nervous basis of the disorder is, further, well exemplified by the laryngismal attacks alternating, in some cases, with slight paroxysms of apnœa and rigidity alone, without glottic symptoms, such attacks being referable to diaphragmatic spasm. In some few instances death may occur in prolonged spasm; and, as a result, the brain and membranes may be seen gorged with blood after death.

It is probable, though not certain, that some disturbance of the stomach or bowels, or occasionally dentition or other peripheral excitants, may occasion this and other convulsive phenomena; but such causes are not necessary for the due explanation of the nervous disorder. Dr.

Gee has made the valuable observation, which most experience confirms, that laryngismus is most common in the winter months, and attributes it to the nervous conditions encouraged by close confinement and bad ventilation.

The main difficulty in **diagnosis** is the possible confusion of laryngismus with the results of a foreign body in the air-passages; but time soon removes the doubt. After the attack the voice is clear, and all symptoms vanish. The spasmodic dyspnœa which may result from recurvation of the epiglottis or other infantile peculiarity is not severe; but it is more continuous, or is at least more frequently excited by anything that hurries the breathing. This condition appears always to right itself with increasing age.

For **treatment** the sufferer from laryngismus should be placed in a hot bath. Should the attacks be frequent, the bromides, and preferably the ammonium salt, should be given; and the more recurrent they are the greater is this indication. In general and repeated convulsions the inhalation of chloroform is sometimes useful between the paroxysms; and morphia may be cautiously given with good effect. The rectal injection of two or three grains of chloral hydrate, with or without some bromide of potassium, has been often found very successful. Artificial respiration should be tried when the period of apnœa is prolonged; and the finger should be passed down to the epiglottis in case the tip of that body be imprisoned between the posterior wall of the larynx and the pharynx, as has been pointed out by some observers. For the rest, the most important part of the treatment is that of rickets generally; and special care should be taken to furnish the child with plenty of light, fresh air, and good nourishment, so indispensable for the adequate supply of nerve-force. A series of cold or cool baths is a good prophylactic; and cod-liver oil and iron are often advisable. All known or supposed exciting causes should be avoided or counteracted, and especially chills, which, by causing catarrh of the air-passages, may also occasion this particular spasm.

Glottic Spasm with Laryngeal Catarrh.

In this class of cases, frequently styled " false croup " or "stridulous laryngitis," the symptoms of laryngeal spasm predominate, and the attacks, which often begin quite suddenly, are apt to be confused with pure laryngismus. But there is always evidence, however slight, of some preceding catarrh; and vocal hoarseness, perhaps with some disturbance of breathing, is noticeable in the intervals of the attacks. The subjects of this affection are usually children from about one and a half to five or six years old, their liability to it almost always showing itself quite early; and there are frequently symptoms of nerve-disorder other than this tendency to spasm. The laryngeal affection is to be traced, however, to a strictly catarrhal origin, which can be demonstrated laryngoscopically in older patients, and is evidenced by the hoarseness and cough which may remain after the spasmodic symptom has disappeared. The attack as a rule begins suddenly and most often, as in laryngismus, at night; the symptoms of bronchial or nasal catarrh, which always accompany and often precede it, being generally unobserved or neglected. It may be of great severity and cause much alarm. It is very apt to recur, sometimes night after night; and there may be several closely successive attacks in one night. Sufferers from this disorder are said by their mothers to be "subject to croup." In almost every case some catarrhal symptoms can be established by the careful observer, and there is often a considerable degree of fever. The cough is loud and hoarse, unlike the whispering cough of severe obstructive inflammation of the larynx; and the noisy croupy breathing is almost wholly inspiratory, affording thus another distinction from membranous laryngitis. The urgent symptoms soon subside, and the breathing becomes comparatively or quite quiet. Examination of the fauces shows either nothing abnormal or, more frequently, a slight redness or swelling of the mucous membrane with some tonsillar enlargement, or, it may be, ulceration. It would be a mistake, however, to suppose that such attacks as are here described never indicate any more serious affection of the air-passages; for a child predisposed to spasm may thus suffer at the outset of an ultimately severe laryngitis or even of diphtheria. It is always well to wait awhile before pronouncing any given case to be of the class we are describing, and before giving, in consequence, the very favourable **prognosis** which this affection *per se* usually justifies. We should also remember that, the cause of this disorder being catarrhal, it is not uncommon to find the catarrh spreading to the bronchial tubes or the lungs, especially in its younger subjects; and that thus, although the incidental spasmodic phenomena which were first noticed may have led to a hopeful forecast, the case may soon be grave or fatal.

As regards **treatment,** all cases should be confined to bed and surrounded by a hot moist atmosphere, if possible by means of a bed-tent and steam-kettle. A hot mustard bath, which alone may promptly relieve the most urgent symptoms, should be given at the outset. Counter-irritation, in the form of a mustard poultice to the upper part of the chest, is strongly advisable; and warm drinks should be frequently given. An emetic of antimony wine, given early, seems sometimes to be of use, possibly as diminishing spasm; and small doses of this drug in a saline mixture may be continued every three or four hours. If our diagnosis be correct that the case is not one of membranous laryngitis, the child will in all probability be much better under this treatment in from twelve to twenty-four hours.

Prophylactic treatment is very important, and all care, therefore, should be taken to protect from cold the child who has once suffered. With this object I recommend cool or cold baths daily, and thoroughly warm clothing.

Acute Laryngitis.

Under this heading I shall consider the various grades of inflammatory affections of the larynx, whether of simply catarrhal or of membranous nature. Laryngitis is characterised by alteration of the voice and by cough, and, in its severer forms, by interference with the breathing, often amounting to serious obstruction. Usually it is accompanied by tracheitis and more or less bronchitis; the membranous form being especially marked by a tendency to extensive involvement of the bronchial tree.

It is usual to class acute laryngitis in children as " simple " and " membranous." During life, however, it is more by inference from the results of previous experience than by direct observation that we make the diagnosis between these forms; for, with the exception of the comparatively few cases where membranous casts may be coughed up spontaneously or after the action of an emetic, the existence of membrane can be established only by tracheotomy or by post-mortem examination. Further, the question of the diphtheritic nature of cases which we either suspect or have proved to be membranous adds another problem to the diagnosis. In the matter of acute laryngitis, then, we have clinically, at one pole, the simple or catarrhal cases with many degrees of symptomatic gravity, as best typified by the examples met with in the early stage of measles; and, at the other, the confessedly diphtheritic laryngitis, as evidenced by the presence of faucial or nasal membrane and other symptoms of diphtheria. Between these two extremes there is an aetiologically questionable class of cases, which are characterised during life and also after death by membrane in the air-passages' alone, and are unmarked by any other symptom or sign of the universally recognised diphtheria.

The term **catarrhal laryngitis** denotes such cases as are directly due to catarrh of the laryngeal structures, are usually associated with signs of catarrh elsewhere, generally recover, and are, even when fatal, unmarked by membranous exudation in the air-passages. Of this the best example is seen in *measles* before the rash appears. The symptoms are loud brassy cough, hoarse voice, rasping inspiration, and more or less dyspnœa which is markedly increased during sleep and when agitation induces crying. At this stage the affection may remain, and may soon disappear, with its accompanying fever, under the treatment referred to in the preceding section, or sometimes without any treatment at all. Other cases are aggravated by great dyspnœa with stridor, sometimes of an expiratory character, and marked by the working of the accessory muscles of respiration and even by recession of the soft parts of the thorax. For a time it is impossible to differentiate these cases clinically from those with membranous exudation, whether diphtheritic or not; and those who recognise a membranous laryngitis apart from the results of the diphtheritic poison must probably admit that a catarrhal inflammation may become membranous. It is mainly in the milder cases, with clear evidence of general catarrh, that the diagnosis of simple catarrhal laryngitis can be made with a fair amount of confidence; but in all severe cases, beginning suddenly, even where there is marked expiratory stridor, we must wait until the fifth day for the possible appearance of the measles rash before we diagnose a membranous laryngitis. In connexion with catarrhal laryngitis we must remember some cases which are due to *traumatism*, such as the inhalation of irritant vapours or of steam from attempting to drink from a kettle-spout, the severer examples of which are not seldom marked by a membranous exudation in the upper air-passages; and also, possibly, some instances of membranous tracheitis which are apparently the result of irritation by a tracheotomy tube in cases with no previous evidence of the existence of membrane.

In the absence of evidence of membrane, of signs of diphtheria in the nose or fauces, or of other symptoms of that disease, and of any history of concomitant cases or of marked epidemic prevalence, the provisional diagnosis of catarrhal laryngitis may be made. The child should be placed in a bed surrounded by a steam-tent; a small mustard leaf should be applied over the sternal notch and followed by continuous hot fomentations; and repeated small doses of antimony wine may be given in a saline mixture. In cases where the dyspnœa is great, vomiting should, if possible, be produced by emetic doses of antimony wine or of the sulphate of zinc or copper. Failing these, $\frac{1}{35}$ to $\frac{1}{30}$ of a grain of apomorphia may be subcutaneously injected. Under this kind of *treatment* most cases of catarrhal laryngitis will recover or at any rate improve; but if the affection be membranous there will be no improvement, or at most but very temporary. When the pulse intermits with each inspiration, when the retraction of the soft parts of the thorax continues, when hardly any air enters the bases of the lungs, and when the child is becoming drowsy and apathetic, operation is demanded; and in the immense majority of cases membrane is thereby revealed. "Scores of times," says Mr. Scott Battams, with his very extensive experience during many years at the East London Hospital for Children, "I have had to face this problem and my decision to operate has nearly always been followed by the discovery of membrane."

It will be seen here that we encounter at this point some difficulty both as to diagnosis and treatment—a difficulty presently to be dealt with in connexion with membranous laryngitis; but the practical lesson of cautious and expectant treatment at any rate is to be learned from the greatly probable recovery of cases of catarrhal laryngitis, however severe, at the onset of measles, which so frequently suggest tracheotomy to the inexperienced; and from the fact that, however treated, by tracheotomy or otherwise, cases of non-traumatic membranous laryngitis under two years old are extremely often fatal.

The term "**membranous laryngitis**," or its often-used equivalent "true croup," denotes, from the clinical point of view, cases of severe and dangerous laryngeal dyspnœa with frequent involvement of the lower air-passages, evidenced by physical signs during life; and marked post-mortem by false membrane in the larynx or trachea or both, and very often by a continuation of the exudation in the bronchial ramifications. The false membrane consists of fibrin, pus, and dead epithelium; and becomes less tenacious and adherent to the underlying tissues, the lower it extends.

The *causes* of universally acknowledged membranous laryngitis are the diphtheritic poison and traumatism; and many believe that there is a more or less extensive class of cases which is unconnected with these agencies and probably due to a severe degree of simple inflammation. There is, however, a very wide-spread belief at present among pathologists, clinicians, and systematic writers generally who love finality, (with the notable exception of several who have had prolonged experience of disease in childhood) that, apart from occasional cases of traumatism, membranous exudation in the air-passages is always diphtheritic in origin. This almost threadbare question is still unsolved, and will probably remain so until we are able positively to test the diphtheritic process by an absolute histological or bacteriological criterion, and, still more, to apply such criterion to cases where membranous exudation is confined to the air-passages. As yet, in spite of the strongest probability of diphtheria being due to the action of the Klebs-Lœffler bacillus, we have mainly to trust to other clinical observations, and to post-mortem phenomena, as the basis of our opinion on the question of the universally diphtheritic origin of membranous laryngitis; and it would appear that, thus far, the presence of the specific bacillus has not been by any means established in all cases of either membranous pharyngitis or membranous laryngitis in which it has been searched for. Baginsky especially has stated [1] that there are two forms of tonsillar

[1] See Archiv. für Kinderheilkunde, Bd. XIII.

and pharyngeal disease, marked by membrane, which are clinically indistinguishable except by the presence or absence of this bacillus as established by cultures; and that the one form is true diphtheria and highly fatal, while the other is not dangerous to life. It must, however, be freely conceded that cases of membranous exudation, strictly confined to the air-passages, as evidenced by post-mortem examination, may occur in such close connexion with cases of confessed diphtheria as practically to remove all doubt of their diphtheritic origin, and that they may probably give rise by infection to the ordinary faucial disease; and, consequently, that in many individual cases of membranous laryngitis it may be impossible to deny their diphtheritic nature, either in life or, sometimes, even after death. But in my opinion the fact remains that there is a very considerable number of cases of membranous laryngitis in children, which, in their whole course to recovery, with or without tracheotomy, or to death, have no sign in common with recognised diphtheria other than the presence of membranous exudation. As long, then, as membrane alone is not explicitly regarded as sufficient evidence of the working of the diphtheritic poison, this clinical question awaits further solution; and it is perhaps now of greater importance than ever since the official recognition of diphtheria as an infectious disease by the Local Government Board, and the consequently possible treatment side by side with diphtheria of cases of membranous laryngitis, which so many consider, in my opinion erroneously, as *ipso facto* diphtheritic. I add briefly the following reasons which seem to me to oppose the doctrine of the necessary unity of diphtheria and membranous laryngitis. *First*, a very considerable number of cases of membranous laryngitis are sporadic, and neither in life nor after death show any sign of the faucial or nasal involvement which is an integral element in the original conception of diphtheria. This fact is established by many cases in my experience, both clinically and by post-mortem evidence; and I state this, fully recognising that false membrane which has escaped observation during life is frequently found after death in the upper part of the pharynx. *Second*, most of the severest cases of membranous laryngitis have laryngeal symptoms from the outset of their illness, and can but seldom be regarded as instances of extension from the pharynx; a small minority only of cases admitted to hospital as evident faucial diphtheria showing subsequently very marked laryngeal symptoms. My own experience negatives the prevalent dictum that there is a very frequent tendency for the faucial diphtheritic process to spread after many days to the air-passages. Confessedly diphtheritic laryngitis as a rule, though not always, sets in early in the course of the pharyngeal affection. *Third*, the cases where the symptoms and signs are limited entirely to the respiratory tract are mainly sporadic; epidemic diphtheria numbering but few purely laryngeal cases. Further, they very rarely infect others. In all the many cases of direct infection I have known there has been a marked membranous appearance in the fauces or nose. Albuminuria, too, is seldom present, and then only in a slight degree, in purely laryngeal cases. *Fourth*, the frequent occurrence of membranous laryngitis after measles, without any other sign or probable source of diphtheritic infection, favours the view of its non-specific nature in many instances. It must be held by those who regard membrane anywhere as always of diphtheritic nature that measles establishes a special proclivity to diphtheria ; but, if this be so, the proportion of cases of measles marked by a pharyngeal membranous deposit would surely be considerable. It is, in truth, but very small.

I have already said that the early diagnosis between diphtheritic and non-diphtheritic membranous laryngitis is not always possible. In the absence of further evidence of diphtheria we are largely influenced by the matter of epidemic prevalence. The sequent paralysis of diphtheria is too late to be of diagnostic value ; and the old argument of the asthenic type of diphtheria and the sthenic type of croup is of little practical use. With regard to the treatment of individual cases this diagnostic difficulty is not very important; but I believe that, in face of present evidence, we are not justified in grouping or treating together all cases of membranous laryngitis. We should, when possible, isolate the sick from the healthy, but never subject them to the contagion of confessed diphtheria while doubt as to the nature of the case remains.

In membranous laryngitis we find the *symptoms* of laryngeal obstruction in accentuated form. Besides the hoarseness of voice and cough, and the noisy respiration, the expiration is markedly strident ; and great difficulty of breathing is shown by the actively dilating nostrils, the backward movement of the head, and the inspiratory retraction of the soft parts of the thorax and the lower ribs. This condition is emphasised by sleep. As the affection progresses the head is persistently thrown back ; the face becomes cyanotic; the expression anxious; the voice whispering ; the pulse feeble, and intermittent with inspiration ; the cough silent; and the child often clutches at its throat. The respiratory frequency is not very great until there is considerable involvement of the bronchial tubes ; nor is the temperature generally high, though there is always some fever. The prevalent restlessness greatly hurries the pulse-rate.

It is the severity and ingravescence of these symptoms that point to the *diagnosis* of the case as one of membranous laryngitis, especially when the probability of early measles and of " œdema glottidis " can be excluded. Although this condition may be met with in a marked degree in purely catarrhal cases, we know as a fact that it is usually associated with membrane. If evidence of diphtheria be given by faucial membrane, great nasal obstruction, markedly enlarged glands at the angle of the jaw, or albuminuria, the diagnosis of membranous laryngitis is practically certain; and, even without such symptoms, it is at least probable in times of epidemic diphtheria. Very sudden development of symptoms of laryngeal obstruction is not specially indicative of the membranous affection, which is often ushered in for a day or two by the milder signs of laryngeal and often bronchial catarrh, and, in confessedly diphtheritic cases, by a short period of constitutional disturbance with some fever. Expulsion of the membrane by coughing is a definite sign but rarely met with. I have already said that the diagnosis of membranous laryngitis is of greater practical moment, from the therapeutical point of view, than the decision as to whether any given case be diphtheritic; and would add that, when we are in doubt on the latter point, as we so often are, we should provisionally diagnose diphtheria. Undoubted diphtheritic involvement of the fauces is occasionally observed subsequently to symptoms of a laryngitis which would not otherwise have suggested diphtheria. It is, however, in my opinion certain, as has been ably insisted on by Fagge, and, moreover, recently evidenced by the bacteriological observations of Baginsky above-mentioned, that not all membranous-looking exudations on the tonsils, in cases of severe laryngitis which may prove to be membranous, are evidence of diphtheria. Pharyngitis in various degrees is frequent in clearly catarrhal cases of laryngitis; and we have already seen that, without laryngitis, extensive exudation limited to the tonsils and erroneously styled diphtheria by many, is exceedingly common. One point, bearing on the diagnosis of membranous laryngitis, I would, from frequent observation, regard as of some practical value. In cases of established laryngitis where inspection shows the pharynx to be quite free from morbid appearances, membrane is more likely to be present than when the pharynx is red and swollen, and covered with dirty mucus.

When the probability of membranous laryngitis has been established, the *prognosis* is always grave. Under the age of two years an immense majority of cases die, however they may be treated. Without treatment, the obstruction in the larynx and trachea may be soon fatal; the membrane rapidly increasing although some of it may be artificially or, sometimes, naturally expelled. The frequent extension of the exudation to the smaller bronchi and bronchioles, causing signs of bronchitis and broncho-pneumonia which are evidenced post-mortem, is a still more frequent cause of death, and too often frustrates the relief attained by intubation or tracheotomy. Increased pyrexia and frequency of breathing are the best clinical evidences of this extension, when auscultation may fail us owing to the predominance of the laryngeal sounds over the râles and rhonchi of the smaller tubes. In these cases the child sinks into somnolence and coma, the strident sounds subside, and death is often preceded by convulsive twitchings.

The *treatment* of cases of supposed or demonstrable membranous laryngitis, as of those of acute laryngitis generally, must be promptly undertaken. If hot mustard baths, counter-irritation to the laryngo-tracheal region, or the abstraction of blood therefrom by means of leeches fail, and emetics be inoperative, all depressant remedies should be avoided; for the child needs all its strength for the chance of recovery. A steam-tent is advisable from the first, nourishment should be frequently given, and alcoholic stimulation is almost always indicated. With persistence or increase of symptoms of obstruction, and especially if there be marked inspiratory intermittence of pulse and recession of the soft parts of the thorax, operative interference should not be long delayed. There is in my opinion but little reason to believe, as Henoch does, that the simply inflammatory membranous cases give greater hope for recovery under operative treatment than the confessedly diphtheritic; for although, as he insists, the tendency to death in diphtheria is multiform, while in simple "croup" it is the local affection which kills, my experience of the great mortality of all cases forbids me to adduce such an argument in support of my belief in the non-identity of croup and diphtheria. The immediate prognosis of membranous laryngitis, from whatever cause arising, must depend mainly on the severity and rapidity of ingravescence in each individual case. When there is evidence of involvement of the lungs, there is but little hope of ultimate success from the artificial admission of air. The effect of intubation or tracheotomy being mainly or only the entry of air into the lungs, the steam-tent and general treatment must be persevered with after the operation. In spite of much doubt and even opposition on the part of several authorities as to the efficacy of hot moist atmosphere, I am well convinced that it is necessary for the best chance of success. Of the relative merits of tracheotomy and intubation in acute membranous laryngitis, as well as of local after-treatment and sundry complications, I shall say but little, leaving these important details to the province of the surgeon. Intubation may be tried at first in cases requiring operative aid, and often gives complete relief to the laryngeal obstruction. Tracheotomy instruments should, however, be at hand (as insisted on by Dr. Hastings, from his extensive experience at Shadwell, and by others), owing to the risk of the membrane being pushed down by the intubation tube and thus blocking the trachea; and, as there is some danger, at least with many tubes, that ulceration may be rapidly set up by irritation, the tube should not be allowed to remain for more than twenty-four hours without removal. There is occasionally some difficulty in its re-introduction. In very bad cases, with marked expiratory stridor, tracheotomy on the whole is indicated in preference to intubation; and it must be said on behalf of tracheotomy that the after-treatment is thereby better carried out. Evidence of diphtheritic disease of the fauces or nose contra-indicates intubation; for only by tracheotomy can the inspired air be prevented passing over the infected surfaces.

It is on the whole very unlikely that intubation will supersede tracheotomy (at least in England, where neither operation is lightly undertaken in cases which experience shows will probably recover if left alone), owing both to its frequent failure in really acute cases and to the much greater expense and difficulty of management of the instruments. I would further observe that the great objection often made to tracheotomy, on the ground of the frequently sequent broncho-pneumonia found in fatal cases, is practically almost baseless; for this morbid phenomenon is really the result of the extending disease. From many post-mortems that I have seen I feel sure that there is no material difference in this particular between the cases which have died, whether or no they have been tracheotomised.

Chronic Laryngeal Disease.

There is but little special to childhood to be dwelt on under this heading. Chronically altered or absent voice and greater or less impediment to breathing may be due to *papillomatous* or, sometimes, other tumours, both benign and (though very occasionally) malignant; to *syphilitic* ulceration in the larynx; or to *tubercle*. Sometimes, after a long course of laryngeal symptoms with recurrent exacerbations, tracheotomy, necessitated by an exceptionally bad attack, may reveal the presence of *membrane*. I have seen two instances of this where there was no other evidence of diphtheria. It is probable that the membranous inflammation here was an instance of acute disease arising out of chronic, though some may maintain that the diphtheritic poison was at work on the tissues already made susceptible by the chronic lesion. Simple chronic laryngeal catarrh in childhood, arising from the acute form, is not common. It occurs mostly in ill-nourished children, and after measles. In cases which recover from membranous laryngitis, whether diphtheritic or not, laryngeal trouble, both vocal and respiratory, may persist, even after it is possible to dispense with the tracheotomy tube which may have been worn. But in most of such cases tracheotomy has been necessary, and is probably causal. I have known but few instances of recovery without tracheotomy where the breathing and voice have not been soon perfectly restored.

The **diagnosis** of the various forms of chronic laryngeal disease in childhood rests mainly on inference from the history of the case and on concomitant signs and symptoms. Laryngoscopic examination, when possible, is very important. Warty growths, especially when occurring in a child with a marked history of syphilis, may be overlooked when laryngoscopy is impracticable, as in a case of my own which was ultimately tracheotomised. In another case of long standing, where the history pointed to a quite sudden onset of the laryngeal symptoms, the diagnosis was chronic pharyngo-laryngitis, the tonsils and pharynx being much swollen. Laryngoscopy was several times unsuccessfully attempted. Tracheotomy, necessitated by an exacerbation of symptoms during an attack of broncho-pneumonia, relieved the breathing; but when, several months afterwards, the child died from broncho-pneumonia with measles, still wearing her tracheotomy tube, abundant warty growths were found on the cords and elsewhere in the larynx.

Suspected syphilitic cases are best **treated** with iodide of potassium or a combination of that salt with the bichloride of mercury. I never trust to mercury alone in any syphilitic condition calling for prompt treatment. In all instances local treatment of the interior of the larynx should be tried, such as the application of the strong solution of perchloride of iron with glycerine, of the strength of one drachm to the ounce, once or twice a week. In cases due to catarrh this method is frequently satisfactory. Every attention should be given to ensure good nutrition, healthy surroundings, and avoidance of chills. Chronic affections requiring operative treatment have no special mark in childhood.

CHAPTER III.

TRACHEO-BRONCHIAL CATARRH AND CHRONIC BRONCHITIS.

Catarrh limited to, or at least mainly affecting, the **trachea** and **primary bronchi**, without prominent laryngeal symptoms, although often accompanied by slight hoarseness, is common after infancy; but is not very frequently met with in the first few years of life. We have seen that such a catarrh often immediately precedes a laryngeal attack; and the consideration of general bronchitis and broncho-pneumonia teaches us how great is the tendency for catarrh of any part of the respiratory tract to involve the smaller bronchi and air-cells in infancy. It is mainly in children over four years old that we meet with the ordinary and generally unimportant "cold in the chest," with which we are all familiar, marked by pain behind the upper part of the sternum, hard painful cough, some hoarseness, and, at first, by scanty or no expectoration. The attack is attended by little or no fever. Doubtless in many or, perhaps, most cases of infantile pulmonary catarrh such a comparatively limited condition exists for a short time; and there is, in addition to this, a practical reason for shortly considering these cases as a separate clinical group. Early treatment by warmth and confinement to bed, aided by vigorous local counter-irritation applied to the upper sternal region and the administration of the compound powder of ipecacuanha, will very frequently be followed by arrest of the catarrhal process, and probably prevent the involvement of the small bronchi and air-cells. The earliest symptoms, then, of a " cold in the chest " in a young child, however slight, should never be neglected; for in any such case there may be a potential laryngitis or broncho-pneumonia. In older patients we observe the greater frequency of the well-known tracheo-bronchial catarrh, which usually runs its course to recovery in a few days or weeks according to the care or neglect which may be bestowed upon it.

The physical signs of this affection are, at the most, coarse rhonchus, generated in the larger bronchi and alterable by cough, and some lengthening and harshening of the expiratory sound. In many cases there may be no auscultatory signs. The cough, however, may be very loud, frequent and troublesome; and there may be some sense of oppression in breathing. Expectoration is very rarely observed in children under five or six, and but seldom in those under ten, years of age.

This condition is frequently seen in an extended form in whooping-cough, and not seldom in measles; and may pass into chronic bronchitis. When the rhonchi are abundant, accompanied by sibili signifying the involvement of smaller tubes, and, still more, by moist râles, the case becomes one of general bronchitis presently to be considered in close connexion with broncho-pneumonia. I would repeat that acute bronchial catarrh limited to the larger tubes is mainly seen beyond infancy; while primary acute general bronchitis pre-eminently affects infants and quite young children. After this early age acute general bronchitis is most often a secondary affection, and should always suggest the pre-existence of other morbid conditions. Of these, among others, heart disease, kidney disease, tuberculosis, and the zymotic disorders (especially whooping-cough, influenza, measles and enteric fever) should be thought of, as well as chronic bronchitis, out of which an acute attack so often arises.

Chronic Bronchitis.

This affection, frequently seen in quite early childhood, is in my opinion fittingly considered here, from the clinical point of view, before the acute form of the disease; for their relationship is far more often that of acute upon chronic than chronic upon acute; and, when chronic

z

bronchitis does seemingly arise out of repeated attacks of acute or sub-acute bronchial catarrh, these attacks are usually not of the serious form known as acute general bronchitis with more or less fever, but rather of the non-febrile kind previously dealt with, the smaller tubes not being involved.

Chronic bronchitis in childhood, from whatever cause arising, generally begins insidiously, and often fails to attract much attention until it has lasted some months. Many of its subjects appear perfectly well, their only symptoms being cough and some amount of wheezing. The cough is paroxysmal, worse in the morning and evening, and rarely accompanied by expectoration until after the age of five years or later, when it may be very abundant and sometimes offensive. I have, however, more than once seen copious muco-purulent sputum in children of three years old. Physical examination reveals coarse dry rhonchi and moist râles, and sometimes fine râles at the bases of the lungs behind. The percussion note is normal, except in that class of cases marked by very early emphysema. In these severe forms dyspnœa is prominent, the child is lethargic, the face is pale or bluish, the body wastes more or less, the fingers and toes may become bulbous, and the sides and lower part of the chest may be drawn in with inspiration. The inspiratory murmur is short and obscured, the emphysematous sign of a resonant percussion-note over the cardiac area and down the sternum is frequently present, and epigastric pulsation and fulness of veins may give evidence of enlargement of the right heart. In some cases bronchial or cavernous breathing, with or without gurgling sounds, points to dilatation of bronchi.

Every case of chronic bronchitis is liable to *acute attacks* of varying severity and frequency; and in young children such attacks may be broncho-pneumonic and fatal. A very large number of cases of moderate or subacute bronchitis in children, with little or no fever, are merely exacerbations of the chronic disease. Weakly or strumous children, as very many of these are, with a history of wheezing more or less all their life, are subject to sudden attacks of *collapse of lung* which may be rapidly fatal. Witness the following case among others which I have seen. A child, whose mother was bronchitic, had been ailing and wheezing off and on since birth, and began to suffer from increased cough and trouble in breathing a few days before admission. There was coarse rhonchus all over the back with doubtfully impaired percussion-note at the right base and the left inter-scapular region. For nearly twelve days the child remained in the same condition, breathing noisily and coughing, but not appearing much distressed; and the temperature varied between 99° and 101°. On the twelfth day he became suddenly worse, and died after a short struggle; the temperature just after death was found to be 106.5°. Nothing was found post-mortem besides complete collapse of the middle lobe of the right lung and slight signs of bronchial catarrh. I have seen·several instances of a rapidly rising temperature just before death in cases of sudden and extensive pulmonary collapse.

This affection is almost universally worse in the winter; and, in its slighter forms, the symptoms remit altogether in the warmer weather. Some few cases, however, seem to be but little if at all modified by seasonal change. The course and prognosis of the disease are largely affected by its determining conditions and the constitution of its subjects.

A considerable number of cases of chronic bronchitis are not referable to any exciting **cause**, and are connected with hereditary tendency. Such have frequently early developed emphysema and a history of parental asthma and bronchitis. Even in this class there is a good hope of recovery with increasing years, good nutrition, and suitable climatic treatment. I have seen many instances of this variety of the affection, and am convinced of its clinical importance. In neurotic subjects spasmodic asthma is often excited by this condition. It is in this class of cases that we not seldom see the symptoms enduring more or less throughout the year.

Chronic bronchitis is extensively associated with *rickets*, quite apart from the thoracic deformity to which that disease often gives rise; and may indeed, with other pulmonary affections, be regarded as often part and parcel of the rickety state. It springs too out of *measles* and very often out of *whooping-cough*, and in this case is the best example of chronic disease following on severely acute bronchial catarrh. It may indeed be said that a large majority of cases of chronic bronchitis, at all ages, are the result of whooping-cough, and very especially those rare instances which begin in youth and early maturity; and that such an origin much encourages asthmatic tendencies.

The *scrofulous* habit is specially marked by a tendency to prolonged bronchial catarrhs which notably diminish or disappear with improvement of the general condition. I cannot deny that repeated attacks of simple acute catarrh may set up a chronic bronchitis; but, considering the rapid restoration of the epithelial structures, especially in childhood, and the clinical fact that the signs of even chronic and continuous bronchitis of several years' standing may completely disappear, I should not be inclined to regard this *a priori* as a very frequent cause; and I am sure that a large majority of cases of chronic bronchitis in childhood are not sequelæ of simple acute attacks, but either are insidious in origin and frequently hereditary, or result from rickets, scrofula, or one of the zymotic diseases.

Chronic bronchitis may lead in time to *collapse* of parts of the lung, especially at the base, *emphysema*, and more or less *dilatation of the bronchial tubes*, which often contain thick pus. All degrees of the usual post-mortem appearances in the tubes may be seen; but as a rule they are not prominent in childhood; and the lesion does not often extend beyond congestion of the mucous membrane, marked thickening and extension to the deeper structures being rarely observed. It is to be especially noted that, in spite of the large bubbling sounds and other cavernous signs which may have been present, saccular dilatation of the bronchi is scarcely ever observed post-mortem in cases of simple chronic bronchitis in children dying from an acute attack or some other disease. The occurrence of true bronchiectasis, with localised cavernous signs, is probably limited to cases where there is either some chronic disease of the lung itself, especially of the fibroid variety, or pleurisy, or both these conditions. In long-continued cases there will be found post-mortem an enlarged right heart with congested kidneys and liver.

The **prognosis** in the chronic bronchitis of childhood is doubtless much better than in later life, the frequent subsidence of the symptoms with proper treatment pointing to the recovery of the injured tissues. Some cases resist all treatment, and may contract tuberculosis; and in others recurrent and sometimes extensive pleurisy is observed. The frequent intercurrence of acute attacks darkens prognosis.

Foremost in importance as regards **treatment** are climatic conditions. If possible, the child should be sent for the winter and early spring to a place where the climate permits it to be much out of doors. It may be true that most cases do best where the air is prevalently dry, such as it is in Egypt, Algiers, some of the European resorts on the Mediterranean coast, or at Bournemouth. Others, however, improve much at such places as Madeira, Pau, or Torquay. Sea-voyages are often highly beneficial. I do not think that any hard-and-fast rule can be laid down for individual cases; and the frequent presence of an asthmatic tendency often renders prominent the unknown quantity as regards treatment. As to home resorts, I can speak well of Bournemouth, Torquay and Aberystwyth, as all very suitable to certain cases; but the more distant places above-mentioned are in many respects to be preferred. Cases which cannot be sent away from home must be kept in warm and well-ventilated rooms, and allowed to go out on warm and dry days only as a rule, or, with caution, in cold weather when the air is still.

Nutrition should be maintained by generous diet; but we should bear in mind the frequent digestive weakness and gastric congestion which mark chronic bronchitis. Cod-liver oil is undoubtedly of great value in many cases, as well as iron and arsenic. As to remedial measures, especially when there is evidence of much bronchial secretion, opium in small doses is pre-eminently useful, as also may be such drugs as the Peruvian and tolu balsams, benzoin, copaiba, cubebs, or terebene. In other cases, where the cough is hard and there is but little secretion, the carbonate of ammonia should be given, and iodide of potassium or antimonial wine are of good service. Inhalation of steam alone, or of steam medicated by tincture of iodine, tincture of benzoin, creasote or oil of turpentine, frequently gives much relief, and should always be tried in the severer cases. The strength of all these remedies may vary from half a drachm or a drachm to the pint of water, according to the individual case. In proportion to the prevalence of asthmatic symptoms, such remedies as belladonna, lobelia, and æther may be used with frequent advantage. Of squills, senega, ipecacuanha, and the much be-praised strychnia I can but say that after frequent trials I have found them practically valueless in this affection.

CHAPTER IV.

EMPHYSEMA AND ASTHMA.

Emphysema in all its forms may occur in quite early childhood, but is not often symptomatically prominent. In its generalised and apparently primary form, the tendency to which at least is probably hereditary, this affection is often overlooked until some extra stress, perhaps in later life, reveals respiratory weakness. The observations of Jackson of Boston regarding *heredity* in emphysema are quite in accord with my own experience. He found that of 28 emphysematous patients 18 had emphysematous parents, while of 50 non-emphysematous patients the parents of only three were emphysematous. I have several times established by physical examination the existence of undoubted emphysema in young children who have not been the subjects of any prolonged catarrh, or of whooping-cough ; and in many of these I have ascertained that one parent was subject to asthmatic attacks. I believe that there is sufficient clinical evidence to support the view that some vice of lung-structure, leading to loss of elasticity, is transmitted from parent to child ; and that this condition explains more cases of primary and permanent emphysema in children than the theory of premature attempts at breathing during the process of birth.

The **subjects** of this affection have, generally, more or less distended chests, a pale complexion, oldish look, and a spare or even wasted frame ; and an undue excess of elevation over expansion of the thorax is observed during inspiration, which is marked by a short and deficient sound on auscultation. In many cases the heart's dulness is lessened or absent, the pulmonary note is over-resonant on percussion, and there is epigastric pulsation. The *secondary* form of emphysema, usually described as " vicarious," where the lesion is of more partial distribution, is common in childhood.   It occurs in connection with obstruction to the upper air passages; with disease which interferes with the action of parts of the lungs, such as inflammation, consolidation or compression from any cause; and with violent expiratory efforts, often repeated with closed glottis, as typically exemplified in whooping-cough.   Such emphysema is also observed in cases of chronic pleurisy where adhesions prevent the retraction of the lung, and is often associated with dilatation of bronchi. In cases where there is no hereditary predisposition this secondary form of dilatation of the air cells may disappear, when its cause is removed, almost as readily as it came; and the affection is, in many instances, of no great importance.   There is here no textural change in the walls of the vesicles, and therefore, strictly speaking, no emphysema.   Doubtless, however, a permanent lesion may be occasioned by any disease which has long impaired the function of the lung, especially when accompanied by severe coughing, as in whooping-cough.   We must never forget, indeed, that *whooping-cough* is the apparent origin of a large number of persistent cases of emphysema and asthma.   I have known several such instances in children who have all their lives suffered more or less from dyspnœa and palpitation on exertion, the original attack of whooping-cough having been forgotten or regarded as of no importance.   Such children are thin and sometimes markedly emaciated; are often, though not always, bad feeders; and not seldom have chronic winter cough. The chest may be also asymmetrical, or more or less of the pigeon-breast type, especially when there has been much bronchial catarrh.   These patients are often brought to the physician as cases of consumption.

But little need be said of that form of emphysema known as *interstitial*, in which, owing as a rule to violent coughing, there is rupture of the vesicles, and air escapes into the interlobular and subpleural tissue, or, in some cases, if the opening does not soon close, into the subcutaneous connective tissue, where alone it gives rise to diagnostic signs.   This event is not common nor in itself necessarily serious, its gravity being dependent on the extent and permanence of its causative lesion.   It is said to occur only in young children, seeing that in them alone are the lobules of the lung separated by distinct intervals of connective tissue.[1]

Emphysema, giving rise to marked symptoms and to the characteristic expression of face and figure so familiar to us in adults, is not of great frequency in childhood.   When it occurs, it is usually of the secondary form and due to severe and prolonged bronchial catarrh or whooping-cough.   It is not often in childhood that we meet with the consecutive events of demonstrably dilated heart and generalised œdema, with con-

[1] I have, however, seen one case of extensive subcutaneous emphysema in an adult, occurring after the rapid removal of a very large pleural effusion with much coughing, which seemed, after full consideration, to admit only of this explanation.

gestion of internal organs, which are so common in later life. Of the very frequent localised emphysema found post-mortem in the chronic lung disease of childhood and later life it is unnecessary to speak, for it has but little clinical importance.

The **prognosis** of emphysema in children is thus considerably better than in the adult. The tissue change which underlies so many adult cases is less frequent in children; and much improvement or even recovery may take place. It must, however, be remembered that there is ample post-mortem evidence of permanent structural change, even in quite early life. In pronounced and lasting cases there is, as in adults, a liability to asthma. We must ever remember that at all ages the existence of emphysema adds much to the gravity of all acute pulmonary affections, and at least in adults is very frequently a dangerous basis for an otherwise probably harmless pneumonia.

Where the affection is apparently primary the **treatment** must be directed towards improvement of the general nutrition and prevention of those causes which, by favouring catarrh, extend the disease. In the secondary cases, which are at once more frequent, prominent and curable, the treatment is mainly prophylactic and remedial of the commonest causes,—catarrh and inflammation. A warm, dry climate, therefore, where the patient can at the same time enjoy the nutritive effects of sunlight and fresh air, is to be recommended when possible. A generous diet and cod-liver oil, iron, arsenic, or all these drugs, are valuable aids. The bronchial catarrh, which so often is the exciting cause of emphysema, and the frequently incidental asthmatic attacks are to be treated by the methods mentioned under those headings. By these means we can cure some cases, and at least alleviate most.

Asthma.

By asthma I mean that paroxysmal dyspnœa with well-known clinical characters which, whether excited or not by bronchial catarrh or other pulmonary affection, seems always to own a marked nervous or spasmodic element. In considering, however, this affection in childhood it is convenient to associate with it the cases of dyspnœa which seem to be due (whether frequently or not, authorities differ) to direct pressure from enlarged bronchial glands on the lower end of the windpipe.

That asthma is, to a very general extent at least, an expression of *neurotic disorder*, as evidenced by its hereditary character, by its close association with other forms of disturbed nerve function, and by the frequently sudden onset and departure of its attacks, which are not seldom both excited and allayed by so-called mental impressions, is in my opinion sufficiently clear. The contention that bronchial catarrh underlies and excites many if not most cases, as it unquestionably does in childhood, in no way weakens this view. For often in adults, and sometimes in childhood, the attacks are purely spasmodic, neither heralded nor followed by sign of catarrh ; and all recognise the exceptional occurrence of asthma in the course of childhood's bronchitis, pointing to the probable necessity of some other factor for its production. I have seen several examples of purely spasmodic asthma in children with marked neurotic heredity, where there was absolutely no evidence of any affecttion either bronchial, pulmonary or glandular. Yet in most cases I admit the association with asthma of emphysema, bronchial catarrh, or broncho-pneumonia. There are several cases, indeed, of this latter disease where prolonged respiratory trouble, unquestionably due to an asthmatic element, is for a while mistaken for indications of greater gravity, and is only rightly understood when a fresh review of the case reveals a condition of improved physical signs, and lessened or absent pyrexia, inconsistent with the diagnosis of continued or extended inflammation.

I would call special attention to the influence of *excitement* in determining attacks of asthmatic breathing in young children, as in adults, even in the most ordinary cases arising out of bronchial catarrh. I have seen many instances of sudden onset and almost sudden departure of these attacks in the course of bronchial catarrh of various grades, which have been believed at first to be due to pulmonary inflammation.

Asthma, whatever its nature or exciting cause, is not often very prominent in early childhood ; and it is somewhat rare in hospital practice. Nevertheless the beginnings of typical adult asthma can often be traced as far back as the second quinquennium of life, or even to a still earlier period. Besides its very frequent association with catarrh and emphysema, a *gouty heredity* can not seldom be made out ; and I can speak with certainty of a few well-marked cases which strongly corroborate the views of West, Eustace Smith and others that there is a connection between asthma on the one hand and *ekzema and urticaria* on the other. I cannot go so far as to say that long-continued and extensive ekzema is almost always joined with a tendency to asthma; but I have seen such unquestionable examples of cured ekzema replaced by asthma, which in its turn disappeared with returning ekzema, as to have no doubt of the frequently close pathological kinship of these affections. For the rest, as in adults, *inhaled irritants*, whether demonstrable as dust or smoke or pollen, or less determinate as emanations from animals, and also so-called *climatic* influences may occasion asthmatic attacks. One of the best examples of asthma is that form which is seen as an expression of "hay-fever." Reflex agencies, such as pressure of glands on the "vagus" nerve, or gastric disturbances, seem to play some causal part ; and polypi in the nose and even enlarged tonsils may be included among possible excitants.

Attacks of asthmatic character are undoubtedly often associated with *enlargement of the bronchial glands*, and sometimes in all probability, as shown by Eustace Smith, directly caused by pressure of such glands on the lower end of the trachea or on a main bronchus. He regards as diagnostic of this cause a venous hum which is heard over the upper end of the sternum when the head is retracted; and teaches that this bruit is due to compression of the innominate vein between the sternum and the enlarged glands in the bifurcation of the trachea, which are carried forward with that organ when it is free to move with the retraction of the head. I have, however, so often found this symptom unconnected not only with asthma but also with any other evidence of enlarged glands, and have further seen post-mortem markedly enlarged glands in this region which have been unattended during life by any such signs or symptoms as above recorded, that I cannot but question both the leading significance of the sternal bruit, and the predominant rôle of enlarged glands in the production of asthma.

It must be remembered, in connection with this subject, that a large number of cases of adult asthma begin before the tenth year of life.

There is nothing special in the **symptomatology** of childhood's asthma. In the pure, but rare, cases unconnected with catarrh the chest is nearly fixed in the inspiratory position, the expiration is prolonged and marked by sibilant and sonorous rhonchi, and the inspiratory murmur is much lessened. The attacks, like other spasmodic neuroses, such as epilepsy, most often occur at night. The skin is moist, there is no pyrexia, and often no cough. In the catarrhal cases there may be some prolonged inspiration as well, and all degrees of moist râles may be heard. The attacks here are usually of more gradual onset and departure, and may last, with varying exacerbations and remissions, for several days, weeks, or even months.

The **prognosis** in many cases of bronchitic asthma in childhood is good, the symptoms disappearing with its exciting cause. Many cases, however, last through life, and I am unable to give any criterion by which an accurate forecast can be made. I have known cases of bad chronic bronchitis in quite young children, with frequent and severe asthmatic paroxysms, which have apparently made perfect recovery after wintering for a few seasons in suitable climates, and some who have similarly improved in spite of seemingly bad conditions. Others, again, with all medical care and apparent climatic advantages, continued to suffer for an indefinite period.

The less demonstrable, perhaps, the exciting causes of the attacks are, whether bronchial catarrh be present or not, the more the affection seems likely to endure.

Treatment should endeavour to promote nervous health by the hygienic and medicinal tonics, and to ward off or remove all demonstrable or suspected exciting causes, whether local or general. Climatic influences are certainly very important in many cases, though no special rules on this head can be laid down for all. The more prominent the bronchial catarrh, the more necessary it will be to treat the child on the principles indicated under the head of chronic bronchitis. Iron, arsenic, and cod-liver oil are sometimes invaluable ; and will also, as far as medicine goes, be suitable for those cases which may be referred to glandular enlargement. Iodide of potassium may also be given and is highly spoken of by many ; but I should hesitate to give this drug for any long period unless distinct improvement were clearly established.

During the attack the recognised remedies are all suitable, and have apparently the same shares of success and failure as in the case of adults. I believe the breathing of nitre fumes is more frequently efficacious than most other inhalations ; and I feel sure that lobelia and belladonna are often of very great value. I have prescribed lobelia largely at all ages for all varieties of asthma, especially in the more chronic forms, and have never observed any ill effect from it other than occasional nausea or still rarer vomiting. A child of ten years old will readily take 10 minims of the tincture every six hours. Depressant and nauseating remedies are undoubtedly of service in cutting short severe attacks; and for this purpose full doses of ipecacuanha may be found useful. Pilocarpine, $\frac{1}{10}$th to $\frac{1}{8}$th of a grain, may also be tried subcutaneously for a child between six and ten years old. But very severe attacks are rare, and such treatment is but seldom called for.

CHAPTER V.

ACUTE BRONCHITIS AND BRONCHO-PNEUMONIA.

BOTH clinical and pathological considerations point to the advantage of studying these affections of the respiratory tract together. With almost all cases of broncho-pneumonia, of whatever origin, there is generalized bronchitis, as shown by the examination of fatal cases or by physical signs during life ; and in severe bronchitis of the finer tubes, which lasts more than a few days, some broncho-pneumonia is the rule, even though the physical signs of pulmonary consolidation may be absent. The passage is strictly gradual, though often extremely rapid, from bronchial inflammation to exudation in the air-cells. Acute general bronchitis and broncho-pneumonia, as seemingly primary affections, are pre-eminently diseases of young children, their counterparts in adults being almost always accompaniments or sequelæ of other maladies.

By **acute bronchitis** I denote here the inflammatory affection of the mucosa, and sometimes of the deeper tissues, which involves the bronchial tree in all its ramifications; exclusive of those cases, already considered, where the larynx, trachea or largest bronchi pre-eminently or solely suffer. The main symptoms of acute bronchitis are hurried breathing in proportion to the extent of the bronchial tree involved, feverishness, more or less cough which tends to diminish as the disease progresses, and sometimes, though almost exclusively in later childhood, expectoration of mucus or muco-pus. **Broncho-pneumonia** in its most frequent form, is, as stated, an extension of this inflammatory process to the smallest bronchioles and air-cells, leading to various degrees and extent of consolidation of the lung; and is as a rule symptomatically evidenced by continued and increasing dyspnœa, more drowsiness, and higher fever.

The conditions out of which bronchitis and broncho-pneumonia arise are many. These affections occur both in an apparently independent form and also as clearly secondary to other morbid processes; the observed lesions having for the most part, with the great exception of many tuberculous cases, no differential relation to their various sources.

"*Catching cold*," which is probably a reflex process following on undue exposure of the cutaneous nervous surface and ending in inflammation of mucous membrane, seems to be the only discoverable immediate excitant in many cases; and these affections are unquestionably far less prevalent in the summer and early autumn than in any other part of the year. Very often, however, such exposure can be almost certainly excluded. It is to be remarked that in a very large number of instances the apparently primary attacks occur in children who are *badly nourished*. Severe cases of acute bronchitis and broncho-pneumonia are not very common among the well-to-do, when unassociated with an early tendency to chronic bronchitis or with some of the other conditions now to be noticed. *Rickets*, *heart*-disease, *kidney*-disease, *syphilis*, the *scrofulous* or *tubercular* diathesis, and notably *tuberculosis* itself, all favour or excite the development of catarrh of the air-passages, although in some of these instances the immediate origin may not be known. It must never be forgotten that *chronic bronchial catarrh*, dating from early infancy, is the underlying condition of a very large number of cases of subacute and acute bronchitis of varying extent in children, with or without pyrexia. Most, indeed, of the slighter cases of generalized bronchitis, in infants which recover from the attacks, have a history of chronic wheezing and coughing with exacerbations; and these children are usually anæmic, thin and weakly. The *fevers* are specially apt to occasion it, among which measles, diphtheria, influenza and enteric fever are prominent. I have seen, too, a sufficient number of instances of bronchitis and broncho-pneumonia immediately following or accompanying scarlatina to convince me of closer relationship here than is, perhaps, usually recognised. *Whooping-cough* is almost constantly attended by some bronchitis, and often, even in its earlier stages, by broncho-pneumonia, which in some cases tends to become chronic. It may be said, I think, that bad nutrition, whether due to insufficient alimentation or to definite disease, is one of the greatest predisposing causes of acute bronchitis and broncho-pneumonia, be the exciting cause what it may. These affections are certainly among the most frequent scourges of poverty.

The first symptoms of **acute bronchitis** are often those of catarrh of the trachea and large bronchi, namely cough, wheezing, slightly increased rate of breathing and, it may be, some evidence of pain in the upper sternal region; but the dyspnœa soon becomes marked with the rapid involvement of the smaller tubes, the cough often tends to lessen, and the child becomes alternately restless and drowsy. There is mostly some pyrexia, often as high as 102°; the face is flushed and, later, bluish; and there is usually sweating. The pulse is frequent, sometimes exceedingly; but as a rule the normal pulse-respiration ratio is not maintained, the respiratory frequency being in considerable excess. An extreme case of bronchitic dyspnœa is very similar to that of laryngeal obstruction, showing the cyanosis, the recession of the yielding parts of the thorax, and the auxiliary muscular action, without, of course, the stridor and vocal loss which mark the latter disease. In both cases the bronchial tree is gravely affected at its root and terminal branches respectively. Expectoration but very rarely occurs; for, at the age most subject to bronchitis, children usually swallow any mucus which may reach the pharynx by coughing.

On examining the chest at the earliest stage coarse rhonchi or moist râles may be heard, succeeded soon by finer sounds often audible over the whole chest. The percussion note is not necessarily affected. In cases which recover, without going on to discoverable collapse or broncho-pneumonia, these signs diminish after a few days, and usually disappear within a week or two; but it must always be remembered that it is impossible to mark by physical examination the initial stages of these further developments. Small and irregularly distributed patches of diminished resonance without bronchial breathing may, indeed, frequently be detected by careful and light percussion in cases where neither collapse nor broncho-pneumonia are of great extent; but, on the other hand, as we shall see, there may be many small foci of broncho-pneumonic consolidation scattered over one or both lungs, as proved in fatal cases, where there have been absolutely no other physical signs than those of bronchitis. Much cyanosis is usually the herald of death, which is often preceded by convulsions. It may be believed, however, that death scarcely ever follows on a bronchial inflammation which does not also involve the ultimate bronchioles and air-cells. There is almost always some post-mortem evidence of broncho-pneumonia in cases which, during life, have been styled capillary bronchitis.

Collapse of the lungs of greater or less extent is a very common event in bronchitis, occurring in most fatal cases. The parts involved can be, as a rule, easily recognised post-mortem by their purplish and smooth appearance, their depression beneath the lung surface, and their capacity of being inflated except in some cases of ancient date. Often, however, in close proximity to the collapsed parts there are patches of broncho-pneumonic consolidation, which, according to its proportionate extent, obscures these marks. Rickets adds much to the chances of collapse, and frequently seems to bring it about even when the bronchitis is not very severe. This fact, considered in connexion with what is seen post-mortem, and with comparison of adult cases of bronchitis, strongly corroborates the view of Fagge and others that the immediate cause of collapse is not plugging of the bronchioles but weakness of the inspiratory act, as in the case of simple atelectasis. The younger and weaker the child, the more extensive and rapid is the collapse. When of small extent, collapse may readily recover, disappearing with returning vigour, and may indeed do so even in cases where it is sufficiently extensive to cause diminished percussion- and breath-sounds or some degree of bronchial breathing. I have repeatedly made this out; but doubtless, in the larger number of instances which quickly recover, collapse is not demonstrable by examination but can only be inferred from increase of respiratory trouble.

Broncho-pneumonia, as has been said, whatever its origin, is nearly always preceded or accompanied by bronchitis, but may arise so quickly and progress so extensively as to express almost from the first the physical signs of consolidated lung, namely, more or less impaired resonance on percussion, and bronchial breathing especially marked on expiration. It is these acute cases which are often at first and for some days, or sometimes throughout, indistinguishable from acute pneumonia of the lobar form—a totally different disease. It is doubtless owing to something special in the young child's respiratory organs that inflammation spreads so rapidly from the larger to the smaller tubes and to the air-cells; and the fact of the ultimate bronchioles of infants being considerably larger in relation to the air-cells, which are bud-like dilatations from them, than in the adult, may have some causal connexion with this clinical peculiarity. Besides this, the absence of expectoration in young children readily favours the inhalation of secretion into the lower air-passages. We have, I repeat, very different sets of physical signs and post-mortem appearances in the broncho-pneumonias of young children. On the one hand there are the very frequent cases of *more or less gradual onset*, where fine mucous râles, generated in the smaller bronchi, are the predominant or even often for a while the sole physical signs; but where the accession of lobular consolidation, with increasing fever, dyspnœa and cyanosis, is indicated in many instances by more or less loss of resonance in disseminated patches, over which very fine and high-pitched metallic-sounding râles are heard and, sometimes, varying degrees of bronchial breathing. On the other hand there is the important class of cases, by no means always to be differentiated either as to their conditions of origin, course, or event, where *a more acute* onset and extensive physical signs of consolidation, frequently limited to one side, often render the diagnosis from ordinary lobar pneumonia difficult or, for a while, impossible. Between these two extreme examples we have also cases showing all degrees of intermediate physical signs, according to the extent of the consolidation of one or both lungs.

Concerning the first kind I need add but little to what has been already said under "acute bronchitis," of which it is practically an advanced stage. The post-mortem appearances are those of discrete patches of lobular consolidation and collapse, and intervening areas of crepitant lung, in varying proportions. There is always evidence of bronchitis, and the small consolidated patches are grouped round bronchial tubes filled with secretion. In some cases the smallest bronchioles and air-cells are laden with pus, giving rise to an appearance which at first sight simulates tubercle. This latter appearance, however, may be seen in the other forms of broncho-pneumonia presently to be mentioned, and is said by Sturges and others to be sometimes met with apart from much evidence of general bronchitis.

With regard to the second kind of broncho-pneumonia, which, being in many respects akin in clinical appearance to the pneumonia known as lobar, is treated by some writers, under the general heading of pneumonia, as the "pneumonia of children," it behoves those who class it as I do to point out as far as possible what distinguishing clinical and post-mortem facts there are to justify them. It has already been implied that, as far as physical signs are concerned, the extent of the consolidation in these cases is practically lobar, and post-mortem examination shows involvement of large areas, a whole lobe, or a whole lung. We must rely for diagnosis during life on the history and course of the case as well as or often more than on physical signs. Even general bronchitis, which precedes or accompanies most cases of broncho-pneumonia, is seen in some degree in the true lobar pneumonia of adults as well as children, and therefore cannot by itself be an all-important sign, valuable as it is in combination with others. Some little aid may be gained by remembering the fact that in the extensive broncho-pneumonic consolidations of children the signs are very often first observed at or above the root of the lung, and, further, that the upper parts are quite as often affected as the lower. In most cases, again, there is evidence sooner or later of consolidation of both lungs—certainly not a common event in the true pneumonia of children and therefore a useful positive sign ; but of course one of our chief difficulties in early diagnosis is in those cases where one lung at least apparently escapes. I would lastly say that marked physical signs of pleural effusion, whether plastic or liquid, over the consolidated lung in an acute case, are evidence *pro tanto* against broncho-pneumonia.

Much light can be thrown on many cases by the *previous history*, for in the great majority of acute broncho-pneumonias there is antecedent ill-health of some kind, sometimes chronic bronchitis, and, exceedingly often, a recent attack of some definite disease as previously alluded to. The pulse-rate is often much higher in broncho-pneumonia than in pneumonia ; and diarrhœa is much more frequent, owing, probably, to the co-existence of intestinal catarrh. Again, even when the onset of the attack has been misleadingly sudden, reminding us of true pneumonia, the course is as a rule much more irregular both as to general symptoms and temperature ; recovery when it takes place is gradual ; and the signs of resolution are only slowly progressive. There may be frequently recurring exacerbations, with a rise of temperature demonstrably coinciding with the involvement of a new area of lung ; and the normal temperature line is often reached in the intervals. The breathing at the height of an attack of broncho-pneumonia is usually much more laboured, and cyanosis more marked, than in true pneumonia, where it is most often merely hurried. This is probably due to the much greater involvement of the bronchial tubes and generally wider dissemination of the affection in broncho-pneumonia. The facts of gradual recovery and resolution of the lung are of the greatest importance, and in a considerable majority of cases lead us, though somewhat late in the day, to the correct diagnosis. I need scarcely say, however, that as true pneumonia has occasionally a rather lingering course, and as its physical signs are sometimes obscured towards the end by those of pleural effusion, the slow disappearance of physical signs is by no means pathognomonic. There will, as yet at least, remain a number of cases in which the diagnosis cannot be made during life.

Post-mortem examination shows certainly a notable difference between the lung in broncho-pneumonia, however acute the course and however extensive the consolidation may be, and the well-recognised pneumonic lung of either adults or children. The swollen, dense lung, with the granular appearance and roughish feel of its cut surface, is found but in a very small proportion of the many children who die with lung-consolidation. The broncho-pneumonic lung is less homogeneous in appearance and smoother on section ; and on close inspection the lobular origin of the morbid process can usually be made out. The microscopic appearances, moreover, show the great preponderance of catarrhal over fibrinous material. In most cases there is evidence of disseminated lobular pneumonia in both lungs, which may have been unsuspected during life, and of general bronchitis. Pleurisy in any of its forms may accompany both broncho-pneumonia and pneumonia itself; but in a marked or extensive degree is very much less frequent in broncho-pneumonia, at least in cases of death at an acute stage. I would, finally, repeat here what I have said elsewhere, that true pneumonia, although common enough in quite young children as evidenced by its typical clinical course, is very rarely fatal, unless double or complicated; and that on practical grounds, although I fully acknowledge a frequent difficulty in diagnosis, I strongly deprecate the classing together of all extensive lung-consolidations in children, however acute they may be, under the common term of lobar pneumonia. For the title "lobar pneumonia" is mostly used as denoting the well-known disease which some call "true" and others "croupous," but to which all accord a separate nosological place.

Broncho-pneumonia of whatever origin is far most frequent in children *under three or four years old*, its occurrence, indeed, being nearly limited by that period, with the exception of some cases which are tubercular or are secondary to the acute specific diseases. Taking a consecutive series of over 400 cases from my note-books, which includes none with any diphtheritic connexion, I find but very few over this age (and those nearly all sequent upon measles) which are not certainly or probably referable to tuberculosis, either from definite signs and symptoms, from their chronicity with wasting, or from post-mortem appearances. To tubercular broncho-pneumonia I shall again allude shortly when considering tubercular disease of the lung. I will only say here that the broncho-pneumonia which follows measles is very often the herald of pulmonary and general tuberculosis. Apart from tubercle, influenza, enteric fever and some other specific excitants, the disease is certainly almost unknown in adults. I have never seen any such instances of apparently primary broncho-pneumonia beyond the age of childhood as are reported with great precision by Dr. Fagge. Out of one unclassified series of 43 broncho-pneumonias, I find only 7 over three years old, while, out of 30 cases designated as true pneumonia, 22 are over that age.

With regard to the **mortality** and **prognosis** of broncho-pneumonia generally, statistics are of little value ; for, although, as I have said, the determining conditions from which this lung affection springs can rarely be inferred from its clinical course, its mortality largely depends on its origin. Cases, however, are usually confused together in note-taking, and broncho-pneumonia thus appears in the incorrect position of a substantive disease. In comparing, again, as is often the case, the mortality of broncho-pneumonia and pneumonia in children, we are met by the double fallacy of want of attention to an age-limit and the frequent falsification in the post-mortem room of the diagnosis of true pneumonia made in the wards. Taking, however, the age of four as a limit, I find that the mortality of 350 cases registered as broncho-pneumonia is about 35 per cent. ; while of 42 cases diagnosed as pneumonia four died, and the two that were examined showed marked complications. Speaking very generally of all cases together, the prognosis in broncho-pneumonia is decidedly grave, and I should put the chances of recovery as practically not much more than even ; but consideration both of the probable origin and of possibilities of treatment are no unimportant aids to our forecast. I know, however, no class of cases in practice, with the physical signs of broncho-pneumonia, which I could speak of as tending to recovery in any considerable majority ; for after study of my cases I find but very few of those, regarded by some as simple and as of usually good prognosis, which occur after exposure or chill in children who are the subjects of neither recent and acute nor of chronic or constitutional disease. I have seen more cases following measles or other fevers recover, than of those apparently sudden or "idiopathic" cases occurring in rickety children and others improperly fed and cared for. Nevertheless acute bronchitis and broncho-pneumonia during an attack of measles or whooping-cough, especially in a young child, are very often fatal. Nearly half of my fatal cases were rickety, and most of them suffered from marked intestinal disorder, with or without vomiting. On the whole a persistent or oft-recurring high temperature is of very bad augury, as also, but by no means always, are marked nervous symptoms, such as retraction of the head, convulsions, or great apathy. The extent of the lung mischief, of course an all-important element in prognosis, must be judged of largely by the general appearance of the child, and by the nature of its breathing and cough.

A rare event, according to my experience, in acute broncho-pneumonia is the formation of an abscess cavity of any considerable size. I have notes of only one case which ran a course simulating acute phthisis. The boy, aged four, was brought to the hospital two weeks after measles with cough and fever, and on examination the signs of extensive lobular pneumonia were found. After a slight improvement in the symptoms the temperature rose to 103°, the signs increased, and in a fortnight there were amphoric sounds at the right base. He wasted rapidly, with much sweating, and died four weeks after admission, the temperature having never remitted. Post-mortem there were found marked lobular pneumonia in innumerable discrete patches, abundant points of pus in the bronchioles and air-cells, and a large ragged cavity containing pus at the base of the right lung. The pleura was adherent and the bronchi were generally thickened and dilated.

Tuberculosis exists in many cases where only broncho-pneumonia can be diagnosed with certainty ; and it is more from the family and previous history, the wasting, and the persistently remittent pyrexia, than from physical signs, that we may be able to foretell the post-mortem discovery of tubercle. Some give evidence of abdominal tubercle, and others die with meningitis. Broncho-pneumonia occurring in the course of *diphtheria* is an especially fatal event, as also is that which follows on *cancrum oris*. Each case must in effect be judged carefully from all its conditions, and from the signs of vitality, appreciable by the experienced observer, which may frequently aid a just forecast according to unwritten laws.

Besides the ending of acute broncho-pneumonia in death or recovery there is frequent evidence of a *chronic condition* which is sometimes mistaken for, and sometimes results in, tubercular disease of the lung. When the signs of consolidation persist (especially when most marked at the apex in children beyond infancy) and cough and other symptoms remain, the possibility of caseation of the lung must always be remembered. Without entering into the debatable question of the exact part played by the tubercle bacillus in destructive lung-disease, I am inclined to recognise a recoverable or arrested caseation of the lung which is indistinguishable from that which accompanies tuberculosis. Broncho-pneumonic consolidation may thus be chronic ; some cases certainly ending in slow reabsorption of the inflammatory matter and recovery, and others in ultimate death from persistent lung-disease, with varying proportions of softening and fibrosis. Instances of this are quite frequent as a sequel of whooping-cough. Lastly both acute and chronic *pleurisy*, very often purulent, attend and follow many cases of broncho-pneumonia whether of simply catarrhal or specific origin. The empyemas found in this connexion are sometimes quite small and often variously loculated. Dr. Sturges is inclined to hold that there may be a similar relationship between purulent broncho-pneumonia and empyema to that between pneumonia and plastic pleurisy. However this may be, I have certainly seen many empyemas in children beyond infancy, as well as the familiar cases in adults, which followed directly on typical true pneumonia.

An equable temperature of about 65°, good ventilation, and hot moist air to breathe, are the most important means for the best **treatment** of cases of acute bronchitis or broncho-pneumonia. All other methods in the early stage are either unnecessary or quite subordinate, as I have been amply convinced by experience. A stuffy and over-heated room, with closed windows and lit by gas, offers the surest conditions of the worst result. I have often seen immediate improvement follow on opening windows and forbidding gas-light, without any further treatment. Too much stress cannot be placed on this point. Everything that tends to further impede the mechanical act of respiration or to disfavour the access of pure air to the lung-cells must be rigorously avoided. Thick poultices, which are obstructive to free breathing, and, sometimes, all poultices and applications to the chest other than sufficient to lessen draughts of air, which are in some circumstances otherwise unavoidable, must be tabooed, especially in small weakly children. The mere removal of the favourite "jacket-poultice" will sometimes work wonders. In severe and established cases I never allow poultices at all, nor any wrap to the chest except a thin layer of cotton wool or wadding; for I am well assured of their uselessness. Of all methods of "counter-irritation" I would say the same, except, of course, in early cases of inflammation of the upper air passages and larger bronchi, which we are not here considering.

The child should, when possible, be moved from time to time between two rooms, the unoccupied one being thoroughly ventilated by open windows. At the foot of the bed, which should be inclosed by thin curtains on a tent-arrangement, there should be kept a steam-kettle with a spirit-lamp, made on the principle of the "steam-draft inhaler" intro- duced by Dr. R. J. Lee. I have not found special medications of the aqueous vapour to be of marked value, and after trying many have nearly abandoned all. For medicine, alcoholic stimulants in small doses are necessary, I think, where there is great prostration and embarrassed heart; but this drug should never be given at random in severe cases, and should be carefully watched, for its narcotic effects, when there is much cyanosis and drowsiness. Carbonate of ammonia I give in nearly all cases. In the acutest cases with much fever I have for the most part, until the last few years, adhered to the time-honoured custom of giving small doses of antimonial wine; but I cannot say that I have ever seen reason to regret its omission in several more recent instances, and no longer give it as a matter of routine. It may serve, however, some- times to relieve the symptoms of cough and restlessness. I have never known the ordinary sudorifics act at all in the somewhat few cases where the skin is very dry. So-called "expectorants," such as ipecacuanha, squills, and senega, I have often tried and long ago rejected.

In serious cases, with cyanosis and labouring heart, I have seen some lives saved and others at least prolonged by bleeding. Leeching is perhaps the best method for quite young children, and cupping, either dry

or wet, for older ones; but in severe cases requiring prompt assistance
I usually cup, or bleed by phlebotomy. The process may be repeated
from time to time. From half an ounce to an ounce of blood may be
taken from a child over a year old, but such directions are of little value.
The amount should be fixed by result, the finger being kept on the pulse,
and the general condition of the patient carefully observed. When a
frequent, feeble and, still more, irregular pulse improves markedly in
some or all of these points, the bleeding should be stopped. In these
affections, even of the severest form, there is often some hope; and I
would insist, from experience of some few remarkable instances, that
none should be left to die with cyanosis without some attempt being
made at bleeding.

When empyema is suspected, exploration should be made; and pus,
when found, should be evacuated at once, however ill the patient may be.

To accelerate convalescence, which, from the nature of many cases,
is often very prolonged, with enduring prostration, nutritive and tonic
treatment in the widest sense is called for. All care should be taken to
prevent undue exposure, but fresh air should be secured. Cod-liver oil
should, I think, almost always be given.

CHAPTER VI.

PNEUMONIA.

WITHOUT discussing the various views which have been held as to the
proper nosological position of acute pneumonia, often styled "lobar" or
"croupous," I only state here that I regard it as an independent fever;
not as a lung-inflammation with symptomatic pyrexia. Both its clinical
and anatomical characteristics strongly support this view; and recent
bacteriological research, though failing as yet to justify completely the
reference of this disease to the action of one specific germ, points with
the greatest probability to an origin from the action of organisms introduced from without. For what I venture to deem by far the best account of the nature and ætiology of pneumonia generally, I refer the
reader to the admirable monograph by Sturges and Coupland, and record
here chiefly the results of my own experience as to the conditions in
which pneumonia in children seems to arise. The study of the proximate causes of the disease must always be of great clinical importance,
however definite our future knowledge may be of its essential origin;
and it must be remembered that, while the immediate conditions out of

which pneumonia apparently springs are multiform, the best accredited germ, the bacillus of Fraenkel, is frequently found in the normal saliva.

Chill is doubtless an important element in the causation of many cases which follow directly on definite exposure to suddenly occurring cold, especially when accompanied by wind. Out of 200 cases occurring in my hospital practice rather more than half were admitted in February, March and April, and less than one fourth in June, July, August and September. Several more are recorded in May, October and November than in December and January.

An apparently *epidemic* form of pneumonia not seldom prevails, and several members of one household may be affected. I believe there is sufficient evidence that bad drainage and other insanitary conditions play some part in its production. That the disease is communicable directly from the sick to the healthy I have but little reason to believe from my own experience. Such communicability, however, is taught by some, and appears to be probable from several cases reported by various observers. Dr. W. A. Wills has recorded some instances of apparent contagion from bed to bed in vol. vii. of the Westminster Hospital Reports (1891).

In the majority of cases pneumonia is in appearance *primary* and idiopathic; in others it is clearly *secondary*, as a part of, or addition to, some other demonstrable disease. It may attack, in its primary form, both seemingly healthy and delicate children; although, as we shall presently observe, there is a greater tendency in the latter class to certain unfavourable sequelæ.

As in adults, so in children, pneumonia begins suddenly with the well-known symptoms of fever, runs a usual course of from four to seven days, and ends with a more or less rapid fall of temperature. In some cases the fever lasts but two or three days, or it may be of yet shorter duration; in others, ten or even twelve days; and in yet others, known as " pneumonia migrans," it may be of irregular course and indefinite period. Sometimes there is a sudden descent of temperature, followed by a considerable rise shortly preceding the ultimate critical fall. Some of the physical signs of lung consolidation almost always remain for a while, after the patient's practical recovery with the subsidence of the fever.

Vomiting very frequently marks the onset, and may be recurrent for several days. In no disease of childhood, apart from definite cerebral mischief, is vomiting so frequent as an initial symptom, scarlatina standing next in order in this respect. *Diarrhœa* not seldom accompanies the attack and may persist throughout, but *constipation* is more often marked and obstinate. I have seen two instances where this symptom, with urgent and repeated vomiting, suggested the diagnosis of intestinal obstruction.

Convulsions, one or more, usher in the disease less seldom than in other febrile disorders of childhood; but this symptom is not very frequent, and most often implies rickets or some other condition of nervous instability. *Rigors* are but rarely observed at the outset in children under five years old, although I have seen this phenomenon occasionally at a much earlier age. *Pain* at the epigastrium or at the side of the chest is often met with, probably indicating the local pleuritis, which, whether detected or not, almost always accompanies pneumonia. *A bright flush on the face*, sometimes limited to one cheek, is present in a very large number of cases, occasionally involving the chest as well, and sometimes leading to the mistaken diagnosis of scarlatina. This rash, however, should not often deceive a careful and experienced observer. It rarely, if ever, occupies the arms or legs. *Hurried breathing* marks most cases from the onset, but this symptom is often not prominent at first; and *cough*, which when present is short and hacking, is often infrequent or even altogether absent. Pronounced dyspnœa, or laboured breathing with cyanosis, is not a usual symptom, but is rather a mark of bronchitis or broncho-pneumonia. *Nervous symptoms* are often pre-eminent at the beginning; and great drowsiness and apathy, or delirium even of the wildest character, or severe headache, may throw all other symptoms than pyrexia into the shade. Occasionally, too, there may be temporary strabismus. Such cases have been named by some observers "cerebral pneumonia," and are often mistaken for meningitis. When, however, delirium is prominent this mistake should never be made, for delirium is no marked sign of meningitis. I have notes of a case where definite pneumonia, ending in good recovery, followed on a condition of vomiting, headache and strabismus, which, lasting a fortnight, seemed to indicate cerebral disease. A persistently rapid rate of breathing without irregularity of rhythm is a strong point in favour of pneumonia and against cerebral mischief.

The *physical signs* of the disease are often later in appearance than in adults, and may indeed throughout be sought in vain. Such "latent pneumonia," of which I have seen several unquestionable examples that have been submitted to frequent and searching examinations, must always be borne in mind. It can only be accurately diagnosed when the course and crisis of the disease, with the usual symptoms of altered ratio of pulse and respiration, have completed the picture. This absence of physical signs is probably due, as has been often said, to consolidation of deep seat and small extent, or possibly to its entire absence. It is to be remembered, however, that very careful examination of the chest will in many cases detect early signs of pneumonic consolidation which escape the superficial observer. The earliest signs of all are much diminished breath-sound and very slightly impaired resonance over part of one lung; and these signs may continue for a while unattended by any râles or bronchial breathing. A little later the inspiration becomes harsh. Not seldom has the recognition, in cases with pneumonic symptoms, of but a slight lessening of percussion and breath-sound without any marked evidence of thoracic trouble, led me to the confident diagnosis of pneumonia at the very outset. The physical signs are often very limited in extent, especially when at the apex, and sometimes are evident only high up in the axillary space. Early fine crepitation is perhaps less frequent in the child than the adult; but pleural sounds are very common, especially in the lower lateral region. Increased vocal resonance, as heard with the cry, is often a valuable and sometimes the predominant sign of consolidation. For the rest, the physical signs and symptoms in most cases are very similar to those in adults, with which I assume the reader's familiarity; and, just as in later life, the extent of the one has no certain correspondence with the severity of the other. The characteristic appearance of the blood in the sputum is very rare in early childhood, any expectoration, indeed, being seldom seen in my experience under ten years old. *Drowsiness* throughout the disease is very common; and real dyspnœa is rare, however hurried the breathing may be, except in cases accompanied by pleural effusion or when both lungs are involved. It is not usual to see the large auxiliary muscles of respiration at work, although the dilators of the nose are almost always active; but in severe cases with very extended consolidation there may be inspiratory retraction of the soft parts of the thorax with every sign of urgent dyspnœa. *Herpes of the face* is not so common in young children as in adults, and has no specially favourable prognostic value as is believed by some. It occurred in but 25 cases (two of which were fatal) out of 135 which I have referred to on this point.

The *temperature* rises often to a very high degree, even in cases of no marked severity. I have frequently seen it over 105 and not seldom over 106. It is usually at its highest on the third day, and has but few marked remissions till near the end.

Pneumonia may attack children at a *very early age*. Of the above-mentioned 135 cases 39 were under three. The general *mortality* averages somewhat lower than is shown by this list of cases in which there were fourteen deaths, all under four years old (and mostly under two), with the exception of one case of eight years old which was complicated with mitral disease. *Termination* by crisis marked 65 (nearly half) of these cases, the duration of the disease varying between three and nine days.

The diagnosis in these early cases rested either on a typical course with typical signs, most often with crisis, or on the discovery of true pneumonia in those fatal cases which could be examined post-mortem. It must be remembered here that, considering the rarity of true pneumonic consolidation post-mortem and the difficulty in many cases of making a diagnosis between pneumonia and broncho-pneumonia during life, it is very probable that some of these fatal cases were wrongly classed as pneumonia. In only six out of the fourteen cases was the diagnosis confirmed by post-mortem examination, and in four of these both lungs were extensively consolidated.

Consolidation of the apex of the lung is more frequent in children than in adults, occurring in about one-fourth of all cases, and perhaps more often in those under three years old. Although more frequently accompanied by delirium than the basic kind, it has an equally favourable course. Of the only two fatal cases of apical pneumonia among the above, one had extensive double consolidation, and the other had jaundice, pleural effusion and diphtheria. The right lung suffered at the apex far more, and at the base considerably less often than the left. Seven cases were double, of which four were fatal. *Jaundice* occurred in two cases, the lung-consolidation being left-sided in one.

Of *meningitis* in pneumonia I can say but little from personal experience, and I am in accord with those authorities who consider the meningitis primary in the case of concurrence of the two affections as demonstrated by an autopsy. It is further probable, as taught by Sturges and Coupland, that these cases are of septic origin; and they are probably allied to the epidemic form of cerebro-spinal meningitis which is often accompanied by pneumonia. One case that I have seen began with convulsions, followed by well-marked symptoms of meningitis which lasted three weeks before apical signs of pneumonia developed, and persisted for a while after resolution of the lung. The temperature was but slightly raised and often not above normal, a vesiculo-pustular eruption appeared on the face and limbs, and there was discharge from one ear. This case ultimately recovered perfectly. The well-known nervous symptoms at the onset of pneumonia, including delirium, occasional squinting and headache, are not justly to be referred to meningitis. I would call attention here to the fact that we very occasionally meet with pneumonia accompanied by extremely slight and evanescent pyrexia.

In much the larger number of cases there is a rapid cessation of symptoms with a rapidly falling temperature, the critical sweating being very often present, but sometimes substituted by a critical diarrhœa. A more gradual fall of temperature and gradual improvement (lysis) is, however, of no less favourable import, nor is there any material difference in average duration between these otherwise similar sets of cases.

A certain proportion of pneumonias merge into *pleurisies,* serous or purulent; the signs of fluid being superadded to those of consolidation, or following on them at a shorter or longer interval. Marked dulness, remaining more than a week or so after the proper symptoms of pneumonia have disappeared, should always be regarded as a likely indication of fluid; and a small exploratory syringe should then be used for diagnostic purposes. Such exploration should be made even if there be but slight or no rise of temperature. Although persistent pyrexia, persistent dulness, and diminished breath-sounds are doubtless the most frequent signs of empyema, I have seen a large number of cases where pus was plentiful and fever absent. Termination in empyema is, I think, more common than is generally suspected, although it occurs in a small minority of cases. The history, however, of many empyemas points to an onset exactly like that of pneumonia. It is moreover well worth remembering that Fraenkel's diplococcus, the microbe which has the most claim to be regarded as causal in pneumonia, has been found in many empyemas; and that Netter (as quoted by Sturges and Coupland) infers from his observations that the greater number of purulent pleurisies in childhood are "pleurésies à pneumocoques." A certain amount of coarse *bronchitis* occurs in the pneumonia of children more often than in adults; and *pericarditis* is from time to time observed, as evidenced either by friction or by the signs of liquid effusion. Pericarditis, however, of any extent or duration should excite the suspicion of rheumatism, of which pneumonia at all ages is from time to time an expression.

The **diagnosis** of true pneumonia in early childhood is often a matter of some difficulty, owing to the frequently lobar distribution of the physical signs, with a sudden onset of symptoms, in cases whose concomitants and results show that they are of broncho-pneumonic nature. This difficulty is especially met with in the broncho-pneumonia which accompanies or follows measles, whooping cough, and other acute diseases. Nevertheless in most instances the abundant râles which usually precede and throughout accompany the signs of consolidation in broncho-pneumonia (not only over the consolidated part where they are usually fine and metallic in character, but also over the chest generally on both sides), and the more prolonged, indefinite and remittent course of the pyrexia, generally give valuable help towards an early, if not an initial, diagnosis; while the termination of the disease will often clear up the doubt which may remain. That there are, however, severe cases of acute broncho-pneumonia which begin with comparative suddenness, and end in as short a time as pneumonia, must be fully recognised; and the true diagnosis cannot here be arrived at until the case is well or dead, and sometimes not at all. Broncho-pneumonic and pneumonic consolidation, moreover, may co-exist, as evidenced by post-mortem examination; and in such cases the clinical and physical signs must of course be confused. But with all this admitted difficulty I would insist on the fact of true pneumonia, as diagnosed from its typical symptoms and course, being very common, and for the most part an easily distinguishable disease, in young children. It should never be confused in thought with broncho-pneumonia, whatever diagnostic difficulty we may meet with in practice; nor should the rare discovery of the typically "pneumonic" lung at post-mortems in childhood induce us to hesitate in our diagnosis, for we have seen that pneumonia in early years in its ordinary form is an essentially benign disease. I need not repeat in detail the widely different physical signs of a typical pneumonia and a typical broncho-pneumonia; but content myself with emphasising the fact that in the one they are as a rule unilateral and prominent, in the other bilateral and obscure; and that, while râles are mostly absent in the one during the height of consolidation and sometimes in the stages both of ingravescence and resolution, in the other they are, as a rule, continuously predominant.

Of the diagnosis between pneumonia and *meningitis* I have already spoken. The predominance of cerebral symptoms often creates a suspicion of primary brain trouble; but suddenly occurring high fever with delirium and burning heat of skin in a previously healthy child should always suggest pneumonia, and this diagnosis should not be abandoned even when the physical signs are long delayed or are ultimately of the slightest kind.

Enteric fever in children especially often begins suddenly with considerable pyrexia and headache, and is thus liable to be diagnosed in place of pneumonia, especially when there is diarrhœa in addition. The flushed face, and some signs or symptoms of pneumonia which as a rule reveal themselves to the careful observer, generally clear up the difficulty; but we must recognise occasionally the impossibility of being positive in our diagnosis at first, however closely we study the case. Delirium is very rare quite at the outset of enteric fever.

The **mortality** of primary pneumonia in children is, as we have seen, but small. Much involvement of both lungs is unfavourable, as also is a very frequent and irregular pulse. A large pleural effusion is both rare and of bad prognosis. In short, all complications are elements of more or less gravity. Especially is this the case when extensive bronchitis or broncho-pneumonia co-exists with the pneumonic attack. The ultimate result of those cases which end in empyema will be treated of with the subject of pleurisy. Secondary pneumonias, occurring in the course of other affections, such as Bright's disease, acute rheumatism and, very rarely, enteric fever, are all, in varying degrees, of more serious import than those which are regarded as primary; as also are the cases which are distinctly epidemic in character, or of probable septic origin. It is among this class of cases, and perhaps especially the nephritic ones, that we occasionally meet with the grey hepatisation of lung well known in adults.

The prevailingly favourable course of pneumonia in early life renders all attempts at active **treatment** as unnecessary in principle as they are demonstrably useless in practice. The patient's diet should be liquid, and the drink copious, slightly acidulated with lemon. In the height of the fever a simple saline mixture may be given whenever it is deemed desirable to prescribe something. Acetate of ammonia with spirit of nitrous ether will sometimes relieve thirst and possibly may promote diaphoresis; although, when the temperature is high and the skin dry and hot, this result is but rarely attained even with very large doses. If the fever be excessive, quinine in full doses, from one to five grains according to age, and frequent sponging with tepid or cold water, should be ordered. I entirely disapprove of cold baths in those severe cases which are supposed by some to indicate the use of this remedy; and, having had but little occasion to try ice-compresses to the chest, can say nothing as regards this alleged remedy. The diarrhœa which is not uncommon in pneumonia is of no grave import, and scarcely ever requires to be checked by art. If it be very profuse, we should empirically use astringents or opium, if not otherwise contra-indicated. When dyspnœa and cyanosis are great, and the case is therefore grave, leeches, from four to eight in number, should be applied to the chest walls, where pressure can be used to arrest bleeding, if necessary; or dry cupping, which is more prompt in action, may be prescribed. When there is much co-existent bronchitis, the case should be treated with a steam-tent, as advised in broncho-pneumonia; and ammonium carbonate and alcohol should be given. If pain be great and there be no marked signs of impeded aëration of the blood, I always advise occasionally repeated doses of Dover's powder, regulated according to age; and hot fomentations or poultices should then be applied to the chest. I regard pain as the chief or only indication for poultices, and deprecate their general use, especially in severe cases, as tending to hamper the breathing.

CHAPTER VII.

PLEURISY.

ALTHOUGH it is the purulent form, or "empyema," which pre-eminently concerns us in discussing the diseases in early childhood, pleurisy in both its plastic and serous varieties is frequently met with.

The question of the **ætiology** of this affection demands some notice here, for the prevailing belief in idiopathic acute pleurisy, or in simple pleurisy due to chill or "cold," is mainly based on superficially apparent cases of this nature which are not at all infrequent in childhood.   Of this origin for the disease, however, I have the gravest doubts ; and long experience and study of numerous cases have caused me to teach for many years that what is called simple pleurisy is almost as rare in the child as it certainly is in the adult, and is practically of but little account. Pleurisy is almost always a secondary affection, occurring in the course or as a sequel of some antecedent or more general disorder; and I am convinced that the more this practical truth is borne in mind the better our prognosis will be.

The first generalisation which will probably be made by an experienced observer of pleurisy in children is, that apparently simple cases arising out of no definitely demonstrable disease are very rare among the children of the well-to-do, though at first sight common enough among those of the poor ; and the next, that a very large majority of cases of all kinds and in all conditions are distinctly referable to some antecedent illness, either local or general.   Out of one series of 108 patients with liquid pleurisy, serous or purulent, taken consecutively from my hospital case-books, 35 were admitted as pneumonia or broncho-pneumonia; and, of the remaining 73, only 12 were stated to have been well before the pleuritic attack for which they were admitted.   An acute onset with pain on one side or in the abdomen was noted in 31 of these 73 ; while in the rest (42) the origin of the affection was symptomatically indefinite, most of them having been ill for several months, usually with cough, and many having never been well since measles or whooping-cough months or years before.   Of the 31 with acute beginnings, 4 were definitely rheumatic, 6 demonstrably tubercular, and 2 undoubtedly traumatic; leaving only 19 which could with any show of probability be put down to the score of primary acute pleurisy.   By far the larger part of these 108 cases were empyemas.

In another series of 18 cases, all plastic or serous in nature, ranging from 1½ to 13 years old, I find two only with a clean bill of health before attack.   Seven had either tubercle or a very strong tubercular history, and five had never been well since measles or whooping-cough some months before.   Two were rheumatic and had heart-disease, and in the remaining two the onset was insidious and origin undiscovered.   These patients, with the exception of two who died with tuberculosis, completely or partially recovered from the pleural attack, which began with pain in the side or belly in nearly half the cases.

Among liquid effusions occurring in childhood, and following on or connected with broncho-pneumonia or pneumonia, empyema is about twice as frequent as serous pleurisy.

In yet another series of 21 cases, all empyemas, in children varying from one to nine years old, 3 were definitely pneumonic in origin, 2 broncho-pneumonic, 3 scarlatinous, 2 tubercular, and 2 traumatic. In six there was a history of long illness with cough, beginning, in half of these, after measles; and in the remaining three the symptoms are reported to have set in acutely with fever, pain in the "chest" or "stomach," vomiting, or rigors. It would appear that in only these last three cases could a primary origin be claimed at all; and it is certainly possible, and in my opinion, most probable, that they were pneumonic. An initial rigor points in all likelihood to coexistent pneumonia, as is also taught by Fraentzel. Moreover, in the cases quoted above as of indefinite origin, and especially those following on measles, a broncho-pneumonia is very likely to have occurred.

I am well assured of the *usually secondary* character of pleurisy, from my experience both of children and of adults; and desire to lay special stress on this point, owing to the little emphasis or occasional doubt expressed thereon by most authorities, and to the frequent and sometimes disastrous mistakes I have seen made in practice by those who hold the popular view that this affection is of but trifling import. The harm arising from such mistaken prognostics is not confined to the patients and their friends, but often seriously damages professional reputation.

To sum up:—*Pneumonia and broncho-pneumonia* account for a very large number of pleurisies, both purulent and serous, and *tubercle* is a very frequent cause. *Acute rheumatism* gives rise to many plastic and serous effusions, and *measles* and *scarlatina* head the list of those numerous fevers or septic diseases of which empyema is an accompaniment or sequel. The affection is not uncommon in *Bright's disease*, and is seen occasionally in *syphilis*. Direct *traumatism*, including falls or blows which may bruise the pleura with little or no external injury, accounts for a certain number of cases of both serous and purulent pleurisy; as also does *extension of inflammation* from neighbouring structures or abscesses, including those below the diaphragm. Other occasional causes are not few, but need not be detailed.

The **onset** of pleurisy in childhood is very often insidious and therefore unobserved. Although a considerable proportion of plastic and serous cases begin with the classical symptoms of pain in the side or epigastrium, and with a hacking, short cough, yet these are more often wanting than in adults; and by far the larger number of empyemas are unattended at first by thoracic symptoms, and often unsuspected, until prolonged illness with pronounced wasting and pallor and more or less dyspnœa at last incites some one to examine the chest.

It is not possible to draw a hard and fast line between either the symptoms or signs of serous and purulent pleurisy.

Empyema, especially among the poorer classes, is, as we have seen, one of the common wasting diseases of children. It is much less often unaccompanied by some pyrexia, continued or remittent, than the serous form; but the temperature may be scarcely raised or even normal. Diarrhœa is often seen with chronic empyema; and a waxy yellow colour of the skin is very common, even in cases of short duration. Clubbing of the fingers and toes is frequent and, as a rule, a mark of chronicity. I have not observed the unilateral occurrence of this phenomenon. In cases where there is evidence of fluid in the pleura the exploring needle only can decide definitely on its nature, and it is therefore useless to detail or discuss the varying observations and mere opinions which have been recorded with reference to differential diagnosis on this point. In all pleural effusions, where the liquid is free or nearly free in the cavity, marked displacement of organs is more common in children than in adults. This is chiefly evidenced by the altered position of the heart's impulse; or, where this cannot be established, as not seldom happens, by the absence of the impulse from its proper place and the detection of the heart's new position by auscultation. As in adults, so in children, the displacement of the heart is greater and more likely to cause disturbed action, both as regards regularity and frequency, when the effusion is on the left side. At the same time I would observe that I have but rarely seen marked cardiac embarrassment in the pleurisies in children; and never a fatal issue, as recorded by some, and attributed to twisting of the vena cava. Clotting in the pulmonary artery is perhaps the most frequent cause of death in those extremely acute cases of pleurisy, more often seen in adults, with rapidly increasing effusion which seriously compresses the lung. On the whole, too, dyspnœa is less marked in children than in adults. I have seen several instances in young children of extensive effusion which scarcely interfered with their playing or running about.

The usual symptoms and physical signs of *non-purulent pleurisy* in children are as follows. There is generally hurried breathing in proportion to the amount of effusion, a hacking cough, and pale complexion. With those rare effusions which rapidly occupy the whole of one pleura, the dyspnœa may be extreme and the face cyanosed, the opposite lung being in a condition of great hyperæmia, and the right heart much embarrassed. Such cases call for instant aspiration to relieve urgent symptoms or, possibly, to save life. The temperature is not often high, rarely much over 101°. Pain in the side is often felt at the onset, and may be almost the only symptom when lymph alone is effused. When the inflammation is confined to the diaphragmatic pleura, epigastric pain, occasional vomiting, and, though very rarely, abdominal tenderness will probably mark the case. More or less impaired movement of the affected side is usually seen; and friction sound is generally heard, if examination be made at the outset, but is often missed from its short duration in cases which go on to effusion. It is always detectable during the period of absorption, and in dry pleurisy may persist for an indefinite time. It is especially frequent in the axillary and infra-mammary regions. The friction sound, heard often over the lower ribs on one side in diaphragmatic cases, is not seldom an important diagnostic sign, aiding us in detecting the thoracic nature of an apparently abdominal attack. In free liquid effusion there is dulness on percussion, extending from the posterior to the anterior base, unlike the dulness of basic pneumonia, which but rarely extends in the anterior direction. The dulness is more intense, and percussion causes a far greater sense of inelasticity or resistance, than in pneumonia. Above the level of the fluid, when the effusion does not fill the pleural cavity, there is almost always a subtympanitic note on percussion. But little is usually learned from examination for vocal fremitus, owing to the high pitched voice of young children, and to the ready transmission of the vibrations, when present, from the unaffected lung. The heart may be felt beating at various distances from its normal position by the hand placed on the chest, the most marked example of this being found when a large left-sided effusion pushes the heart's impulse far over to the right.

Auscultation detects much-exaggerated breath-sounds on the healthy side (giving rise sometimes to the diagnosis of lung-affection in that position), and diminution or complete abolition of vesicular breathing on the affected side. Very often bronchial breathing, sometimes in a most marked degree, is heard all over the area of effusion, and exaggerated or puerile breathing above the level of dulness.

A very large number of effusions are circumscribed, especially when *purulent*. They may be very small, occasioning much difficulty in physical diagnosis, and often escaping even frequent exploratory punctures. There may also be more than one collection of fluid completely isolated by adhesions. I once drew off serum by one puncture and pus by another on the same side of the chest, at the same sitting. In cases of some standing the diagnosis between thickened pleura alone, and the same condition with pus-collection somewhere, is quite impossible without exploration. In general, when the exploring syringe or trocar, well thrust in, seems to be firmly held and is withdrawn unstained by pus or blood, the diagnosis of thickened or gelatinous pleura at that spot may be made with confidence; while the appearance in the syringe or aspiration bottle of pure blood in any quantity would probably indicate puncture of the lung. In cases where combined fever, wasting and other symptoms point to active disease, repeated exploration should be made at various points of dulness.

The **events** of pleurisy are very different according to the nature,

extent and duration of the effusion. Some plastic pleurisies may be
entirely absorbed ; but not seldom they leave more or less adhesive
thickening, with or without physical signs. Serous effusions, even of
magnitude and many weeks' duration, may ultimately be absorbed, but
very often result in thickened pleura, marked by dulness and lessened
breath-murmur, with or without friction sounds either single or double ;
and the affected side of the chest may fall in when there are adhesions,
causing more or less deflection of the spinal column, with the concavity
towards the lesion. Similar results to these last, but more marked,
extensive and frequent, may follow empyemas which are spontaneously
absorbed or which recover after one or more aspirations. Empyemas,
however, far more often become chronic ; and, if unrelieved by opening
and draining the pleural cavity, lead to profound wasting accompanied,
as a rule, by fever of varying degree, and to death from exhaustion.
Amyloid disease may be another result of unrelieved or chronic empyema ;
and tuberculosis may, as in other diseases, supervene, encouraged, it
may be, by the morbid condition induced by the suppuration. In some
cases the pus points outwardly at the surface of the thorax, or, after
burrowing the tissues, at other parts of the body, both internal and
external ; in others it is freely expectorated, with much coughing, after
penetrating a bronchus, while occasionally it is discharged in both of
these modes. Fibroid induration of the lung, with or without marked
dilatation of the bronchial tubes, is also in some cases a result of chronic
pleurisy, both purulent and non-purulent. A careful study of the natural
history of empyema in childhood leads to the conclusion that the ultimate prognosis is for the most part bad if the cases be not actively
treated ; and the practical lesson I have learned from experience is to
aspirate at once and fully when the diagnosis of pus is established,
regardless of the absence of symptoms of pressure on the lung, which is
so frequent in localised empyemas. Should the general health improve
and there be no fever, local treatment must then yield to general, until
there be further indication for interference from increase or reappearance
of physical signs.

I have seen several cases of empyema recover well after one aspiration. But, if there be hectic fever or any sign of deterioration of the
patient, I invariably have the cavity opened and drained antiseptically,
without losing time by a second aspiration. I have ample evidence,
from the comparative results of my earlier and later cases, that incision
of the pleura without delay is completely successful in many instances ;
and am further convinced that the modern plan of sub-periosteally
excising a portion of one or more ribs is almost always advisable, as
conducing at once to better drainage in all cases (whether the lung
can re-expand itself or not), and to the earlier closing of the pleural

fistula by aiding the chest-walls to fall in when the lung is per-manently crippled. Difficult as it undoubtedly is in some instances to drain and close the anfractuous cavities of a long-standing loculated empyema, I have the records of numerous cases, where the chest has been opened, resulting in perfect expansion of the lung, with no deformity of the thorax, nor any other trace of the disease than some comparative dulness and the surgical scar. In many cases, indeed, the scar alone has remained to mark the side which suffered. Among these successful cases several are included where the pus was so thick that it could not be aspirated, where there were many adhesions, and where the pleura had to be freely and extensively scraped, being entirely occupied by a gelatinous effusion.

Between this complete success, and as complete occasional failure with fistulous opening and wasting unto death, there may be all grades of thoracic deformity with impaired health. I have sometimes seen a very good result from the spontaneous evacuation of an empyema either through the chest wall or through the lung. Nevertheless "Natura medicatrix" scores but few successes here, and her efforts should always be prevented by surgical art. I would mention here that I believe very many empyemas are such from the beginning, purulent conversion of a serous effusion being certainly rarer than is generally believed. A serous effusion may remain serous for an indefinite time.

As to the **diagnosis** of pleurisy in children, we must remember that it is often very difficult or impossible to distinguish between pleural and pulmonary sounds by auscultation alone, and that pleural sounds are often heard at a spot where puncture may prove the existence of fluid. The great frequency of loculated effusions must remind us to examine for such collections not only at the base but also at other parts of the chest; for a circumscribed empyema may be situated even at the apex of the lung. Careful examination will almost always lead to a correct diagnosis between lobar pneumonia and pleural effusion. I therefore refrain from repeating the physical signs of these well-known conditions, and omit the familiar but useless and generally misleading disquisition of the text-books on a difficulty which is only occasional. If in any case of chest affection there be persistent local dulness with absent, muffled, or tubular breathing, and even accompanied by additional sounds of undetermined character, an empyema is at least probable; and an explora-tory puncture should be used for diagnosis when possible. It is in the chronic cases that the difficulty of distinction between fluid effusion and, for instance, a solid fibroid lung or thickened pleura may arise; and it is just in these cases that a puncture, which so often removes all diagnostic doubt, can be performed without exciting objection or causing risk. In some cases however of small loculated empyemas even frequent exploration is unsuccessful. Change of level of dulness with change of the patient's position is a sign of fluid at all ages ; but the diagnosis in such cases is usually already certain, and the absence of this phenomenon has no importance, considering the frequency in childhood of thickened pleura with adhesions and loculi. The diagnosis between tubercular or other consolidation of the lung and a pleurisy confined to the apex must be made more from general consideration of the history and symptoms of each individual case than from physical signs alone.

The **prognosis** in pleurisy largely depends on the causation of the case in question, being generally bad in very early infancy, when it is usually purulent, and in cases of marked diathetic disease, such as the "scrofulous" condition. Serous pleurisy usually recovers, at least approximately, even in tubercular subjects, provided the underlying disease be not advanced elsewhere. It is, however, neither in children nor in adults to be regarded without apprehension as to its ultimate meaning, for it is frequently but one indication among others of serious and general disease. I have seen several cases discharged as cured, which were subsequently the victims of phthisis.

In all empyemas the prognosis is better in proportion to the completeness of the evacuation of the pus and the early institution of surgical treatment. In few diseases does the ultimately perfect or partial cure depend so much on skilful and assiduous treatment, both surgical and medical. By far the larger number of the many complete cures of empyema occur in those cases which are the result of pleuro-pneumonia.

For the **treatment** of pleurisy we seek to relieve pain and discomfort, to antagonize the conditions out of which the disease seems to arise, and to evacuate pus, when present, out of regard not only to the local trouble it may cause but also to the general and ulterior mischief it entails. In the comparatively rare cases of acute dry pleurisy discovered at the outset, counter-irritation, with Rigollot's mustard leaves or a blister, quickly lessens or removes the pain, and may perhaps arrest the inflammation and prevent liquid effusion. When, however, liquid effusion has occurred, no special treatment is called for or is in the least efficacious unless the symptoms of pressure on the lung are so great as to demand instant aspiration. I have seen on the one hand, in a sufficiently large number of serous effusions, such a rapid re-absorption in their natural course as to preclude belief that this process can be hurried by any drug treatment; and on the other hand my failures in attempting to reduce chronic effusions by means of so-called absorbent medicines, among which potassium iodide is still in the best repute, have been too frequent to warrant the teaching that the occasional disappearance of fluid under this treatment is to be regarded in the light of an effect.

In all pleurisies, other than serous cases due to traumatism or those

occurring in the actual course of a rheumatic attack, a diet as generous
as the patient can take, and iron, cod-liver oil, arsenic, or, in fact, any
effectual aids to good nutrition, are to be recommended, not forgetting
the dictates of general hygiene. Cardiac stimulation either by alcohol
or ammonia is often indicated by the state of the circulation. I have
no reason whatever from experience to believe that curtailing the amount
of liquid in the diet promotes the absorption of a pleural effusion; and,.
further, a dry diet is practically out of the question when any fever is
present which abolishes appetite and impairs digestion.

I have already spoken of the imperative necessity of evacuating pus,
in whatever quantity it may be present; and must leave most of the
details of this treatment of empyema to the province of the surgeon. I
would but add here that Potain's aspirator with a medium-sized trocar
should be used in the first instance; that the puncture should be made,
not at the extreme base, but preferably in the sixth space; that if the
fluid re-appear, or the general symptoms demand it, the thorax should
be opened with, as a general rule, sub-periosteal resection of a piece of
one or two ribs according to strict antiseptic methods; and that a drainage tube, shortened from time to time, should be inserted, securely fixed,
and retained until the discharge ceases and the wound is healing. In
all cases, unless otherwise indicated, the chest should, in my opinion, be
opened either in or behind the mid-axillary line, in spite of the preference
of some surgeons, on various accounts, to make the incision further forward.

No washing out of the chest is required unless the discharge be fœtid,
or there be reason to believe that there is much adhesion and inspissated
and retained pus, which is often indicated by persistently raised temperature, scanty discharge, and retarded healing of the fistula. In this case,
as also when the usual marked improvement fails to follow on the
operation, another opening may be necessary, and the chest may be from
time to time washed out with an aqueous dilution of tincture of iodine
(half a drachm to the ounce) or a solution of boracic acid. In aspirating
a large effusion I can, from experience, strongly recommend the precaution
of occasionally suspending the flow of fluid for a few minutes in all
cases. Positive indications for this are fits of coughing, and any evidence
of embarrassed breathing or heart action.

As soon as the child is well enough to be moved, which is often long
before the discharge has ceased, he should be encouraged to take exercise
short of fatigue, and placed in conditions where he can obtain as much
fresh air and sunlight as possible.

Pleurisy with effusion, especially empyema, is sometimes, though not
very often, double. The prognosis is then graver than in the ordinary
unilateral cases, but the principles of treatment are the same. There is
no reason to wait for the healing of the incision wound on one side

before opening the other pleura. I have lately had two cases of double empyema, one recovering perfectly after a single aspiration on each side, a second exploration finding no fluid ; the other showing an equally good result after a preliminary double aspiration, and subsequent opening and draining of both sides of the chest, with a fortnight's interval between the last two operations. There is, however, I believe, no reason for more than a day or two's interval, or perhaps even a much shorter one, between opening the two pleuræ.

As a valuable illustration of the excellent result obtained in empyema after resection of rib, I add the following short account of several cases examined at various periods after discharge from the children's hospital. The observations were made in 1891 at my suggestion by Dr. Hastings, then Resident Medical Officer, who kindly took the trouble of trying to find out all the cases which had been in hospital during a period of eight years. Out of a very large number, only twenty-four came up for examination. Of these, two cases were examined seven years, four between four and five years, two between three and four years, seven between two and three years, six between one and two years, and three less than one year, after discharge from hospital. Nineteen cases were under six years old, including two of one year and six of two years, the remainder varying from seven to thirteen years. Most of them had either been noted, or with great probability regarded, as arising out of an acute attack of pleuro-pneumonia. As regards symptoms at the time of the examination referred to, in two cases there was stated to be occasional pain in the affected side, and in eight there was some cough, severe in only one. Of these eight, four had slight bronchitic signs, one granular pharyngitis, two shortness of breath on exertion, and in one, where the discharge had continued for two years, there was some evidence of dilated bronchi. In the rest there was no complaint at all. The general nutrition was good in nineteen cases, and fair in five. Not one looked wasted or ill. As regards physical signs, inspection of the chest in the majority of the cases gave no indication of disease beyond the presence of the scar. The spine was straight in nineteen cases, distinctly curved in only two, and the shoulders were of the same height in fifteen. In fourteen the movements of both sides of the chest were equal ; in two only was there distinct deficiency on the affected side. Percussion showed no dulness at all in eight cases ; localised dulness in the region of the scar in twelve ; and distinct dulness of more extensive area in four. To auscultation the breath sounds were quite normal in ten cases, and more or less weakened over greater or less areas in the rest. In only one were there any adventitious sounds limited to the affected side.

The position of the heart's apex beat was little, if at all, altered in a large majority of these cases.

It must be remembered that in almost all of the patients deformity of the chest, impaired movement, dulness, and weakness of breath sounds were conspicuous on discharge from hospital, even when the wound was, as was usual, quite healed. This series of cases, however, amply proves the great frequency of good recovery of the lung after greater or less lapse of time.

CHAPTER VIII.

ON PHTHISIS AND MEDIASTINAL GLAND DISEASE.

Most that is special to our subject in the clinical and pathological aspects of destructive pulmonary disease, or "phthisis" in its widest sense, is mainly confined to cases in children below the age of about six years. After this age we meet with more and more instances of "consumption of the lungs" of the patterns familiar to us in adults, where the symptoms and signs of chest-mischief are greatly predominant, and where anatomical examination shows that the destructive process has begun in the lungs and is mainly or sometimes almost entirely localised therein. Until the age of about three years the largest number of cases of tuberculosis of the lungs show also more or less generalised tubercle in the other great cavities and in the lymphatic glands, evidenced often by special symptoms; and thus many more instances are met with than in adults which have the prominent clinical aspect of either cerebral or abdominal disease. Again, as we have already seen when considering the subject of tuberculosis generally, examples are frequent of advanced tubercular disease, in the lungs as well as in other parts, where both the symptoms and physical signs during life have been indistinctive of their true cause, consisting often of wasting and fever with but slight evidence of pulmonary catarrh; and we also meet with other cases where, still with only general symptoms and but slight cough, extensive disease in the lung during life may be established by physical examination.

From the clinical point of view, therefore, I shall regard as phthisis or pulmonary consumption only those cases in young children where the symptoms and signs of lung disease are prominent; and shall treat shortly of their chief points of difference from their counterparts in later life Concerning the anatomical basis of phthisis in children, it may be said here once for all, without entering into discussion as to whether all cases are strictly tubercular in origin or not, that there is practically the same variety of appearances in the lung at all ages. We meet with tubercle in all its states and combinations, whether caseous, grey, or miliary, softening, obsolescent or obsolete ; and, though less frequently, with all degrees of fibrosis. There is also, as in adults, a great variety of signs and symptoms. As a general rule, with the exception of those cases of somewhat doubtful origin hereafter to be noticed under the name of "fibroid disease," phthisis in children, with all varieties of lung appearances, tends to run a much shorter average course than in adults; and hence we meet with fewer cases of advanced cavitation or of much fibrosis as a sequel of tubercular mischief. After a careful review of my experience, and reference to my case-books, I am constrained to say that it seems as impossible to make any clinically valuable classification of the phthisis of childhood according to the anatomical appearances found post-mortem, as it is of that in later life. I cannot recognise any important clinical differences either between the so-called "pneumonic" and the so-called "tubercular" cases of acute onset and rapid course, or, again, between the more chronic forms which are similarly differentiated by some authorities. There are doubtless many cases of broncho-pneumonia in young children, where the signs of consolidation, usually in one lung, endure for long with perhaps some slight fever, cough, and wasting. Many of these, as we have seen, recover; while more go on to phthisis, with softening and cavitation of caseous matter as shown post-mortem, and with both lungs affected. Whether the consolidation of lung that recovers be caseous, or if caseous, already tubercular or not, that which ultimately breaks down and is soon followed by a similar affection of the other lung is assuredly and confessedly tubercular. I cannot therefore but regard the clinical facts of phthisis, in children and adults alike, as quite corroborative of the pathological tenet that almost all destructive disease of the lung is ultimately tubercular; although it seems at least probable that the chronic inflammatory or perhaps even caseous process may last for some time (as for instance in the case of many broncho-pneumonias) unaffected by tubercle, and ultimately recover perfectly. Such cases may indeed be said to be almost the monopoly of early childhood, a period when tuberculosis is as a rule so fatal. For some time, however, both as regards physical signs and general symptoms, they may be clinically indistinguishable from those which prove to be destructive tubercular disease.

The apparent differences in the various clinical forms of phthisis are probably due to constitutional and environmental circumstances which cause the infective tubercular process or bacillary activity to be more limited and more readily checked in some instances than others. In children, as in adults, some cases run a symptomatically severe course with disproportionately small extent of lung-disease ; while others may have considerable local mischief for long, with but few grave symptoms. In the former class of cases, however, we usually find evidence of some

more generalised tubercle ; while in the latter the tubercular process probably spreads, and spreads slowly, from its primary seat alone. The
only classification of pulmonary phthisical affections that I regard as in
any way useful is very general, and is much the same as that adopted by
Professor Jacobi. (1.) **Acute miliary tuberculosis**, without marked or
any signs of consolidation, which may be almost confined to the lungs
and begins usually in caseation of the bronchial or mediastinal glands.
These cases are not very common, and are as a rule rapidly fatal. (2.)
Ordinary phthisis, or the " caseous pneumonia " of authors. This often
arises out of broncho-pneumonia or out of caseous disease of the bronchial glands ; and may run either an acute or somewhat chronic course,
with various degrees of signs of consolidation, sometimes slight and
mainly shown by bronchophony with little dulness, at other times
extensive with all the usual characters. In many cases there is temporary, and sometimes apparently permanent, recovery ; but there is
always a liability to fresh outbreaks, and to acute miliary tuberculosis
of the lung or other parts, especially of the brain and pia mater. (3.)
Chronic phthisis, of various antecedents and often very slow course ;
marked by considerable fibrosis of lungs, or, in some cases, by extensive
fibrosis of one lung only, and by a tendency to symptomatic quiescence.
This third class, however, as we shall see, is one of heterogeneous content,
and is to be considered apart only from reasons of practical convenience.

Of the **acute form of miliary tuberculosis** almost confined to the
lungs I need say little more here, the affection being not common in
childhood, and even rare in young children. The physical signs are
usually only those of catarrh and are therefore not distinctive ; and the
diagnosis is to be made from a careful consideration of the case in the
light of a knowledge of the general characters of acute tuberculosis. We
must remember that, practically, a large majority of cases of tuberculosis
of the lungs of any form is included in the category of general tuberculosis. In suspected cases, ophthalmoscopic examination for choroidal
tubercle should be remembered ; and due note made of the fact of high
but remittent temperature, often showing the remittence in the later
part of the day.

In the second and **ordinary form** of pulmonary phthisis common
to all ages, acute or subacute, and characterised post-mortem by grey
or caseous tubercular deposits, with or without miliary tubercle, there
are some points to be noticed as more or less special to children. In the
majority of cases below the age of four or five, and in some as old as five
or even six years, the lung-disease begins, not at or near the apices of
the lungs, as is the rule in older children and adults, but either *at the
root or in the lower lobes*, thus depriving us of one of the chief elements
in the physical diagnosis of adult phthisis, based on our knowledge of

the usual progress of tubercular disease from above downwards. The infrequency of cavitation, moreover, of such extent at least as to give rise to unequivocal auscultatory phenomena, is a further hindrance to the right interpretation of the varying pulmonary signs we discover, which, considered by themselves, are easily confused with those of non-phthisical affections such as pneumonia, broncho-pneumonia, or pleurisy. Tubercular disease of the lung, often beginning at the root, is here like broncho-pneumonia, which is so frequently an extension of bronchitis and at the same time not seldom a forerunner of tubercle. I must nevertheless insist on the fact that at the age of about four years, when pulmonary tuberculosis begins from its somewhat longer course to assume gradually the familiar symptomatic characters of phthisis as seen in adults, there is a very notable minority of cases where the disease does start at the apices and proceed downwards. Of this I have seen many examples.

Expectoration is decidedly a less marked symptom in young children than in adults, and we are frequently thus left without the diagnostic help of microscopical examination for bacilli and lung-tissue. It is none the less true that in phthisis we meet with a great exception to the general rule of the absence of expectoration in the pulmonary diseases of children under seven or eight years old; for I have seen numerous cases as young as four, and some younger, where expectoration was considerable or even profuse. *Hæmoptysis* is certainly not so frequent as in adults; but, though seldom excessive, is by no means rare. Rapidly fatal hæmoptysis, arising from aneurysms or erosion of vessels in cavities, is very rare, owing, probably, to the usually shorter course of juvenile phthisis; but instances of both may be met with. The temperature charts show nothing very peculiar; progressive tuberculosis, with little or sometimes no pyrexia, of which I have seen some instances like those reported by Henoch, being mostly confined to wasted infants of low vitality. One more point to be insisted on, at the risk of some repetition, is that a considerably greater liability obtains in phthisical children generally, than in adults, to tubercular disease elsewhere, as marked by symptoms of both head and abdominal mischief. The younger the patient—and this applies as well to the years beyond childhood—the more frequent are cerebral tuberculosis, meningitis, peritonitis, and intestinal ulceration (with enlarged mesenteric glands) as often evidenced by obstinate diarrhœa.

The following case shortly illustrates, both in its history, signs and course, the form of phthisis frequently seen in young children. A boy of nearly four years old was admitted with an account of suffering from sickness, headache, diarrhœa and noticeable cough for a few weeks previously. There was no family history of lung or other disease; but the boy had never been well, and had coughed slightly since measles followed by whooping-cough two years before. At first scattered râles only were heard at the bases; but later on some dulness and rather fine crepitation were discovered, mainly over the upper half of both lungs. The fingers were clubbed; there was much diarrhœa and night sweating; and the temperature was hectic throughout, varying between normal and about 103°. The glands in the neck and axilla were enlarged. After five months, with progressive physical signs, cough and emaciation, the boy died, having developed considerable œdema of his feet and a small purpuric eruption. The post-mortem showed extensive caseous tubercle, softening into small cavities, in both lungs; large and caseous bronchial and mesenteric glands; tubercular ulceration of intestine, and numerous small tubercles in the liver.

With regard to the frequent difficulty in *diagnosis* of the acuter forms of phthisis with caseous pneumonia from non-tubercular broncho-pneumonia, nothing can be said from the point of view of physical signs; but we should remember that an acute broncho-pneumonia following on measles, especially when extensively involving both lungs, is very apt to be tubercular; and we may be further aided by the previous and family history of the case. In tuberculosis, especially of the pulmonary form occurring in children beyond infancy, there is a considerably larger proportion of cases with a marked hereditary history of phthisis than in infantile tuberculosis taken as a whole. The presence of glandular affection or other evidence of struma favours the diagnosis of tubercular lung-disease; and I think that it is especially and perhaps only in pulmonary phthisis, among all the other forms of tuberculosis, that the so-called "tubercular appearance," with fine features and complexion, good stature and slender bones, is of any importance as showing a predisposition to tubercular disease. In practice we should certainly beware of giving any diagnostic weight in individual cases to the somewhat vague descriptions of the scrofulous, tubercular or phthisical "habits." We must always hesitate long before pronouncing any case of apparent lung-mischief in children to be unquestionably phthisical; and should remember that not only broncho-pneumonic consolidation at the apices, but also pleural thickening and empyema localised at the upper part, may closely simulate phthisis as regards both physical signs and many symptoms. In all doubtful cases the exploratory syringe should be used; and repeated examination should be made of the sputum for bacilli, whenever possible. It must also be noted that physical signs of excavation, quite indistinguishable from those heard over tubercular cavities, may often be found over one or more regions of lungs which have been the subjects of prolonged attacks of broncho-pneumonia. These signs are due to marked dilatation of the bronchial tubes, which in some degree is usually observed in fatal cases ; and they often vanish completely, though gradually, owing to the contraction of the tubes, with the subsidence of the catarrhal process and general improvement of symptoms. I have over and over again, especially in past years, been obliged to renounce a too hasty diagnosis of phthisis and a consequently erroneous prognosis, in cases of pulmonary mischief in young children. The truth impresses itself on me more and more that at least in childhood and in the absence of proof of the existence of tubercle bacilli, we must rely for a diagnosis of phthisis much more on the symptoms than on the physical signs. As regards the symptoms, with the exception of the point already alluded to, they are identical with those of the adult disease.

Under the general heading of **chronic phthisis** I shall consider, for practical purposes, the very mixed class of cases which have, as points in common involving chronicity, both *fibrosis* of lung, of varying extent and distribution, and a tendency to more or less *symptomatic quiescence*, with stationary or slowly progressing physical signs. In this class we find some cases where a tubercular origin cannot be questioned ; others where a tubercular connexion, although existent, is nevertheless obscure in its nature ; and still others where no tubercular relationship at all can be demonstrated. In all chronic cases of tubercular disease of lung there is more or less secondary fibrosis in the neighbourhood of those deposits which are in a state of retrogression ; and thus fibrosis may be regarded as a process of repair and as part and parcel of the chronic, retarded or "cured" cases of the ordinary phthisis that we have already considered. Extensive fibrosis, in cases of this kind, is infrequent in childhood, owing to the usually more rapid course of the disease than in later life. We often, however, meet with cases of chronic consolidation of the lungs, and more especially of one lung, in children beyond the period of infancy, where the symptoms and physical signs are of insidious origin and slow progress, and where after death, which is usually from some intercurrent attack of acute disease either inflammatory or tubercular, extensive induration of the lung is found, due to a development of nucleated fibroid tissue in the interlobular septa, the alveolar walls and the bronchial tubes, and leading to extensive destruction of the lung-cells. In some of these cases caseous deposits are seen in the lungs or bronchial glands or elsewhere ; and tubercular cavities are sometimes found, as well as dilatation of the bronchial tubes, which in some degree accompanies most cases of chronic fibrosis. It is in this class of cases, where some evidence of tubercle is found post-mortem with a predominant amount of fibroid change, that the difficulty meets us as to whether we should regard the disease as primarily tubercular with secondary fibrosis, or as chronic inflammation of the lung, or pulmonary "cirrhosis," arising out of a catarrh and subsequently infected by tubercle. There is no doubt that in some instances, after a long course of more or less illness marked by the symptoms presently to be noticed and by the signs of consolidation limited to one lung, there is ultimately evidence of breaking down of the lung, with some affection of the other; and the symptoms of ordinary phthisis supervene, tubercle bacilli being found in the sputum. In another class, however, the fibrotic affection apparently undergoes little change, there are no signs or symptoms of softening, and, whether the disease be limited entirely to one lung, as it very often is, or involves both to some extent, no trace of tubercle of whatever kind or condition is found post-mortem in the cases which die from the intercurrent attacks of bronchitis or broncho-pneumonia to which they are greatly subject. It is not within the scope of these remarks to discuss the question as to whether all fibrosis of the lung which ends in phthisis is primarily tubercular; but it may be said that, as far as this affection as found in children is concerned, there is at least considerable clinical evidence of some tubercular connexion in many instances, even when apparently one lung only is affected; and, further, that we have no practical means of distinguishing during life between tubercular and non-tubercular cases. We are on firm ground, however, in stating that extensive pulmonary fibrosis in children arises mainly out of one or several attacks of broncho-pneumonia, or follows on chronic bronchitis, often with intercurrent acute attacks; and many hold that it may result from a true "lobar" pneumonia or a pleurisy. Morbid anatomy favours the view that the inflammatory process begins in most instances in the lung itself, invading the smallest bronchial tubes, the interlobular septa, and the walls of the air-cells. According to the stage of the disease the fibrosis may be seen in the form of streaks of various sizes intervening between healthy portions of lung, or involving the whole of a lobe or the entire lung. In many cases the pleura is much thickened throughout, and so closely adherent that the lung can be removed only by cutting. There has, however, in my own experience been little or no clinical evidence or suspicion of a pleurisy being the starting-point of pulmonary fibrosis as seen in children; and out of a large number of cases of pleurisy which I have been able to examine long after the attack I cannot remember a single case of this form of chronic lung-disease, often as an apparently simple pleural effusion has been the first overt sign of ordinary tubercular phthisis.

In a very considerable number of my cases with the signs and course of extensive fibroid disease of lung there has been a markedly phthisical family history, a fact which may have some bearing on the vexed question of the possible causal part played by a tubercular process which may have become obsolete even in cases with no evidence of tubercle post-mortem. The subjects of the disease are mostly children beyond

four or five years old with, as a rule, a history of frequent chest attacks
not seldom dating from measles or pertussis, and gradually increasing
dyspnœa on exertion. Cough may be very slight at first, but in time
is usually prominent, with much expectoration and a tendency to paroxysmal attacks closely simulating whooping-cough. Emaciation is often
slight and always gradual, and in some instances, where the disease
seems stationary, the child may be in fairly good condition. In proportion, however, to the respiratory difficulty there is pallor or blueness of
the face, the fingers are clubbed, the veins of the upper part of the body
are full, and there are the usual signs and symptoms of a dilated and
hypertrophied right heart.

In most cases in childhood the *physical signs* of disease are limited to
one lung, at least for a long and indefinite period; the so-called cases of
" fibroid phthisis" with signs in both lungs, arising out of inhalation of
irritating substances, being almost confined to adults. We therefore find,
after a while, contraction of one side of the chest with much impaired
movement and percussion note, and the heart is displaced towards the
affected side. This displacement is especially notable when the right
lung is diseased, the heart being sometimes found beating in the right
axillary region. The auscultatory signs vary according to the activity of
bronchial catarrh and the degree of bronchiectasis, or to the presence of
cavities arising from ulceration through the bronchial tubes or from the
tubercular processes which may be present. There may be all combinations of dry and moist sounds, from a slight degree up to well-marked
signs of one or more cavities. The sputum varies much in amount and
character; in some early or almost stationary cases there is none, or it
may appear only during intercurrent attacks of bronchial or pulmonary
catarrh; while in most, where bronchial dilatation is established, or where
there are excavations in the lungs either from bronchial ulceration or a
tubercular process, the sputum is both profuse and markedly fœtid.
Pyrexia is often absent for long periods, and in some cases for an indefinite time; but in many, even when there is no other evidence of progressive disease, pyrexia in a slight degree and of a remittent character is
observed. When indeed pyrexia, and still more when signs of excavation as well, are noted, accompanied by increased wasting and cough, the
case is one which is called by some "fibroid phthisis," and is only distinguishable from ordinary phthisis by the limitation of the physical
signs to one lung. In some of these cases the other lung may then show
signs of apical mischief, and the disease may progress; while in others
the active signs may disappear and the disease resume a chronic and
apyretic condition. A practical lesson I have learnt, from studying fibroid disease of lung in children, is that we must not pronounce a positive
prognosis, or exclude the possibility of ordinary phthisical events and a

fatal issue, on the ground that these one-sided cases appear to arise from a chronic pulmonary inflammation of non-tubercular origin.

Whatever their source may be, and although evidence of tubercle post-mortem, as we have seen, is sometimes undoubtedly absent, death from an outbreak of tuberculosis is by no means rare. We must, however, remember that there are many cases where the disease, confined to one lung, becomes after a while apparently stationary, and the patient may live for several years with somewhat impaired nutrition and breathing power and with normal temperature. The physical signs as well as the symptoms are here those of past, rather than of present, mischief, consisting often of contraction of one side, with intense dulness and bronchophony and but few, if any, additional morbid sounds.

The two following cases shortly illustrate this interesting form of lung disease.

A girl, æt. thirteen, subject to attacks of " bronchitis " since the age of three months and treated at Shadwell Hospital for rickets when two years old, was admitted with the history of cough, fœtid sputum often tinged with blood, and general weakness, of two years' duration. Her mother, father and several brothers and sisters suffered from chronic cough ; and one sister, æt. eleven, died in Guy's Hospital with "chronic bronchitis." Several near relatives had died of consumption. The patient was somewhat thin, with a fine clear skin and long eyelashes. The left chest was markedly contracted and dull all over ; and abundant extensive crackling was heard, with loud bronchophony and whispering pectoriloquy near the angle of the scapula. There were no abnormal signs on the other side of the chest. At first the temperature was occasionally 100° in the evening, but soon fell to normal. The girl rapidly improved with ordinary hygienic treatment, and left after some months in apparently good health.

(2.) A boy of three, with a history of much family phthisis, and of " bronchitis " when a baby, but of good health since, was admitted with cough, wasting and night-sweating of six months', and fœtid expectoration of six weeks', duration. He was thin, with bluish complexion ; coughed spasmodically ("like whooping-cough "), and spat profusely. The right chest was dull all over, and much contracted at the upper part, where the breath sounds were cavernous and attended by large crepitation, and there was distinct cracked-pot sound. The left side appeared normal to examination. During six weeks in hospital the child improved much, and left with a marked gain in weight.

The two subjoined cases are inserted to exemplify difficulties often met with in diagnosis between *consolidation of the lung* and *pleurisy.* Of the first it may be questioned of which kind the affection originally was. A girl of seven had had cough all her life. The right side was contracted and dull all over, and the breathing for the most part was loudly bronchial; but there were no added sounds of any kind, and the left side was apparently normal. The heart's position was found on the right side of the sternum. A little pus was seen on exploration at the angle of the scapula, but none on four subsequent trials. In the course of some weeks coarse metallic râles were heard over a large area on the right side, and slightly, probably from conduction, on the left. There was never any expectoration or fever. It was thought that the case was one of fibroid disease, and that the pus might have come from a dilated bronchus. On the whole, however, considering the absence of expectoration, the universality of the dulness, and the probably normal condition of the left chest, the case may almost equally well be regarded as pleuritic in its entirety. The child left the hospital after some months in the same state as on admission.

The next case, in a child one year old, was thought at first to be pleural, but proved to be due to caseous consolidation. There had been some cough and wasting for several weeks, but the breathing was not much embarrassed at first. Absolute dulness was found over the lower half of the left chest, with somewhat bronchial breathing but no added sounds. No fluid appeared on exploration with a needle. In three weeks the child died with increasing debility and some pyrexia, but with no change of physical signs. The left lower lobe was completely adherent to the chest wall and universally caseous. In the centre, rather towards the base, there was a small cavity containing creamy fluid. The upper lobe was very slightly involved, and the right lung was distinctly hyperæmic. The mediastinal glands were large and caseous, the mesenteric glands much swelled, and there were some tubercles on the surface of the spleen. This case may be classed with those referred to by Dr. Goodhart under the name of cheesy solidification of the lung.

Treatment.—The clinical study of phthisis in children teaches us that, comparatively rapid though the course of most cases may be towards the usually fatal issue, yet there is reason to believe that arrest, or perhaps recovery, occurs somewhat more frequently than in later life. There is, moreover, perhaps some indication for possibly hopeful treatment in the fact that it is especially in young children that catarrhal processes in the air-passages so often seem to prepare the ground for the entry and development of the tubercle bacillus. Besides, therefore, giving all attention to the details of general hygiene and nutrition, which are of the highest importance in cases of suspected or demonstrated phthisis at all ages, we must ever be careful to protect children from all avoidable sources of catarrh. The good effect of this precaution is perhaps best seen in cases of chronic phthisis, or those described under the head of fibroid disease, where an accession of catarrh so frequently aggravates the disease, and where residence in a warm climate often insures quiescence of symptoms and a considerable degree of health. For the rest there is little to be said regarding the treatment of phthisis in children as distinguished from adults. The most nutritive diet consistent with individual digestion must be insisted on, including abundance of cream ; and cod-liver oil should be taken as persistently as possible. In early cases the hypophosphite of lime or soda should I think be given regularly and for long, although I cannot say that I have ever seen any definite good results from this drug in the well-established disease.

Arsenic is a very useful drug in the numerous cases with an element of hope in them ; as also is iron, which I have no reason to believe is harmful in any, although stated to be so by some writers. A dry climate is to be insisted on ; but whether moderate heat or cold is most suitable will entirely depend on individual cases. The best climate for any case is that which renders it possible for the child to have plenty of sunlight, to be much out of doors, and as active as its strength permits. In England there is clearly not much choice of winter residences, all of which are unsatisfactory ; but, as I have said when treating of struma, there are perhaps no better spots than Bournemouth, or the west coast of Wales. Among foreign places, many of which are highly beneficial in arresting disease by affording the above-mentioned conditions, Egypt is strongly to be recommended, and all the more since accommodation is now provided in the desert close to the pyramids, and thus within a short distance from Cairo. In some cases a sea-voyage to the Cape or Australia is of great use. Failing the possibility of going so far, Arcachon, Biarritz, or some of the resorts on the Riviera may be tried. Patients should not return to England until the month of June, and should usually leave home for some more suitable climate not later than the end of September. In the summer many patients do as well in England as anywhere. Both the numerous watering-places on the East coast, and the Yorkshire moors, supply favouring conditions in the hot months.

In advanced cases we must of course depend on the usual symptomatic treatment, which cannot be detailed here. We shall often find much benefit from frequently repeated small doses of opium which check cough and allay irritability ; and from the regular and sometimes free administration of alcohol, either as wine or brandy. Some recommend, apparently on antiseptic principles, the internal use of creasote in doses of one or two drops two or three times a day ; and, for checking diarrhœa, the occasional administration of two to four grains of naphthol or naphthalene, according to age. These latter drugs appear to be of use in some cases ; but my experience of them, as yet but slight, has had no uniform result. Their nauseous taste is difficult to conceal.

I would add that in the class of cases above-mentioned, where there may be doubt whether we have to do with a recoverable broncho-pneumonia or with tubercular disease, and in many of which the signs and symptoms ultimately disappear, there is no harm but possibly good to be expected from the early and repeated use of vigorous counter-irritation to the affected part by means of the croton-oil or the iodine liniment. In spite of the absence of positive evidence of benefit therefrom, I follow and recommend this now somewhat antiquated line of treatment.

I omit all discussion of antiseptic methods of treatment, by inhalation or otherwise, based on the bacillary origin of tuberculosis, in the belief that, however practically valuable for prophylaxis and public hygiene our recent ætiological knowledge may well prove to be, it has as yet had no influence on individual therapeusis, with the exception, perhaps, of the lesson, taught by the infective qualities of phthisical sputum, that we should protect patients from the possibly additional ill effect of their own discharges by directing them to expectorate only into vessels provided for the purpose and containing some germicide.

Disease of Mediastinal Glands.

It is especially the tracheo-bronchial glands, so often the subjects of enlargement and caseation in childhood, which will concern us here. Clinically this enlargement of the glands, if excessive, may give rise to a set of symptoms, due to involvement of neighbouring nerves and organs, which justifies separate notice. Similar results of course may follow, and with much greater proportionate frequency, on other and rarer forms of mediastinal disease, such as abscess or growths of lymphomatous or sarcomatous nature, or to some extent on the chronic mediastinal thickening which is sometimes met with in connexion with pericarditis. It must be remembered that the bronchial glands are almost always enlarged and caseous in tubercular lung-disease, and that we but rarely meet with any special symptoms due to their enlargement in the course of recognised phthisis. There is, however, a well-known group of cases where enlargement of bronchial glands, with or without definite symptoms, precedes the pulmonary mischief, as evidenced by post-mortem examination where the disease is seen to have started at the root of the lung. In most cases, including this latter class, there are no symptoms or physical signs, even when the enlargement of the glands is found post-mortem to be considerable. Of this I have several times satisfied myself. In others there may be symptoms of pressure on the pneumo-gastric nerve, such as hoarseness or spasmodic cough ; œdema of the face and upper part of the body, due to pressure on the superior cava or jugular veins, may appear ; or infarcts and hæmoptysis may result from involvement of the pulmonary vessels. There may also be stridulous breathing from pressure on the trachea. Even when many of these symptoms are present, it is often impossible to discover any special physical signs by percussion or auscultation; but in some few there is dulness in the interscapular space with, perhaps, tubular or bronchial breathing and bronchophony, and in others more or less dulness behind the upper part of the sternum or on one or both sides of it as well, with or without similar auscultatory signs. In addition to these signs there is no doubt that, listening with the stethoscope placed just below the sternal notch, we may hear from time to time a venous hum, produced by retraction of the patient's head and ceasing on its replacement. From observation I can say that this particular humming sound is perhaps very frequent in cases where obvious pressure signs exist; that it is very often absent throughout in cases where the bronchial glands are found much enlarged post-mortem; and that it is not seldom heard where there is no other symptom or physical sign of bronchial or even pulmonary trouble at all. In the absence, therefore, of signs of pressure I am of opinion that we are not justified in attributing diagnostic weight to this phenomenon as an early sign of bronchial-gland disease. Enlarged tracheo-bronchial glands may give rise to one-sided signs by pressing especially on one bronchus, and causing impaired resonance and diminished breath sounds, or marked bronchial breathing. They may also break down into an abscess, and burst into the trachea or a bronchial tube. On the other hand they may shrink and become calcified, as we often find post-mortem; but in all probability this result is confined to caseous glands which are neither detectable by examination nor evidenced symptomatically.

SECTION VI.

DISORDERS OF THE HEART AND CIRCULATION.

SECTION VI.—DISORDERS OF THE HEART AND CIRCULATION.

BEFORE discussing affections of the heart in childhood, which may be mainly classed under the heads of congenital disease and inflammation of the heart and pericardium, it is well to glance at those anatomical and physiological peculiarities of the circulatory system in early life which have a bearing both on the physical examination of the organs concerned and on certain symptoms of their derangement.

The apex beat in infants is placed further outward and higher up than in later life, being generally in the vertical nipple line or beyond, and often in the fourth interspace. The usual adult position of the apex beat is frequently not observed until the child is three or four years old or more, when the width of the heart in relation to the chest becomes less, the thorax becomes less barrel-shaped, and the diaphragm takes a lower position. The apex beat is also more diffused in early childhood, and the sounds are shorter in relation to the silences and sometimes reduplicated. Owing to these peculiarities the exact localisation of sounds and murmurs is often somewhat difficult. Again, the ratio of the volume of the heart to that of the ascending aorta is much less in early than in adult life, being only about one-fourth greater in infancy, while after puberty it is nearly quintuple. The most rapid increase of the heart takes place about the time of puberty ; and it follows from these facts that the blood-pressure is much less in childhood than afterwards, contributing in all probability, as has been frequently observed, to the tendency to take cold and to the ready failure of circulatory power so often noticed in young children, and well exemplified by the familiar chilblain. Anomalous and temporary murmurs are often heard over the apparently healthy infant's præcordium. These are difficult of accurate explanation and are probably best termed hæmic.

The great variability of the pulse-rate in infants and young children is probably accounted for by the imperfect development and consequent instability of the regulating nervous mechanism. The pulse-rate at birth generally ranges between 130 and 140, and may be higher, sinking to little over 100 in the second year, and only very gradually reaching the normal of 75 to 80 about the time of puberty. Very slight movements and mental excitement are accompanied by great rises in the pulse-rate ; and irregularity of force and rhythm, often observable during the first year while the child is at rest, is easily produced and much magnified by the same causes. It is important to bear in mind this normal frequency, irregularity and ready excitability of the healthy infant's heart when looking for the symptoms of any disease such as, for instance, meningitis, or the convulsive state, which such phenomena often indicate. The younger the child the less is the symptomatic importance of irregularity of pulse alone. After the first year or two, however, this irregularity is often of the greatest diagnostic weight in early cerebral disease and some other affections. In a case, for instance, of apparently simple jaundice of about three weeks' standing in a healthy child of six years old, marked irregularity of pulse was one of the earliest symptoms which ushered in the delirium, convulsions, and fatal coma of what proved to be acute yellow atrophy of the liver.

Disorder of the heart's working may occur in childhood, as in adults, from various causes other than anatomical disease, but is neither of much importance nor accompanied by that distress in the sphere of feeling which so often fills older patients with dread. Palpitation, perverted rhythm, and intermission of beats are not rarely observed ; pain, but seldom. Such irregularities appear to be connected with disorder of the stomach and bowels, or bad nutrition with anæmia, and are sometimes marked during dentition. I have seen several cases which seem to correspond closely with those described by Da Costa. These are marked by irregular rhythm and infrequent action of the heart without any discoverable condition or excitant to account for them. Da Costa terms such cases idiopathic, and states that they mostly begin after three years of age, and generally last till puberty, but are sometimes permanent. He notes further that during febrile attacks the cardiac rhythm in these cases becomes irregular. I am inclined to attribute these phenomena to disturbed innervation, which is usually part of a more or less wide-spread nervous disorder.

Functional disorder of the heart is to be treated by attacking the conditions out of which it seems to arise, and the avoidance of all demonstrably exciting causes. The diet must be duly regulated when gastro-intestinal symptoms accompany the cardiac trouble ; and, for the rest, we should strongly advise moderate exercise, mental distraction, and all attainable sun-light and fresh air. I do not think that digitalis or similarly acting drugs are of any avail ; but the nutritive effects of arsenic and iron and, in some cases, cod-liver oil are not to be doubted and are sometimes very soon observable, especially in patients who are anæmic.

CHAPTER I.

CONGENITAL HEART-DISEASE.

Cyanosis, or the " blue disease," characterised by varying degrees of purple discolouration of the skin and mucous membranes, hurried breathing, surface coldness, and clubbing of the fingers and toes, is usually associated with congenital malformation of the heart or large vessels. Cyanosis, however, may occur without heart-disease as the result of long standing pulmonary affection, such as extensive emphysema or thick fibrinous pleural adhesions, and, in moderate degree, in ill-developed children with or without mental deficiency. Congenital heart-disease on the other hand, marked by physical signs and some symptoms of disturbed circulation, may be found unattended by cyanosis.

The *symptoms* of cyanosis in connection with heart-disease are observed, in about two-thirds of all cases, either at birth or within the first week. In the rest they arise at varying intervals after birth, sometimes after many years, their retarded appearance being probably due to extra stress on the imperfect heart, associated with physical exertion, mental excitement, the onset of some acute disease, or anything which may induce disturbance of compensation. The longer the interval after birth when the symptoms appear, the more often a definite exciting cause can be established. The depth of discolouration varies much in different cases and in the same case at different times, some patients when at rest showing almost a natural colour, while in others the nose, cheeks, lips, tongue, fingers and toes may be almost black and the whole body dark purple. Excitement always deepens this colour and increases the frequency of breathing.

The axillary temperature is almost always below the normal, and likewise often, though by no means always, the temperature taken in the rectum. The body surface is felt to be markedly cold, especially at the extremities. The heart-beats are generally hurried, irregular or intermittent, the breathing is frequent, and there may be cough and occasional hæmoptysis. Bad digestion often, and hæmatemesis sometimes, occur as signs of congested stomach, and from time to time there may be bleeding from the gums, nose or bowel. Anasarca is but rarely seen, and only towards the end. The patients are as a rule very small, emotional and especially fretful; they are subject to drowsiness; and their intellectual functions are often torpid, though not seldom apparently unimpaired.

Physical examination shows frequent malformation of the thorax, pigeon-breasts being very numerous, as evidenced by my own case-books and the more extensive collections of many others ; and in most instances there are the signs of enlarged heart and a murmur or murmurs, generally loud and systolic, over the præcordium.   It is almost always impossible to pronounce definitely upon the exact nature of the cardiac malformation which underlies the signs and symptoms of cyanosis ; for, apart from the fact that malformation of various kinds may exist without physical signs, the possible lesions are so numerous and multiform, owing, among other causes, to the varying periods of fœtal existence when arrest of development takes place, that most attempts at diagnosis must be confessed to be mere guess-work or more or less probable inference from averages based on anatomical records.   The classification and explanation of congenital malformation of the heart are of great anatomical importance ; but in this clinical work I must content myself with observing that the most frequent lesion in cases with cyanosis is narrowing or complete obstruction of the pulmonary artery with some consequent communication between the two sides of the heart, a pervious ductus arteriosus and an imperfect ventricular septum having been more frequently recorded than a similar auricular defect, or than a patent foramen ovale.   It follows from this that most of the cases are due to faults of development occurring quite early in fœtal life ; for the ventricular septum is normally complete by the end of the third month.   For a most lucid exposition of this subject I would refer the reader to a paper by Dr. Sharkey in the *Lancet* (1880, vol. ii.), while, to those who would further pursue the study, the classical work of Peacock remains of unrivalled value.

The *prognosis* in cyanotic cases is uniformly bad.   Improvement scarcely ever occurs, and very few survive to middle age, one-third of all dying in the first year and two-thirds in the first decade.   Cerebral hæmorrhage, convulsions and coma are often the immediate causes of death ; and bronchial catarrh with collapse of the lungs and bronchopneumonia is a frequently fatal event.   I can corroborate, too, from my own cases the modern teaching that cyanotic patients, in spite of the pathological dictum of Rokitansky to the contrary, show a certain tendency to suffer from pulmonary tuberculosis.   Among other illustrations of this in my case-books there is recorded an instance of well-marked phthisis in a cyanotic of thirteen years old.

But little can be said with regard to the *conditions* which cause or favour congenital malformation of the heart and large vessels.   Some few cases are referable to fœtal endocarditis, especially in the right heart. Other defects of development sometimes co-exist, such as visceral transposition, anencephaly, spina bifida, and umbilical hernia.   Such cases are doubtless far more common among the working population of large towns, where bad hygienic conditions largely contribute to the production of ill-formed children, than in the country or among the children of the well-to-do. Of this Dr. Lewis Smith of New York gives interesting evidence. Nervous disturbances which accompany maternal impressions may be reasonably suspected, though perhaps scarcely proved, to be causal in many instances. There seems on the whole to be a slight numerical predominance of male cases, as is the fact with other congenital deformities.

The cyanosis itself we must, with Morgagni and Peacock, regard as the result of long-continued capillary stasis and obstruction at the centre of circulation, and but slightly, if at all, to the commingling of arterial and venous blood.

Where there is no cyanosis, congenital disease of the heart may always be suspected when, with or without marked symptoms of disturbed circulation, the physical signs of murmur and enlarged heart point to stenosis of the pulmonary artery; as also in cases where there are paradoxical sounds and signs unattributable to rheumatism or other affections known to be productive of heart-disease. We often meet with such cases at any age, which may be with all probability credited to some congenital defect, and where the frequent absence of any symptom or other sign of heart-disease will usually, at least after the lapse of time, completely justify a good prognosis. A systolic bruit, and nothing more, in the region of the pulmonary artery is very frequent in this class of cases; and I would lay stress on a systolic murmur, heard loudest at the base of the heart, and very clearly over the back of the chest, as often indicating a congenital affection which is symptomatically unexpressed.

The *treatment* of patients with cyanosis may be summed up under the heads of continuous warmth to the body (not forgetting the nocturnal hot bottle to the feet), digestible diet, stimulation from time to time by alcohol, and rest for the body and the mind. Digitalis may occasionally be useful in cases where dropsy may have set in and when the heart-beats are frequent and irregular. Blood-letting is certainly indicated, when dyspnœa is urgent, for the relief of the probably overloaded heart.

CHAPTER II.

CARDIAC INFLAMMATION AND VALVE-DISEASE.

BEFORE discussing the clinical aspects of valve-disease in childhood, and of the affections usually described under the heads of endocarditis and pericarditis, it is practically important to consider the part played by involvement of the muscular substance of the heart itself. Now, although localised abscess or diffused myocarditis, readily recognised post-mortem and sometimes leading to cardiac aneurysm, may occur in childhood more often than in later life either as the result of pyæmia in connexion with bone-mischief or of endo- or peri-carditis or other causes, it is nevertheless very rare and need not be dwelt on except in a monograph or large systematic work. Its proper symptoms are those of rapid heart failure, lessening or absence of the first sound, smallness and irregularity of pulse, difficult breathing, pallor, palpitation, and often præcordial pain ; and acute cardiac dilatation may be evidenced by percussion. When it occurs in association with endo- or peri-carditis, as it occasionally does in severe rheumatism, the physical signs of these affections are of course predominant ; and, generally speaking, its existence, from whatever cause arising, is rather to be inferred from the observation of excessive heart failure in the light of pathological knowledge, than directly demonstrated by physical examination.

It is mostly in its less recognised and less severe forms in connexion with rheumatic heart-disease that myocarditis becomes of great clinical importance. There is probably some involvement of the underlying muscle in every well-marked case of recent endo- or peri-carditis, and this condition of the heart, with its necessarily impaired function, should be prominently before us in the mental picture we form of any acute case of heart-disease. In spite indeed of the probably almost constant primary lesion of the endocardium in rheumatic heart-affection, our clinical conception of endo-carditis, with the exception of its ulcerative form which has its own special symptoms and dangers, should be mainly that of its causal relationship to mechanical valve lesions and their chronic effects on cardiac structure and function. Wherefore, when we think of acute rheumatic endocarditis or pericarditis with marked cardiac disturbance, we must never forget the deeper affection of the heart itself which neither murmur nor friction sound can reveal or explain ; and, when we meet with the not very rare cases where in acute rheumatism there are continued symptoms and signs of heart-trouble with no physical signs of valvular or even of pericardial mischief, then, by keeping a probable myocarditis before our eyes, we shall avoid the common but grave error of treating them lightly because there is "no murmur." In close association with this matter we must remember the numerous cases of heart-failure in children, as well as in adults, which are referable to more or less acute dilatation and weakness of the muscle either from inflammation or degeneration, of which the most familiar examples are seen in the course or sequel of acute specific diseases such as diphtheria, enteric fever, scarlatina and several others. Fatty degeneration of the heart at least often co-exists with or follows on these infective conditions, and the symptoms, generally, are those of more or less dilatation, with difficult breathing, failing circulation or rapid collapse. There may or may not be a soft blowing murmur at the apex; but there is always a feeble or absent first sound, and percussion often gives evidence of enlarged cavities.

Bearing in mind the above remarks we can now proceed to the consideration of **endocarditis and valve-disease** in childhood.

Endocarditis in fœtal life is often connected with some developmental defect, and attacks the right heart especially though not exclusively, the greater liability of the right heart being doubtless due to the much greater circulatory stress sustained by its valves. Union of the segments, both of the auriculo-ventricular and semilunar valves, is frequent. The chief interest attaching to these conditions is the probable explanation afforded thereby of certain otherwise inexplicable cases of heart-disease. The causes of fœtal endocarditis are obscure; rheumatism may possibly be one; and it has been suggested that some cases are due to absorption from hæmorrhagic foci in the placenta.

Endocarditis from rheumatism is still more frequent in childhood than in later life. A larger proportion, too, of valve-diseases, in every way comparable to demonstrably rheumatic cases, is at first sight less clearly attributable to rheumatism in young children than in adults; and, if the occurrence of marked arthritis were regarded as necessary for the diagnosis of rheumatism, this difference would be of considerable importance. I have, however, dealt with this matter in sufficient detail under the heading of rheumatism, and will only say here that endocarditis in childhood, as evidenced by valve disease, is rheumatic in a vast majority of cases. Of 98 instances taken consecutively from my ward-books I find 77 are attributable to rheumatism, either from definite history or highly probable inference. Of the remaining 21, some may be almost certainly referred to scarlatina or measles, the origin of the others being obscure. There are, however, no grounds for believing in idiopathic endocarditis; and we must remember that enteric fever, small-pox and other septic disorders may all be credited with some cases, as also may undiscovered rheumatism.

Respecting the relative frequency of the different *valve affections*, these 98 cases show that under the rheumatic heading there was clinical evidence of mitral regurgitation alone in 39, of mitral stenosis and regurgitation in 25, of mitral and aortic disease in 7, of mitral stenosis alone in 4, of tricuspid and mitral stenosis in 1, and of tricuspid disease alone in 1, the last case, æt. 12, with a history of perfect health in early years before rheumatism, being examined post-mortem and showing normal mitral and aortic valves. I would here, however, remark, from my experience of necropsies generally, that tricuspid valvulitis from rheumatism is much more frequent than is usually taught; and would refer to the publication by Dr. Hebb [1] of 14 cases observed during four years in the post-mortem room of Westminster Hospital.

Of the 21 cases, where no probable rheumatism could be established, 10 had double mitral murmurs, 7 a systolic apex murmur only, and 4 a præsystolic murmur only. I can find no record in the above-mentioned list or among many other instances of heart-affection registered under the headings of rheumatism and chorea—amounting to 260 in all—of a single case of aortic valve-disease with unaffected mitral, nor of any aortic case which was not clearly of rheumatic origin. In one instance, however, which was under my observation very frequently during eight years, a double aortic murmur only was heard during life, though the necropsy demonstrated marked disease of the mitral valve as well. This negative observation illustrates well the clinical truth that aortic valve-disease *by itself* is not connected at any age with rheumatism or the ordinary causes of endocarditis, but is rather an affection arising out of vascular disease or strain in adult life. Mitral stenosis alone seems to be especially connected with the less marked forms of rheumatism, and is conspicuous among valve affections of uncertain origin. It is also specially likely to cause embolism, cerebral and otherwise.

Valve-disease, notably of the rheumatic kind, seems to be rare in children under six, and to become prominent at about the same age as when articular rheumatism is usually first observable. Out of 150 cases of definite valve-disease, registered either as such or under the head of rheumatism, only six were under this age, ranging from 3 to 5½ years old ; and the youngest one was in all probability not rheumatic. It is at least certain that signs and symptoms of valve-disease are but seldom shown under the age of six. On the whole, however, children are far more liable to endocarditis than adults, and most cases of rheumatic heart-disease arise in the early stage of the earlier attacks of rheumatism. Few persons suffer from the fever for the first time after 35 years old, and rheumatic subjects who escape heart-disease in early life have a fair chance of subsequent immunity.

[1] See vol. iv. of the Westminster Hospital Reports (Churchill).

Ulcerative endocarditis is not frequent in childhood, and has no clinical peculiarity at this period. I have seen but three probable cases in children, which were suspected during life, one only having been submitted to post-mortem examination, which alone gives positive evidence of this disease. Ulcerative endocarditis is sometimes found post-mortem, as often in adults, where it has not been suspected; but it is still more often suspected where it is not found. It is, however, of great importance to bear in mind the most probable signs of this affection, with its often markedly intermittent pyrexia and its tendency to cause emboli in various parts. It may exist with few or no prominent symptoms referable directly to the heart, and may closely simulate the febrile condition of some forms of ague or of retained collections of pus in various parts of the body.

The phenomenon of *fibrous nodules* in acute rheumatism as described under that heading has a bearing on the subject of heart-disease; for it is rarely, if ever, unattended by valve-mischief of a usually progressive character, and, as shown by Dr. Coutts and others, is frequently synchronous in appearance with fresh attacks of endocarditis. It is true that these nodules are almost always accompanied by evidence of arthritis; but it would appear that, in the absence of such evidence, nodules are a very strong indication of rheumatic heart-disease, and should give rise to at least a very guarded prognosis even in cases apparently not otherwise severe. Large nodules are of especially bad augury according to Dr. Cheadle, who urges the microscopical similarity of the process of formation of nodules and of valvular vegetations.

The *symptoms* of extensive valve-disease in childhood are much the same as in adults; but, owing probably to more ready and complete establishment of compensation in children, we more often find the symptoms very slight. It is remarkable how often, especially in cases of well-marked mitral disease with enlarged heart, both dyspnœa and dropsy are conspicuously absent, even under the stress of considerable exertion. On the other hand, probably a large majority of children with valve-disease quickly deteriorate and die as the age of puberty, with its greater heart-stress, is approached or reached. I have seen many examples of this apparently rapid failure of compensation; and, although we not seldom see quite elderly persons with valve-disease dating from rheumatism in early youth, the cases dating from early childhood are, I think, but few.

The proportion of cases of rheumatic valve-disease in which the aortic valve is involved is considerably less, according to my experience, in childhood than in later life. It must be also remarked that children with mitral disease are especially liable to become thin, and are usually pale, this peculiarity being probably explicable by the general effect of deficient circulation on the nutrition at a period when the body's wants are great; while in the adult, with less perfect compensation, the mechanical results of the valvular failure are more prominent, and are evinced in the face by capillary congestion.

The symptoms of the onset of valve-inflammation in childhood are often, like other rheumatic symptoms, but slight, or may be altogether absent, the condition being recognised only by the discovery of the murmur; and, even when the primary rheumatic affection has been well marked, both definite murmur and symptomatic evidence of the valvulitis are often long delayed. The heart, therefore, should be carefully and repeatedly examined in all rheumatic cases and, indeed, in all febrile attacks of doubtful nature. If no murmur be heard, altered sounds and actions may attract the attention of the experienced observer and materially influence the treatment and subsequent course of the case.

It must, however, be remembered that the beginning of acute endocarditis is often accompanied by other signs and symptoms probably referable to involvement of the heart-muscle and pericardium. Such are præcordial and epigastric pain, signs of dry or liquid pericardial effusion, or bulging of the heart region, without evidence of noteworthy pericardial effusion, owing to the involvement of the myocardium with swelling of the heart from distension of its cavities.

Among other modifications of the heart-sounds, which may point to the beginning of valve-disease in childhood, I may mention two. The first, especially described by Dr. Cheadle, consisting of a doubled second sound at the apex, with or without a diastolic murmur, seems to be an early sign of mitral stenosis. These phenomena are referred by him to asynchronous flapping back of the mitral and tricuspid valves and, when the diastolic murmur is heard, incomplete falling back of the mitral after systolic closure. The second, also a herald of established mitral stenosis, is according to my experience very much more often observable, being a slightly divided first sound which, after exertion, is frequently accompanied by a thrill, and at once develops into a well-marked murmur, immediately preceding and abruptly ending with the heart's impulse, and usually described as "præsystolic."

I have several times observed the complete disappearance of murmurs, in all probability organic, which have arisen in acute rheumatism. In four cases out of the above-quoted list, besides several others that I have seen, well-marked mitral murmurs, some of them being distinctly "præsystolic" with thrill, disappeared soon after the fever, no signs or symptoms of heart-disease being detected on examination at periods varying from two to three months afterwards. Many instances, however, of vanishing systolic murmurs at the apex are probably not due to valvulitis but to a dynamic cause from ventricular dilatation.

No instance of disappearance of a definite aortic regurgitant murmur has hitherto occurred in my experience. Here must be remembered the frequent difficulty of distinguishing between early double aortic murmur, before consecutive heart enlargement sets in, and pericardial sounds of soft and indefinite character; as well as the possible confusion of some double mitral murmurs with pericarditic friction. Examples of such mistakes by experienced auscultators are from time to time revealed on the post-mortem table.

In two cases of well-marked rheumatic mitral disease with a loud venous humming sound in the neck, a double murmur at the base, exactly simulating that of aortic valve disease, was constantly heard; but both the humming and the pseudo-aortic sounds vanished completely during pressure on the cervical veins. Lastly I would record a remarkable case, in a child of eight, where a loud and long diastolic murmur, as well as a systolic one, was always heard at the apex; but the necropsy showed a mitral orifice of fully three inches round, and no lesion other than a few small granulations on the auricular surface. Such a diastolic murmur, as is well known, is usually associated with marked mitral stenosis.

As regards *prognosis* in the valvular diseases of childhood we must remember, in addition to the early compensation and the frequent failure at puberty, the common occurrence of repeated attacks of endocarditis and pericarditis. For the rest, it is especially true in childhood that the forecast is essentially relative to the individual case, and necessitates careful and repeated study of signs and symptoms. Many cases, severe at the outset, make ultimately good progress; while others, seemingly slight, become rapidly worse with more or less acute dilatation of the heart.

The *treatment* of heart-disease respects, first, the early acute disorder or those affections, especially rheumatism, out of which such disorder arises; and, second, established cases of valvular mischief.

In acute cases of cardiac inflammation, whether demonstrated or only suspected, absolute rest and freedom from excitement are of the first importance. All children with rheumatism of any degree should be kept strictly recumbent. When the onset of cardiac or exo-cardiac mischief is suspected or demonstrated by signs or symptoms, the salicylates or salicin which may have been ordered should, as a rule, be discontinued at once; but, if the heart be working well with neither over-frequent nor irregular rhythm, these drugs may be carefully persisted with in severe articular cases, even when a murmur is present. There is no doubt in my mind that both these medicines depress the heart's action and neither prevent nor arrest endocarditis. Acute cardiac trouble and all cases marked by præcordial pain are very frequently much relieved by opium, which I very often give both for its anodyne and probably anti-inflammatory effects, when it is not contra-indicated by marked pulmonary complication, and am inclined to regard as an invaluable remedy. In most cases where the temperature is high, and especially where ulcerative endocarditis is suspected, quinine in full doses should be given; antipyrin, in my opinion, never. Alcohol is, I think, always strongly indicated by symptoms of cardiac failure; and, with a pulse at once irregular and frequent, but usually not otherwise, digitalis. I say nothing special of strophanthus or convallaria, which I seldom use now; for I have found that the sickness which digitalis is often accused of causing is very rare, if not mythical. Such apparent sickness can often be stayed by changing the vehicle in which the drug is given.

In chronic heart-disease absolute rest is often needful; but, when marked dyspnœa and dropsy are absent and there are no signs of pulmonary œdema, the patient may be allowed moderate, though never competitive, exercise, and should enjoy as much light and fresh air as possible. The body-warmth should be sedulously attended to and a hot foot-bottle always used in bed during cold weather. To promote compensation and general nutrition, ample but easily digestible diet should be ordered, with appropriate modifications on the appearance of the gastric symptoms not uncommon in mitral and right heart-affections. Iron, arsenic, and cod-liver oil are valuable aids to nutrition. Hurried breathing and œdema of the feet with a frequent and irregular pulse necessitate digitalis. If diuretics or diaphoretics be indicated by increasing dropsy I prefer a hot-air bath (its effects being carefully watched), to the many nearly inert drugs of this class and to the active, but always depressing and sometimes dangerous, pilocarpine.

With regard to those cases where, although mitral affection co-exists, aortic valve incompetence seems to be the prevailing mischief, as evidenced by predominant enlargement of the left heart, marked anæmia, recurrent headaches, a tendency to syncope, or, more often, by the absence of symptoms of mitral failure and of right heart distension, I would say that the prognosis in childhood is comparatively good, especially, of course, in the latter class of cases; and, as in adult life the subjects of aortic regurgitation with good compensatory hypertrophy can often endure even many years of strenuous labour, so, in childhood, moderate exercise may not only be permitted but also distinctly recommended.

CHAPTER III.

PERICARDITIS.

The chief clinical interest of pericarditis in children centres round the rheumatic cases, which are in a very large majority, whether evidenced by extensive signs and symptoms or only by a localised friction sound. Pericarditis *detectable during life* is not indeed so overwhelmingly frequent as endocarditis in rheumatic children, physical signs of the latter occurring in more than 80 per cent., and of the former in rather less than half, of my cases of rheumatism. It is, however, frequently demonstrated post-mortem although clinically undiscovered; and is rarely, if ever, absent in fatal cases of rheumatic heart-disease. Whether or no pericarditis in some degree always accompanies endocarditis in rheumatic children, as set forth by Dr. Sturges in a most important lecture on "The Rheumatic Carditis of Childhood,"[1] my own cases amply testify that, whenever it does occur, endocarditis is present.

Liquid effusion into the pericardium is, relatively to the form marked only by friction sounds throughout with but slight extension of præcordial dulness, much more frequent in children than in adults; a large majority of cases at first characterised by friction sounds soon showing evidence of distended pericardium. As regards **symptoms** there is but little special to our subject, the dyspnœa, pain and distress being generally proportionate, as in adults, to the amount and rapidity of progress of the effusion. With the generally more rapid progress in children the initial symptoms are often well marked and severe; and epigastric pain may be prominent. Many children, however, very soon accommodate themselves to even a large effusion; and I have seen some cases, quite unparalleled in my experience of adults, where the patients were able to walk quickly or even run with no apparent distress. ·

In many cases, especially of the rheumatic kind, slight signs of pericarditis without marked cardiac symptoms may be very early accompanied by a rise of temperature. Bearing this in mind, as well as the constantly associated endocarditis which may not be apparent, and the frequent slightness of arthritic symptoms in childhood, we should at once examine the heart-region for friction or other abnormal signs in all cases of pyrexia; avoiding by this precaution many discreditable blunders.

Rheumatic pericarditis as well as endocarditis in children not seldom

---

[1] See *Lancet*, August 27, 1892.

2 D

appears to pass off rapidly and entirely, well-marked signs clearing off with no remaining evidence of either adhesion or of impaired working of the heart. When, however, heart-symptoms continue, unaccompanied, it may be, by abnormal sounds or even by marked increase of cardiac dulness, pericardial adhesion, often the result of repeated slight attacks of inflammation, should always be suspected. Such suspicions are certainly justified by the presence of retraction of the præcordial region during systole, indicating adhesion of the pericardium and the pleura.

Other than rheumatic causes of pericarditis are somewhat more frequent in children than in adults. Such are *septic infection* in the new-born, from phlebitis or absorption from the umbilical vein, and also from osteitis and periosteitis. It may occur, too, in many of the *fevers*, especially scarlatina; and I have seen one case in the course of a severe attack of mumps. *Tubercular* pericarditis, though very rare as the only tuberculosis of the thorax, is not seldom seen in connection with similar affection of the lung, pleura or bronchial glands, or of the cranial and abdominal cavities. *Bright's disease* may be accompanied by pericarditis which is very often rapidly fatal; and some of the scarlatinal cases make their appearance with signs of renal failure. Lastly, pericarditis is often seen in connection with severe *pleuro-pneumonia*, and purulent pericarditis occurs with some cases of *empyema*.

As in pleurisy, so in pericarditis, the proportion of purulent to serous effusions from all causes is greater in childhood than in later life. Even in rheumatism suppurative pericarditis may, I think, very occasionally occur—an exception to the almost universal rule of serosity in rheumatic effusions, as also is the still rarer purulent pleurisy. A chronic effusion into the pericardium may be purulent and point at the surface. In such cases there may be the greatest difficulty in making a positive diagnosis.

Out of a series of 26 consecutive cases of pericarditis registered under this heading as the most prominent affection, apart from the far more numerous cases of pericarditis as a subordinate symptom of rheumatism or other diseases, 17 were distinctly rheumatic and were also associated with valve-disease; four followed on scarlatina (two of them showing also mitral regurgitation); two on pleuro-pneumonia; one on Bright's disease; one on mumps; and one on enteric fever. This list, however, but very imperfectly represents the proportion of pericarditis in pleuro-pneumonia, for many slight and some severe attacks occurring in this category are noted as complications of the pulmonary affection. A very large number of cases of pericarditis, especially of the tubercular variety, other than those of rheumatic origin, are not revealed by physical signs, and their existence is established only on post-mortem examination. Some of these may be suspected from hampered heart action; and we must always bear in mind that acute pericarditis generally connotes some involvement of the cardiac muscle, which may be indicated only by irregular action of the heart and altered or muffled sounds.

The **diagnosis** of pericardial effusion in childhood sometimes presents peculiar difficulties, owing to the heart being nearer to the front of the chest than in adults, and thus causing both impulse and sounds to be less obscured, even in cases of considerable effusion. When, therefore, friction sound is absent, we have to depend almost wholly on percussion for a correct diagnosis. The difficulty of accurately deciding on the causes of increased præcordial dulness, even when there is no concomitant affection of lung or pleura, is of great practical importance in cases of previous heart-affection with enlargement; and all experienced observers will admit that very often the ordinary rules of percussional distinction between an enlarged heart and pericardial effusion, according to the alleged squareness of the one and pyramidal shape of the other, most signally fail us in cases of any age, but especially in children. I have myself known two cases of the right heart and one of the left being tapped, with nevertheless no resultant harm but rather benefit to the patient, in the expectation of finding a pericardial effusion; and I doubt not that similar diagnostic mistakes are more frequently made in practice than reported in print. Another error, not seldom made, of mistaking a pericardial effusion for a left-sided pleurisy is not of much importance, and is more easily avoided.

In respect of the difficulty of diagnosis between enlargement of the heart, especially on the right side, and pericardial effusion, the teaching of Dr. Rotch[1] may prove to be of value. He concludes from many experiments and observations that absolute dulness of any considerable extent in the fifth right intercostal space means pericardial effusion, provided that other complications outside of the heart and pericardium, such as pulmonary consolidations and pleural effusions and adhesions, can be excluded. Since reading Dr. Rotch's paper I have found this dictum diagnostically helpful in one case, which seemed to many to be pericardial effusion until the necropsy disproved this view; but in another very doubtful case, which for several reasons, although diagnosed by others as pericardial, was regarded by me as cardiac enlargement and was almost proved to be so by the withdrawal of a drachm and a half of pure blood with marked relief of severe symptoms, there was unquestionable dulness in the right fifth interspace. I could not, however, quite exclude in this case the possibility of some pleural complication on the right side; and the child, who was the subject of previous heart-disease, soon recovered sufficiently to leave the hospital.

[1] Article "Diseases of Pericardium," vol. ii. of Keating's *Cyclopædia of the Diseases of Children*.

It may be hoped that further observations may test the value of this possible diagnostic aid. For the rest, the matter of detection of peri- carditis is much the same in children as in adults, though large effusions more often cause prominence of the heart region. Double friction-sound of no valvular localisation, but chiefly marked at the base in early cases, is the most certain sign; but it must be remembered that pericardial sounds, when not rough, may be sometimes mistaken for a double aortic murmur. When the sound is systolic only it may be more easily mistaken as valvular, especially when localised at the base and devoid of that prevalent quality which has led to the somewhat misleading use of the terms "friction" and "pericardial" as synonymous in their ap- plication to sounds. In all doubtful cases of this kind we should defer our forecast as to permanent damage until repeated examination has been made.

The immediate **prognosis** in acute pericarditis is as a rule good in children as regards approximate recovery, with the marked exception of cases occurring in Bright's disease, which, as in adults, are most often rapidly fatal. It is very rare for a young child to die from the imme- diate effects of a rheumatic pericardial effusion, however large it may appear, provided there be not much pre-existing heart disease; but a fresh pericarditis, even without much liquid effusion, supervening on old heart mischief, and especially when accompanied by fresh pleurisy, is of very grave and sometimes fatal import. Doubtless many of the severe symptoms of failing circulation which occur with pericarditis of a seemingly small extent are due to a greater or less involvement of the heart muscle; and, seeing that pericarditis in children is almost always attended by valve-disease, the prognosis largely depends on the consecutive damage to the heart from this cause. In all chronic cases a knowledge of their course from continued observation is necessary before a useful forecast for any lengthened period can be given.

Cases of what has been termed *mediastino-pericarditis* have been reported by Drs. Ashby and Hutton, consisting for the most part of an extension of pleuro-pericarditis to the mediastinal connective tissue, sometimes involving the mediastinal glands, and leading to matting together of the pleura, pericardium and great vessels. This inflammatory process may be of various extent, and the indurated tissue may be very thick. Such cases are often tubercular and associated with caseous mediastinal glands, but they are certainly sometimes unconnected with tubercle. The symptoms are those of labouring heart, imperfect filling of the lungs, and pressure on the large veins entering the chest. In some cases which I have observed there has been an extensive but ill-defined area of dulness above the base of the heart. In these, as in those described by the authors above quoted, there was hurried breathing and blueness of face on exertion with a tendency to enlarged veins on the neck and chest; and, in one, clubbing of the finger-tips. In the graver cases there may be general œdema with ascites and enlarged liver. I had one post-rheumatic case under observation for several years, where the less grave symptoms and considerable dulness in the upper sternal region were well marked. The boy, otherwise healthy, was always somewhat distressed and slightly cyanosed on violent exertion, and occasionally had slight syncopic attacks which were certainly not of an epileptic character.

In the **treatment** of pericarditis the chief attainable objects are rest for the hampered heart and relief of pain, for we cannot greatly hope to arrest the inflammatory process. At the onset, however, I usually blister over the præcordium, and give full doses of opium, which, besides relieving pain, may have an anti-inflammatory effect. In severe cases leeching over the præcordium may be useful. If the heart-action indicate it, by a combination of frequency, irregularity and feebleness of contraction, digitalis may be given; but this drug and its clinical associates are, I think, quite useless and possibly harmful if the pulse, although frequent, be of good quality and rhythm. The salicylates and salicin are to be avoided if heart-trouble be marked, as already insisted on in connexion with endocarditis, but are not contra-indicated in rheumatic attacks if the heart be working easily.

Paracentesis pericardii is very rarely necessary. The question will arise only in cases of acute and rapid effusion with urgent symptoms where relief may reasonably be expected, or when we believe the effusion to be purulent. After a diagnostic puncture a free incision may then be made and a drainage-tube inserted. In rheumatic pericarditis indications for paracentesis are rarely offered.

If paracentesis be decided on, a fine aspirating trocar should be used and any forthcoming fluid slowly withdrawn. Should a mistake in diagnosis have been made the abstraction of a drachm or two of blood from the heart will not be harmful; it has, indeed, been in some cases positively beneficial. I have once, as already stated, almost deliberately tapped the right heart in a case where some had made the diagnosis of pericardial effusion; and I can see no reason why future experience may not establish the operation as an easy, safe and speedy means, especially in the case of children, of relieving those symptoms for which we usually practise venesection or cupping. At present, however, I am not in a position to discuss in detail or definitely advocate this procedure, which only further experience could duly accredit.

If the contention of Dr. Rotch, regarding the positive diagnostic value in pericardial effusion of considerable dulness in the fifth right intercostal space, be corroborated by general experience, his further suggestion that this is the right spot for paracentesis is a practical one. The usual place chosen at present is the fourth or fifth left intercostal space, midway between the left sternal border and the nipple-line.

All patients who have had pericarditis with symptoms should be kept in bed after the febrile time for a period to be defined by careful physical examination and observation of the heart's functions.

CHAPTER IV.

RAYNAUD'S DISEASE.

IT is perhaps best for clinical reasons to mention this curious malady here, rather than relegate it, according to prevalent custom, to the domain of skin-affections. It appears to be at least proximately due to local failures of circulation, and the tendency to it would seem to be congenital in some cases. But little or nothing of practical value can be said concerning the ætiology of this remarkable form of circulatory stasis or gangrene, which must often, perforce, be called idiopathic. It has been attributed to spasm of arterioles, to peripheral neuritis, and to functional nerve disturbance; and, from some of its associations, especially with hæmoglobinuria and to some extent with the rare affection in new-born infants known as Winckel's disease, it might perhaps in some instances be plausibly referred to microbic origin.

The affection is most often though not always of symmetrical distribution. It usually attacks the toes or fingers, and sometimes the ears, scrotum, vulva or other parts. The affected region is at first cold and yellowish white or livid in hue, and may be painful, tender and hard; but sometimes it is quite anæsthetic. The arterial pulsation in the suffering limbs is not usually affected, but in some cases has been notably lessened. There is frequently some fever, with headache and anorexia. In the slighter cases the parts soon return to the normal condition by means of warmth, but in others gangrene often sets in and more or less sloughing-away is the final result. In almost all instances there is a tendency to recurrence of attacks at different intervals, at least for a time, and in many a previous history of liability to suffer from cold extremities. Taking into consideration the cases of this disease which are associated with paroxysmal hæmoglobinuria, it seems clear that exposure to cold is at least a frequent exciting cause of attacks. Raynaud teaches that the affection is due to spasm of arterioles excited by cold, and himself observed contraction of the retinal arteries. The various degrees or, in some cases, stages of the affection are denominated local syncope, local asphyxia, and gangrene. In the most advanced form the disease is rare ; but attacks of chilly pallor or blueness of the extremities, neither very intense nor proceeding to gangrene, are not very rare in children, at least among the hospital classes. The blueness sometimes extends up the arms and legs. Such cases usually recover very soon with continuous warmth and stimulative treatment. There may, indeed, be some association between this affection and the ordinary chilblain.

I have seen a case in a boy of eight, who was in hospital for right hip disease, contracted diphtheria later, and then, after two months' mechanical extension of his right leg, showed signs in the toes of the right foot of commencing gangrene which subsequently spread as far as the metatarso-phalangeal line. At first the appearance was provisionally attributed to the bandaging ; but after a fortnight the toes of the left foot became affected as well as the pinnæ of both ears. In the course of the next six weeks, with various fluctuations of condition, all the parts recovered, except the right foot, from which most of the phalanges sloughed away.

Dr. W. Pasteur has kindly shown me his notes of a case in a boy of nine, where there were several attacks of blueness and coldness and pain in the fingers following soon after a dog-bite on the back of the hand. Some of the attacks were very evanescent. The blueness was very deep and extended sometimes up the fore-arm, but usually only as far as the wrist. Artificial chilling of the hand brought on an attack. The left radial pulse was frequently noticed to be smaller than the right during the attacks, but not at other times. The boy recovered completely from the affection after a fortnight. He was, however, the subject of hereditary syphilis, and subsequently suffered from pharyngeal ulceration of apparently syphilitic character. In another case of my own, in a boy of five, there were repeated attacks of pain, with deep blueness of extremities, amounting to blackness at the finger-tips. In one attack the left hand and right foot were affected ; in another, on the following day, the left hand, both ears and both feet ; and, one day later, one foot only. Each attack lasted about three or four hours. In a month from the first they ceased entirely. The urine was normal throughout.

I have referred, under the heading of urinary affections, to two further cases under the care of Dr. W. Pasteur, where there was hæmoglobinuria, always excited by chill, in association with blueness and chilliness of the extremities. These cases seemed to me to be in every way similar to the usually, but not always, malarious examples of paroxysmal hæmoglobinuria with circulatory stasis or gangrene of the ears or extremities, which are from time to time met with in adults.

Warmth, tonics, and the application of the interrupted galvanic current to the affected limb by placing it, with one of the electrodes, in a mixture of warm water and salt, after the practice of Dr. T. Barlow seem to be the best modes of *treatment*, and, when established in time, may prevent "local asphyxia" proceeding to gangrene, as appeared very likely in one case under my own observation.

INDEX.

THE END.